Natural Hazards - Impacts, Adjustments and Resilience

Edited by Ehsan Noroozinejad Farsangi

Published in London, United Kingdom

IntechOpen

Supporting open minds since 2005

Natural Hazards - Impacts, Adjustments and Resilience
http://dx.doi.org/10.5772/intechopen.77525
Edited by Ehsan Noroozinejad Farsangi

Contributors
Md. Shahin, Sanjida Akter, Prome Debnath, A.K.M. Mostafa Zaman, Bhushan Mohan Raisinghani, Chittaranjan Subudhi, Keyoor Pathak, José Antonio Aparicio Florido, Sebia Mutasa, Edson Munsaka, Jared Bly, Louis Hugo Francescutti, Danielle Weiss, Frank Jerome Glago, Peter Bikam, Yusuf Musa Jimoh, Hector Chikoore, James Chakwizira, Emmaculate Ingwani, jestina chineka, Nthaduleni Samuel Nethengwe, Carolyne Ogutu, Antony Rono, Jaime Santos-Reyes, Leandro Torres Di Gregorio, Gustavo Guimarães, Fernando Danziger, Graziella Jannuzzi, Eduardo Qualharini, Aline Pires Veról, Assed Naked Haddad, Marcelo Gomes Miguez, Virgílio Cruz, Daki Bianchessi, Muneyoshi Numada, Arisa Yasui, Chaitanya Krishna, Diana Dimitrova, Wilfrid Gbenankpon Gbenankpon Djima, Abdulah Can Zulfikar, Cüneyt Tüzün, Hong Yan, Yusen Ye, Mahmood Ahmad, Xiaowei Tang, Feezan Ahmad

First published in London, United Kingdom, 2021 by IntechOpen
IntechOpen is the global imprint of INTECHOPEN LIMITED, registered in England and Wales, registration number: 11086078, 5 Princes Gate Court, London, SW7 2QJ, United Kingdom
Printed in Croatia

British Library Cataloguing-in-Publication Data
A catalogue record for this book is available from the British Library

Additional hard and PDF copies can be obtained from orders@intechopen.com

Natural Hazards - Impacts, Adjustments and Resilience
Edited by Ehsan Noroozinejad Farsangi
p. cm.
Print ISBN 978-1-83962-305-9
Online ISBN 978-1-83962-306-6
eBook (PDF) ISBN 978-1-83962-307-3

For EU product safety concerns:
IN TECH d.o.o., Prolaz Marije Krucifikse Kozulić 3, 51000 Rijeka, Croatia, info@intechopen.com or visit our website at intechopen.com.

Meet the editor

Dr. Ehsan Noroozinejad Farsangi is currently an assistant professor in the Department of Earthquake and Geotechnical Engineering, Graduate University of Advanced Technology, Iran. Dr. Noroozinejad is the founder and chief editor of the *International Journal of Earthquake and Impact Engineering*, engineering editor for *ASCE Natural Hazards Review*, associate editor for *ASCE Practice Periodical on Structural Design and Construction, IET Journal of Engineering, Frontiers in Built Environment: Earthquake Engineering, Open Civil Engineering Journal,* and *Journal of Applied Science and Engineering,* and editor of the *Journal of Reliability Engineering and Resilience.* His main research interests include structural and geotechnical earthquake engineering, structural dynamics, computational dynamics, smart structures, solid and fracture mechanics, resilience-based design, reliability analysis, artificial intelligence, SSI, and vibration control in structures and infrastructure under extreme loading.

Contents

Preface

Resilience is increasingly used as an approach for understanding the dynamics of natural hazards. This book presents research in the field of resilience and provides an overview of its development to date, which draws on the wide literature on engineering aspects, ecological science, social science, social-environmental system, and natural hazards. The book is addressed to researchers and practitioners on the subject as well as undergraduate and postgraduate students and other experts and newcomers seeking more information about the state of the art, new challenges, innovative solutions, and new trends and developments in these areas.

The first section provides recent research on risk and crisis management toward a more resilient society. It includes chapters on disaster management; disaster hazards, causes, and impacts; disaster-resilient rescue; risk management; disaster risk reduction; and disaster relief.

The second section focuses on more detailed simulations and modeling in terms of natural hazards and includes chapters on mitigation of earthquakes using the latest design tools, performance-based design for healthcare facilities, evaluation of liquefaction-induced settlements, modeling extreme damages from natural disasters, and effects of earthquakes on buildings.

Dr. Ehsan Noroozinejad Farsangi
Department of Earthquake and Geotechnical Engineering,
Graduate University of Advanced Technology,
Kerman, Iran

Section 1

Risk and Crisis Management

Chapter 1

Disaster Management: A State-of-the-Art Review

Jared Bly, Louis Hugo Francescutti and Danielle Weiss

Abstract

Disaster management involves the pillars of emergency management: planning and preparation, mitigation, response, and recovery. Emergencies are serious events that threaten health, life, and property and can be managed within the capabilities of the affected organization. Disasters, on the other hand, are hyper-complex emergencies, requiring resources not immediately available. Disaster management follows the principles of emergency management, and emphasizes flexibility, collaboration, and teamwork. Lack of resources will challenge people and organizations both in effects of disasters and the ability to manage them. Poverty, climate change, governance, and education are foundations to improve capacity. Hospitals play an important role in disaster response and can prepare accordingly. Plans, to be effective, must be implemented through appropriately-targeted exercises. Building on an all-hazards approach, to more hazard-specific considerations can improve disaster preparedness as well as day-to-day efficiency. Disaster management is complex and crucial. These principles are explored through the fictional tale of Tucci[1], a coastal city in the worst flood anyone can remember. Well, almost anyone...

Keywords: natural disasters, emergency management, disaster management, disaster training

1. Introduction

Sunday:

"Well, this is a disaster" said Jojo, the 19 year old apprentice to his mentor, Raj. They were pulling in the fishing nets near the usually beautiful seaside village of Tucci, now dull and grey and partly under water. The nets were heavy with debris from the churning sea. Raj grunted a mirthless laugh. "No. This is just a hard day of work. Tomorrow will be the disaster."

---

[1] *I went with a fictional disaster to demonstrate the principles of disaster management for a number of reasons. 1. Any current disaster would soon be overshadowed by one more recent. 2. There are many people that would have a much greater understanding than me of any historical event. 3. Any real event risks being 'foreign' to people in other places. The story of Tucci belongs to no one, and so applies to anyone. I agree with Robert Fulghum who wrote" ...myth is more potent than history" (The storyteller's creed in All I needed to know I learned in kindergarten). JB*

Then the old man added, "unless it stops raining, the bridge stays above water, the power line's fixed, and we have enough sandbags for everyone to keep their houses from washing away."

It wasn't totally exaggerated. The rain had been the worst in decades. Many homes in the low-lying village were already flooded. Those that were a little higher than the rest were already overcrowded with friends and relatives who's houses were in a foot or two of water. And the bridge, the only land access to the village, was visible only as rail posts marking a dotted line through the sea between the village and the green foothills.

Disasters require both a potentially harmful event and a component of vulnerability [1]. If an event overwhelms local response capacity, whether by insufficient material resources or by inadequate social systems or structure, outside help is needed. This is a disaster. Thus the magnitude of an event that causes a disaster will vary by organizational capacity. Many of the natural events described elsewhere in this textbook (earthquakes, tsunamis, etc.) create disasters. An earthquake in a remote, uninhabited area might be called a natural disaster, but it is not truly a disaster if people are not severely impacted. Disasters occur at the interface of nature and civilization [2].

Emergency management is usually described in terms of planning, mitigation, response, and recovery. As we move along the spectrum of severity, from emergency to disaster, the same principles apply, with an emphasis on adaptability and collaboration. Specific to hospital disaster management, contextual issues such as triage, decontamination, and patient care are built upon a general and pervasive approach to disaster readiness. In resource-poor environments, the challenge is magnified as the impacts of natural disasters are greater, and the ability to respond and recover less. Education and training will be most effective if methods match the objectives. With all the uncertainty therein, training for disaster must include establishing relationships between organizations and allowing for flexibility in the face of events that can be predicted but never fully anticipated.

Not every windstorm, earth-tremor, or rush of water is a catastrophe...So long as the ship rides out the storm, so long as the city resists the earth-shocks, so long as the levees hold, there is no disaster. It is the collapse of the cultural protections that constitutes the disaster proper. ([3], p. 211)

2. Definitions

From crisis to catastrophe, emergency to disaster, there is a spectrum of events that may threaten people and organizations. Not just the event, but the characteristics of the affected population define disaster. Risk and resilience are opposing forces that must be considered with disaster management.

2.1 Emergencies

Disasters and emergencies differ in quality and magnitude but are often and inaccurately used synonymously. "Disasters are not just 'big emergencies'" ([4], p. 293). Emergencies are time-sensitive, potentially harmful events that put life and well-being at risk. Resources are available at the local level to prevent, mitigate, or minimize the harm, and a single responding organization is responsible [5, 6]. Local resources, as a variable in the equation, can affect what constitutes an emergency, and what goes beyond. An event of the same magnitude, in locations or situations with different capabilities and resources, may be managed within the organization (emergency) or need outside help (disaster). An example in a health care context might be a car crash involving one or two seriously injured people requiring prompt medical investigations and treatment, assuming the facility is equipped to deal with such an event.

2.2 Disasters

Disasters are sometimes considered "hypercomplex emergencies" or "major emergencies" involving multiple people at risk of harm, multiple jurisdictions responding, and resources that are not immediately available locally ([5], p. 8; [7]). Coordination between agencies, many of whom have no prior relationships, becomes a challenge [5]. Plans for resource utilization must change when those resources are overwhelmed [8]. Preparations, planning, and training at the local level, within the abilities and available resources of a single agency, do little to prepare for disaster.

2.3 Crisis

Crisis is a more generic description. A crisis is a "critical event or point of decision which, if not handled in an appropriate and timely manner (or if not handled at all), may turn into a disaster or catastrophe" [9]. We use the word *crisis*, then, nonspecifically, as an emergency event that has potential to evolve; *emergency* as time-sensitive event with potential harm; and *disaster* as an event larger and more harmful than an emergency, with many people at risk, and where management requires resources outside of the responding organization or department.

2.4 Catastrophe

The word *catastrophe*, more severe than a disaster, completes the spectrum [5]. Many variations of the definition exist, but all suggest a magnitude of harm and inadequacy of response capabilities beyond what would be considered disaster [1] (**Figure 1**).

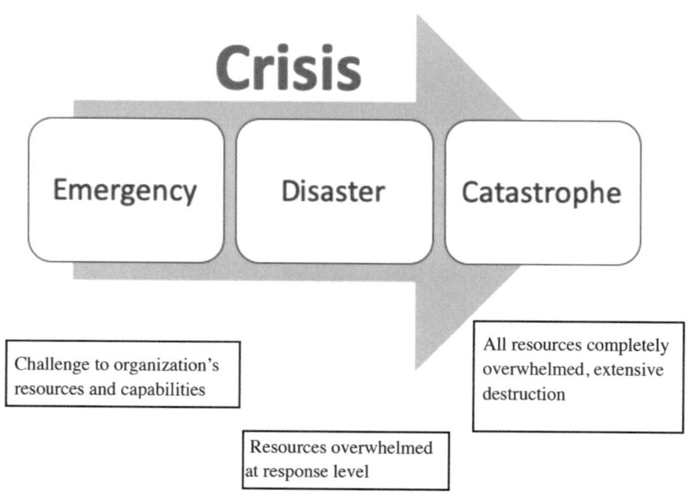

Figure 1.
Spectrum of crises.

Monday:

The school gymnasium was packed with wet bodies. A kind of bored panic filled the air. After all, what more could they do but wait for the worst the storm threw at them and then pick up the pieces when it blew itself out?

"Thanks for being here, I know it's been hard for everyone. And there's still lots of work to do to clean up after yesterday's catastrophe" said Ros, the town's mayor, referring to the wind that had blown off parts of a few roofs, and torn off a main limb of the biggest cedar in town, crushing a corner of J. Z.'s corner store.

Ian spoke up, "we can't worry about yesterday's fiasco. We gotta think about the crisis we're gonna be in tomorrow if the power's not back. Then it will be a real emergency!"

2.5 Hazard

Our first thought when we think of a hazard will often be an event—earthquake, flood, or fire. But only thinking in terms of characteristics of the event — wind-speed, the size of tsunami wave, the magnitude of an earthquake, etc. — is to neglect a critical component. To become relevant to disaster management, nature must collide with human activity [10]. Hazards can be quantified simplistically as the probability of an event occurring, causing harm [11]. And there is no separating hazard from risk and resilience [12]. So the hazard is the oncoming storm and the potential for harm to the village it approaches (**Figure 2**).

2.6 Risk

Risk is connected choice and probability [11]. Choice by the decisions we make. We build in flood zones, we develop seaside resorts, and we ignore all but the

Figure 2.
Hazard, risk, and vulnerability illustrated.

most active fault lines when looking at real estate. We buy fire insurance or not. We upgrade the old building to comply with seismic billing codes or not. We run disaster drills or not.

Probability is the other face of risk. Risk is an abstract concept, forever in the future, always uncertain.

> *Risk is a complex and, at the same time, curious concept. It represents something unreal, related to random chance and possibility, with something that still has not happened. It is imaginary, difficult to grasp and can never exist in the present, only in the future. ([11], p. 47).*

2.7 Vulnerability

Vulnerability will create harm from the hazard. A predisposition to be harmed, intrinsic to the organization or organism is its vulnerability [11]. Poverty, age, gender, racial identification, geography, and many social, economic, and political factors are all parts. The vulnerability can accumulate until recovery is complete [12].

2.8 Resilience

The ability to adapt is central to an organization's ability to resist and rebound from disaster [13]. Resilience is woven through all aspects of disaster management—from preparation through mitigation, response, and recovery [12, 14]. Resilience alters the disaster threshold. The more resilient a system, the more harm can be absorbed before the system is overcome [13]. More resilience means less susceptibility to disaster.

3. Emergency management

Preparation and planning, mitigation, response, and recovery are the basic principles of emergency management [15]. It is called *emergency* management, but should really be called *disaster* management. Necessarily limited to first responders, the title emergency management gives an illusion of control that makes it both "a misnomer and an oxymoron" ([16], p. 5). Regardless of the size of the event's magnitude, management includes all those efforts before, during, and after to minimize physical, social, and economic damages. Both planned and improvised actions should be included [16].

Preparation occurs before the disaster and includes preventative measures [17]. Disaster preparation, then, can also raise the disaster threshold if the disaster is thus avoided. At least, effects are minimized through planned measures. In our example settlement, prevention of a storm may not have been possible, but prevention of harm was through city planning, weather warning systems, and flood-resistant housing and infrastructure. Food and fuel stores could only be built up before the flooding.

Mitigation also includes a component of prevention but is closer to the event than planning. Anything to minimize harms that are not prevented could be considered mitigation. This can be through the reduction of the effects of the hazard, vulnerability of those affected in harm's way. In Tucci, they could build up walls of sandbags to protect their homes. They could moor their boats securely. They could evacuate, or they may have been able to if they had made adequate plans and preparations. Clearly, all these components are intricately connected.

The response may be what we typically think of when we envision a disaster. This is the responders—firefighters, paramedics, police, military, municipal forces, and volunteers—dousing the flames, treating the wounded, rescuing the stranded, and searching for victims.

Recovery entails returning, rebuilding, restoring. It is regaining a sense of normalcy, if not returning exactly to the pre-disaster state. Tucci will never be the same. The coastline will be altered. Attitudes may change forever. Lives may be lost. Houses will have to be repaired or rebuilt. Few residents will rebuild their houses exactly as they were before the storm. Recovery should focus on learning from the disaster and improving those liabilities made apparent by the wind and waves. This applies not only to the repairs to physical structures but to emotional health and economic stability.

3.1 From emergency management to disaster management

Preparation, planning, mitigation, and recovery are all important management principles for crises of any magnitude. As complexity increases towards disasters, we focus on the response at the front lines. This is because this phase sees the most variation and inconsistency [18]. On the front and back ends, in planning and recovery, the skies are clear. There is time to think. Not so in response. The response is the result of planning and facilitates recovery. To be prepared for an emergency should be routine. Preparedness for a disaster does not automatically follow.

By definition, local resources are sufficient to respond to an emergency. When these resources are overwhelmed, either by supply (nature of the event) or demand (response capabilities), the situation is a disaster ([19], Ch1). Outside help is needed. Intra-agency communication and coordination are required, usually without the benefit of established relationships and protocols. As complexity increases, more emphasis must be placed on flexibility and coordination between teams.

When the crisis moves from emergency to disaster, flexibility becomes increasingly important in planning, preparation, and response. In disaster planning, people should be prepared not to respond to specific circumstances, but to be able to adapt to the unanticipated. Training for disaster, then, ideally trains flexibility, communication, and the ability to work across organizational boundaries [20, 21]. Some structure is necessary to create the ability to adapt the structure to the situation. Brandrud's [22] description of their successful system is excellent: "...[the] written preparedness and response plan was structured just enough to remind the health professional of their role and task, yet flexible enough to enable them to release their creativity to improvise solutions" (p. 811).

Tuesday:

"Anyone got a charger?" The question was becoming a little repetitive. At first, the people that asked this were given sympathetic smiles and apologies. Now, if anyone dared ask, it was only met with grunts and grumbles. Part of 'the plan' involved keeping in touch with people by cell phone. There were only a handful of people who still had any battery life left on their phones, and no one had reception.

All but a few of the townspeople were crammed into the school for the night. It was loud. Fifty quiet conversations, a few crying babies, the howling wind, and the incessant rain added up. And the air was thick with sweat and sewer (the toilets had all overflowed). A dozen people were standing in a circle in the middle of the gym, sorting through a pile of walky-talkies.

> *The side door flew open with the outside coming inside, and a group of bodies in rain gear, dripping from head to toe. It was a crew from Uah, an even smaller town down the coast. They had got their whole village out last week and came here on a few all-terrain vehicles to lend a hand. Apparently, there was a team coming from the city to take everyone out. If the rain ever stopped...*

Crisis standards of care are a reflection of the flexibility needed to respond when resources are lacking for the situation's need [23]. The same standards employed in day to day operations, or even in an emergency (when an organization has the capability to manage it), will consume valuable assets (time, supplies, personnel, cognition) when the system is asked to perform beyond capacity. Awareness of the difference between disaster standards and the standards applied to usual operations will facilitate effective disaster planning and response (**Figure 3**).

3.2 Disaster management: resource poor environment

Natural hazards alone do not result in disaster, but rather the vulnerability of the populations of countries impacted [24]. The complexity and chaos of disasters make management challenging in many ways. Even the best plans will be unable to address each difficulty encountered in a disaster [25].

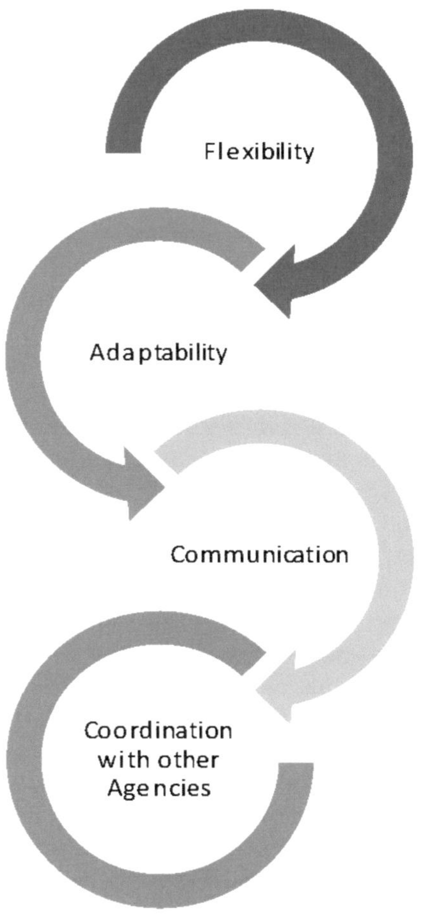

Figure 3.
Principles in management when emergency becomes disaster.

Resources are defined as the organization's fundamental financial, physical, individual and organizational capital attributes [26, 27]. In resource-poor environments, the challenge is greatly magnified. The environments most often impacted by a lack of resources are those of a lower socioeconomic status. Poverty and disasters are strongly associated [19]. Developing countries are repeatedly subject to disasters resulting in reduced or negative development [19].

Wednesday:

There was a lot of talk about fixing houses, repairing roads, upgrading the bridge. People didn't want to talk about the deeper issues. Most would never be able to afford anything more than patching the holes. Someone brought up the idea of building up on the hillside where the waves couldn't reach. But that was so utterly inconceivable. How would they build a new town if they couldn't even build new houses? Some would have to leave. Hard to live in a fishing village if your boat got washed away and you got no other way to make a living.

More impoverished communities are more vulnerable to natural disasters due to a mixture of social, political, cultural and economic factors [28]. Residents within these poorer communities tend to live in environments more prone to hazards such as rural areas with limited access to resources. The reduction in resources results in a more extended reconstruction period and can further delay developmental lag [19]. For example, in 2001, both El Salvador and the United States were hit by earthquakes, resulting in $2 billion in damages [19]. Although the same monetary value, the impact on each country's economy varied drastically. This $2 billion in damages had minimal impact on the U. S. economy, whereas, in El Salvador it resulted in 15% of the countries GDP [19]. These financial setbacks to developing countries can create a cyclical impact of further delayed development lag and economic growth.

Beyond the economic impacts, developing countries also face higher casualty rates. Over 96% of disaster-related deaths in recent years have taken place in developing countries [29]. Disasters may bring about harm to poor, developing countries in many ways beyond death, injury and destruction [19]. Some of the numerous examples include an increase in crime due to poverty and desperation, damage to schools leading to longterm impacts on education and further employment, destruction to hospitals which increase the vulnerability of disease, and the impact to vital infrastructure such as roads, bridges and airports, which may take years to rebuild and further impact resource access [19].

For meaningful disaster preparedness, the focus must be on improving availability and access to resources. This improvement should be a continual improvement effort to implement these resources to the area permanently. This implementation will help to support improvement to the quality of life to those impacted and decrease the inequity of resources and support when faced by disasters. Improved governance, combined approaches on all government levels, empowering communities, assessing vulnerability, ensuring access to quality information, and increasing the resilience of livelihood and infrastructure within these environments will reduce poverty and increase the quality of life [29].

Climate change and sustainable development both also influence the frequency and severity of disasters, particularly in resource-poor countries. Climate change, and irresponsible use of natural resources such as deforestation, make the environment more susceptible to hazards and disaster [30]. Disasters related to natural hazards, such as floods, storms and earthquakes, have significantly risen over recent years [30]. Such an increase in disasters is likely to further the frequency and severity of the impacts on the resource poor countries. Sustainable development is crucial to help reduce this burden.

3.3 Hospital disaster management

Disasters are easily forgotten. The unfortunate truth is that the longer the distance in time and space from disasters, the less influence they have on preparedness and planning [31]. This is especially relevant to hospitals because of a number of other interactions. Perception of disaster preparedness is often quite different between planners and frontline workers, the latter decidedly less optimistic about the facility's state of readiness [31]. And the pressures and problems of everyday operations can easily push aside concerns for an unforeseeable event. The attitude of disaster preparedness needs to pervade all aspects of the organization in the face of so many unseen but real hazards [32].

Specific hospital management principles include, but are definitely not limited to, vulnerability analysis, communications, triage, surge capacity, psychosocial effects, and medicolegal issues [31]. Hospitals must consider the disaster and its effects not only on a massive influx of patients but on existing patients, as well as health care workers in and out of hospital [33]. Patient care may be complicated and compromised by issues of security, chemical or biological exposure, and capacity for definitive care [29, 34].

Typically, an 'all-hazards' approach is employed as a basis of preparation for crises of any nature. More advanced preparedness will be tailored to specific hazards [30, 35]. We cannot plan for every possibility, especially not every extreme and infrequent event covered in this textbook. Plans must be broad enough to allow adaptation as needed [22]. If plans are too narrowly focused the preparation may be ineffective. Flexibility is key.

> *Thursday:*
>
> *Good thing we made it out when we did, although, an hour earlier would have been ideal. The leak that had been dripping constantly in the west corner of the gym turned into a stream, then a river, then the storm outside as the tiles gave way. The sick and the injured were evacuated first, down to Mayor Ros. Raj and me came on the last load. The hospital at Alec wasn't used to a hundred people at all, much less all within a couple hours. It was hard to tell who was who - doctors, nurses, housekeepers — might have been the president of the hospital — who were finding blankets, mopping up the incessant streams of muddy water, handing out bottles of clean water, looking at cuts and bruises and sore throats.*

Hospital disaster planning has important ramifications for capacity-building. That is, the threshold for disaster, an event that overwhelms local abilities, is intricately connected to capacity. "If a disaster is defined as an event that outstrips the organization's ability to deliver healthcare, preparedness is a method of "vaccination," raising the threshold not only in disaster periods but also in normal day-to-day function" ([31], p. xi). Disaster preparation is capacity-building.

Disaster preparedness is also about building networks. Again it comes back to the definition of disaster that requires help outside the immediately-affected organization. Coordination and communication between agencies are important in the success or deficiency of disaster response [23, 31, 36]. Establishing and enhancing relationships between organizations cannot be done in the moment of need. This should be a high priority for any organization in this time of global connectedness. Whether for a hospital, a nation, or a single-family, Alexander's [32] words for current and future emergency managers applies here: "Nothing can substitute for personal relationships" ([32, 37], p. 10).

The worst possible outcome of preparedness activities is to engender complacency. A "paper plan syndrome" refers to passively placing confidence in a document detailing a facility's readiness ([35], p. 3). Written plans do not obviate problems [33, 38]. To be effective, training needs to be continuous, team-centred, and at least as far as disasters go, focused on the non-technical aspects of working in teams [22]. They have to use existing resources and include the possibility of the loss of these resources. The loss of electrical power is particularly important to consider. Our increasing reliance on technology is a modern blessing in times of peace and a serious susceptibility when things are bad [12].

4. Training

Plans are only 'fantasy documents' if they have no real implementation through training ([39], p. 2). Exercises also may only be preparation in fantasy if not implemented conscientiously. When planning disaster training exercises, we should consider our purposes. Is the intent to expose participants to the disaster response plan or their roles in the organizational structure? Is it to test the implantation of the response plan, to expose its weaknesses and oversights? This is often the objective, intended or not ([40], p. 277). Evaluation and improvement of disaster plans may be a useful objective if that is the need [31]. But simply observing shortcomings does not itself remedy them. Lessons "identified" does not mean lessons "learned" ([40], p. 280) Is the intent to learn from or improve collaboration with other agencies? Is the intent to improve decision-making and specific skills? These are all valid objectives and need to be determined to meet the organization's needs, lest any coincidental success be wrongly attributed to ineffective plans [41]. Disaster training should focus on adaptability. "Exercises and training on how to be creative and imaginative under such circumstances would be more useful than detailed disaster plans" ([25], p. 376).

A month later...

"We just need to stick to the plan next time," Jan said, the last part sounding like a question. The storm was a memory like a bad dream. The town meeting, those who were left, was about getting ready for the next one.

The plan was new to almost everyone. Ros dug up some dusty old binder a few days ago. Too bad it made it, untouched, through the storm. It was full of detailed instructions about houses reporting to block leaders, block leaders reporting to councillors, councillors to the mayor, the mayor to the assistance team that was supposed to come from Alec, the capital city. Only thing was, households were all rearranged, trying to find somewhere dry to sleep. The block leaders didn't even know who they were, the mayor didn't have any councillors, and the team, well, not sure there ever was one.

4.1 Barriers to effective exercises

Disaster exercises may not accomplish what is intended during their design [20]. Excessive complexity, targeting the wrong audience, and unforeseen social psychological effects are some of the problems that can impair the efficacy of disaster education.

Complexity. More complex does not mean better when it comes to training exercises [21]. Thinking that testing more skills will improve more skills, stressing more processes will improve more processes, and designing more complex scenarios will enhance a greater repertoire of individual and systemic responses is flawed.

The opposite can occur. Complexity can obscure the purpose of the exercise, lead to passivity among participants, and decrease collaboration [42]. Complexity can also interfere with learning [20]. Complex responses may be better trained by simple exercises. The goal is internal complexity with external simplicity (Loveluck cited in [21], p. 423).

Leaders versus participants. Many exercises benefit the designers and facilitators more than the participants [20, 21]. This may be effective when that is the goal. Some exercises explicitly target leaders and not participants [43]. But often, the intent is to train participants. Even when that is the stated objective, participants may not see it that way [44]. Facilities and educators may not be training who they hope to train. It is important to consider who the exercise is for, and who is actually benefitting.

Social psychological. Recognizing that crisis simulations are meant to evoke some stress in individuals and organizations, some researchers have examined the adverse social and psychological effects of exercises [20, 43]. Sometimes "unintended consequences" of these effects can appear as a failure to participate when trainees fear evaluation from superiors ([20], p. 422). Supervisors giving feedback can reinforce incorrect behaviors if hierarchical relationships are ignored [20].

4.2 Benefits

There is no doubt that planning and training is key to disaster preparedness [41]. Disaster exercises are beneficial when objectives are clear, and debriefing is effective. When objectives are appropriate and align with needs, response capacities improve. Debriefing helps with this and with all aspects of learning and growth. The debrief is one of the most important parts of effective exercise.

Clearly defined objectives. Objectives should identify whether the purpose of the exercise is evaluation or training, individual skills or collaboration, crisis or emergency response. Experts commonly identify the need for objectives to guide disaster exercises [20, 45]. Yet hospital exercises often do not include specific objectives [46] or have not clearly defined them [47]. Objectives help operationalize disaster training. That means we can identify what we wish to improve, measure to see if we have improved, and actually improve in the desired area [20, 21, 43, 46]. In many cases, the method of training and objectives of an exercise is not complementary and do not create the conditions for improvement in operational capacities [46, 48].

Disaster vs emergency, stability vs flexibility, training vs drills. Disasters and emergencies are different events and require different responses [21]. Training for emergencies requires drills, practicing being able to perform planned responses to anticipated events [20, 42, 51]. In a disaster, responses outside an organization's policies and protocols are required [20, 44]. Training for disaster ideally trains flexibility, communication, and the ability to work across organizational boundaries [20, 21].

Collaboration. Disasters require interactions across and within organizations that is outside of usual lines of communication [20]. Collaboration, then, is key. Collaborative communication can help organizations recognize crises in the first place [49] and throughout the event. If there are barriers to effective communication across organizational boundaries, the response will be less timely, flexible, and effective [51]. We should prepare for the need to collaborate through practice working within new relationships and organizational structures [25].

Debriefing. "... the only reason for running a simulation is so that an exercise can be debriefed" (Thiagarjan cited in [20], p. 421). Debriefing is essential in order for learning to occur [20, 49]. Debriefing helps accomplish objectives, be they developing plans, training existing skills, or learning new things [50]. Learning from an

• *Is the exercise overly complex?*
• *Who is the exercise intended to benefit?*
• *Is the possibility of psychological effects addressed?*
• *Are there clear objectives?*
• *Is training for emergency or disaster?*
• *Is collaboration being trained in disaster exercises?*
• *Is there an adequate and effective debriefing?*

Table 1.
Questions to ask to make disaster training effective.

exercise increases with reflection individually and collectively [21, 44, 51]. The utility of an effective and adequate debrief cannot be underemphasized (**Table 1**).

> *Seems like a dream. A dream I'd like to forget. I said as much to Raj, adding "won't see another one like that for a hundred years."*
>
> *He was just shaking his head. "Forget this dream and it might as well be three days till the next one. Be the same dream all over again unless you keep this one in mind."*

5. Conclusion

Disaster management is challenged by the difficulty we have as people and organizations to think about future, uncertain events. The complexity and chaos of disasters further complicate the tasks of planning, preparing, and responding. The more complex the event, the more an organization must adapt and collaborate with other organizations. This frameworks of resource management in disasters will guide organizations in their disaster preparedness activities. We have touched on some applications of these principles to hospitals and resource-poor environments. From an accurate understanding of what constitutes a disaster, education and training will more likely be effective — directed to the right people, developing the right skills in the right places.

Author details

Jared Bly[1,2]*, Louis Hugo Francescutti[1] and Danielle Weiss[3]

1 University of Alberta, Edmonton, Canada

2 Royal Roads University, Victoria, Canada

3 Northern Alberta Institute of Technology, Edmonton, Canada

*Address all correspondence to: jrdbly@gmail.com

References

[1] Blanchard BW. Guide to emergency management and related terms, definitions, concepts, acronyms, organizations, programs, guidance, executive orders & legislation: A tutorial on emergency management, broadly defined, past and present. In United States Federal Emergency Management Agency 2008 Oct 27. Available from https://training.fema.gov/hiedu/docs/terms and definitions/terms and definitions.pdf

[2] De Smet H, Lagadec P, Leysen J. Disasters out of the box: A new ballgame? JCCM. 2007;**20**(3):138-148. DOI: 10.1111/j.1468-5973.2012.00666.x

[3] Carr LJ. Disaster and the sequence-pattern concept of social change. The American Journal of Sociology. 1932;**38**(2):207-218. DOI: 10.1086/216030

[4] Tierney KJ. Businesses and Disasters: Vulnerability, impacts, and recovery. In handbook of disaster research. Handbooks of Sociology and Social Research. 2007. https://doi.org10.1007/978-0-387-32353-4_16

[5] Oliver C. Catastrophic Disaster Planning and Response. CRC Press; 2010

[6] United Nations (2019). UN-SPIDER. Space Based Information for Disaster Management and Emergency Response. Retrieved from http://www.un-spider.org/

[7] McMaster R, Baber C. Multi-agency operations: Cooperation during flooding. Applied ergonomics. 2012 Jan 1;**43**(1):38-47. DOI: 10.1016/j.apergo.2011.03.006

[8] Cannava P, Cicillini D, Higgins M, McGrath A, O'Leary J. Response to H1N1 influenza outbreak in a pediatric children's hospital: Challenges faced and lessons learned. Journal of Pediatric Nursing. 2010 Oct 1;**25**(5):375-381. DOI: 10.1016/j.pedn.2010.03.001

[9] Business dictionary. http://www.businessdictionary.com/definition/crisis.html

[10] World Health Organization. WHO/EHA Panafrican Emergency Training Centre, Addis Ababa. 2002. Retrieved from http://apps.who.int/disasters/repo/7656.pdf

[11] Cardona OD. The need for rethinking the concepts of vulnerability and risk from a holistic perspective: A necessary review and criticism for effective risk management. In: Hilhorst D, Frerks G, Bankoff G, editors. Mapping Vulnerability: Disasters, Development and People. London: Earthscan Publications; 2003. pp. 37-51 Retrieved from http://www.academia.edu/download/54365737/Rethinking_Risk_Cardona.pdf

[12] Etkin D. Disaster Theory: An Interdisciplinary Approach to Concepts and Causes. Butterworth-Heinemann; 2014

[13] Hufschmidt G. A comparative analysis of several vulnerability concepts. Natural hazards. 2011 Aug 1;**58**(2):621-643. DOI: 10.1007/s11069-011-9823-7

[14] Murphy BL, Etkin D. Introuction. In Etkin D, Murphy BL, editors. Disaster and emergency management in Canada. Canadian risk and hazards Network 2011. p. 1-16. Retrieved from https://www.crhnet.ca/sites/default/files/library/Introduction%20Formatted.pdf

[15] Public Safety Canada. An Emergency Management Framework for Canada (Third Ed.) Government of Canada. 2017.

[16] McEntire DA. The status of emergency management theory: Issues, barriers, and recommendations for improved scholarship. University of North Texas. Department of Public Administration. Emergency administration and Planning 2004. Retrieved from http://precisionmi.com/precisionmi/Materials/ManagementMat/Status%20of%20Emergency%20Management%20Theory.pdf

[17] Public Safety Canada. Disaster prevention and mitigation. 2015. Available from https://www.publicsafety.gc.ca/cnt/mrgnc-mngmnt/dsstr-prvntn-mtgtn/index-en.aspx

[18] Van Wart M, Kapucu N. Crisis management competencies: The case of emergency managers in the USA. Public Management Review. 2011 May 1;**13**(4):489-511. DOI: 10.1080/14719037.2010.525034

[19] Coppola, D. P. (2015). Introduction to international disaster management. Retrieved from https://ebookcentral-proquest-com.ezproxy.royalroads.ca

[20] Borodzicz EP. The missing ingredient is the value of flexibility. Simulation & Gaming. 2004 Sep;**35**(3):414-426. DOI: 10.1177/1046878104266221

[21] Borodzicz E, Van Haperen K. Individual and group learning in crisis simulations. Journal of contingencies and crisis management. 2002 Sep;**10**(3):139-147. DOI: 10.1111/1468-5973.00190

[22] Brandrud AS, Bretthauer M, Brattebø G, Pedersen MJ, Håpnes K, Møller K, et al. Local emergency medical response after a terrorist attack in Norway: A qualitative study. BMJ Quality & Safety. 2017 Oct 1;**26**(10):806-816. DOI: 10.1136/bmjqs-2017-006517

[23] Timbie JW, Ringel JS, Fox DS, Pillemer F, Waxman DA, Moore M, et al. Systematic review of strategies to manage and allocate scarce resources during mass casualty events. Annals of emergency medicine. 2013 Jun 1;**61**(6):677-689. DOI: 10.1016/j.annemergmed.2013.02.005

[24] UNDP. Disaster Risk Reduction Governance & Mainstreaming. October, 2010. Retrieved from https://www.preventionweb.net/files/17429_4disasterriskreductiongovernance1.pdf'

[25] Quarantelli EL. Disaster crisis management: A summary of research findings. Journal of management studies. 1988 Jul;**25**(4):373-385. DOI: 10.1111/j.1467-6486.1988.tb00043.x

[26] Hill CW, Jones GR, Schilling MA. Strategic Management: Theory: An Integrated Approach. Boston: Houghton-Mifflin; 2001

[27] Hitt MA, Ireland RD, Hoskisson RE. Strategic Management: Competitiveness and Globalization, 2nd Ed. 1997. West, St Paul, MN.

[28] Benson C, Twigg J, Myers M. NGO initiatives in risk reduction: An overview. Disasters. 2001 Sep;**25**(3):199-215. DOI: 10.1111/1467-7717.00172

[29] OECD. Poverty and climate change: Reducing the vulnerability of the poor through adaptation. Available from www.oecd.org/env/cc/2502872.pdf

[30] De Smet H, Lagadec P, Leysen J. Disasters out of the box: A new ballgame? Journal of Contingencies and Crisis Management. 2012 Sep;**20**(3):138-148. DOI: 10.1111/j.1468-5973.2012.00666.x

[31] Kollek D, editor. Disaster preparedness for health care facilities. In: PMPH-USA. 2013

[32] Health Standards Organization. Emergency and disaster management [guideline]. 2020. Available from www.healthstandards.org

[33] Francescutti LH, Sauve M, Prasad AS. Natural disasters and healthcare: Lessons to be learned. In Healthcare management forum. 2017 Jan.;**30**(1):53-55. DOI: 10.1177/0840470416679338

[34] Baker MS. Creating order from chaos: Part I: Triage, initial care, and tactical considerations in mass casualty and disaster response. Military medicine. 2007 Mar 1;**172**(3):232-236. DOI: 10.7205/MILMED.172.3.232

[35] McCann, DGC. Introduction to disasters and disaster planning. in Kollek, D, editor. Disaster Preparedness for Health Care Facilities. PMPH-USA. 2013. p. 1-9.

[36] Hu Q, Knox CC, Kapucu N. What have we learned since September 11, 2001? A network study of the Boston marathon bombings response. Public Administration Review. 2014 Nov;**74**(6):698-712. DOI: 10.1111/puar.12284

[37] Alexander DE. 2030: Adapting to complexity. Haznet 2020;14(1). http://haznet.ca/haznet-magazine-spring-2020-issue/

[38] Juffermans J, Bierens JJ. Recurrent medical response problems during five recent disasters in the Netherlands. Prehospital and disaster medicine. 2010 Apr 1;**25**(2):127 http://citeseerx.ist.psu.edu/viewdoc/download?doi=10.1.1.176.1952&rep=rep1&type=pdf

[39] Clarke L. Mission Improbable: Using Fantasy Documents to Tame Disaster. University of Chicago Press; 1999 May 15

[40] Skryabina E, Reedy G, Amlot R, Jaye P, Riley P. What is the value of health emergency preparedness exercises? A scoping review study. International journal of disaster risk reduction. 2017 Mar 1;**21**:274-283. DOI: 10.1016/j.ijdrr.2016.12.010

[41] Steigenberger N. Organizing for the big one: A review of case studies and a research agenda for multi-agency disaster response. Journal of Contingencies and Crisis Management. 2016 Jun;**24**(2):60-72. DOI: 10.1111/1468-5973.12106

[42] Berlin JM, Carlström ED. Collaboration exercises—The lack of collaborative benefits. International Journal of Disaster Risk Science. 2014;**5**(3):192-205. DOI: 10.1007/s13753-014-0025-2

[43] Peterson DM, Perry RW. The impacts of disaster exercises on participants. Disaster Prevention and Management: An International Journal. 1999;**8**(4):241-255. DOI: 10.1108/09653569910283879

[44] Berlin JM, Carlström ED. Collaboration exercises: What do they contribute? –a study of learning and usefulness. Journal of Contingencies and Crisis Management. 2015 Mar;**23**(1):11-23. DOI: 10.1111/1468-5973.12064

[45] Hsu EB, Jenckes MW, Catlett CL, Robinson KA, Feuerstein C, Cosgrove SE, et al. Effectiveness of hospital staff mass-casualty incident training methods: A systematic literature review. Prehospital and disaster medicine. 2004 Sep;**19**(3):191-199. DOI: 10.1017/S1049023X00001771

[46] Verheul ML, Dückers ML. Defining and operationalizing disaster preparedness in hospitals: A systematic literature review. Prehospital and disaster medicine. 2020 Feb;**35**(1):61-68. DOI: 10.1017/S1049023X19004187

[47] Jamieson T. Disastrous measures: Conceptualizing and measuring disaster risk reduction. International

Journal of Disaster Risk Reduction. 2016 Oct 1;**19**:399-412. DOI: 10.1016/j.ijdrr.2016.09.010

[48] Beerens RJ, Tehler H. Scoping the field of disaster exercise evaluation-a literature overview and analysis. International Journal of Disaster Risk Reduction. 2016 Oct 1;**19**:413-446. DOI: 10.1016/j.ijdrr.2016.09.001

[49] Lagadec P. Learning processes for crisis management in complex organizations. Journal of Contingencies and Crisis management. 1997 Mar;**5**(1):24-31. DOI: 10.1111/1468-5973.00034

[50] Hermann MG. In conclusion: The multiple pay-offs of crisis simulations. Journal of Contingencies and Crisis Management. 1997 Dec;**5**(4):241-243. DOI: https://doi.org/10.1111/1468-5973.00062

[51] Berlin JM, Carlström ED. The three-level collaboration exercise–impact of learning and usefulness. Journal of Contingencies and Crisis Management. 2015 Dec;**23**(4):257-265. DOI: https://doi.org/10.1111/1468-5973.12070

Chapter 2

Flood Disaster Hazards; Causes, Impacts and Management: A State-of-the-Art Review

Frank Jerome Glago

Abstract

Floods are among disasters that cause widespread destruction to human lives, properties and the environment every year and occur at different places with varied scales across the globe. Flood disasters are caused by natural phenomena, but their occurrences and impacts have been intensified through human actions and inactions. The practice of flood disaster management have evolved over the years from traditional approaches of ad-hoc response measures to integrated approaches involving technologically advanced tools in flood disaster awareness, preparedness and response measures. This chapter proffers understanding into flood disaster awareness, preparedness and management, mitigation and adaptation strategies. Most importantly, the chapter presents a review on the relevance of modern technological tools namely Geographic Information System, Remote Sensing, Internet of Things and Big Data, that are available to flood managers, in the creation of efficient early warnings and Flood decision support systems that elevates the resilience of societies to flood disasters.

Keywords: flood disaster, disaster risk awareness, disaster risk preparedness, disaster management, early warning system, resilience, technology

1. Introduction

Disasters can result from forces of nature which may be aided by human actions. Some disasters build up slowly while others may happen suddenly and unexpectedly. Flood disasters can be classified among the quick and sudden disaster types, but are among few in this category that can be well predicted, anticipated and controlled to a great extent.

Floods, like other disasters, do not qualify to be labelled 'disasters' by the mere virtue of their happenstances. They do become disasters when they cause damage or adverse effects to human lives, livelihoods and/or properties. Floods are probably the widest spread among the various disaster events that occurs in most countries and causes the most deaths [1]. Floods, like other disasters, have the ability to cause widespread disturbances in communities, and alter the way of life of people in the affected areas.

The word flood originated from the old English word 'flod' akin to the German word 'flut' and the Dutch word 'vloed' seen as inflow and float of water [2, 3]. The Oxford Reference Dictionary (ORD) defines flood as an overflowing or influx of water beyond its normal confines. Floods usually happen when the volume of water within a water body, say, a river or a lake, exceeds its total carrying capacity and as a result, some of the water flow outside the normal perimeter of the water body. Floods occur in almost every part of the world with different intensities and effects. Some of the most notable floods that have occurred include the 1981, 1991 and 2002 floods along the Chiang Jiang (Yangtze) river in China, the Mozambican floods in 2000, the 1983 and 1993 floods on the Mississippi river [2].

In the summer of 2005, the remarkable flooding brought by Hurricane Katrina which caused more than $ 108 billion in damages, constitute the costliest natural disaster in U.S. history [4, 5]. Identified different types of floods namely riverine floods, localized and urban floods, normal flood (e.g. 1-year flood), medium flood (e.g. 5-year flood), severe floods, and catastrophic floods. It is indicated that floods can also be distinguished by their style of occurrence [2]. Flash floods occur when water quickly sweeps over an area which is difficult to deal with and it is not easy to predict the amount of rain expected within the spatial area over a short period of time [2].

Regional floods occur when rain falls over a large area for days or weeks causing river levels to rise quickly and fall slowly usually inundating large areas and causing widespread economic losses [6]. Flash floods are also referred to as upstream floods and regional floods, downstream floods [7].

There are varied effects of floods. The primary effects of flooding include physical damage to buildings and weakening of structures [2]. There are instances of loss of human lives and livestock, and the outbreak of disease epidemics. Other effects include instant losses of entire harvest as in the Mozambique flood in 2000 and northern Ghana floods of 2007. Whilst the effects of floods have come to be highly perceived in the negative, it is also true that floods are not entirely of damaging impact on human beings. Flooding can be beneficial such as making the soil more fertile and providing nutrients. Periodic flooding was essential to the development of some of the ancient civilizations especially those along the Tigris-Euphrates rivers, the Nile river, Indus river among others [2].

2. Causes and impacts of flood disasters

2.1 Cause of floods

Floods happen when soil and vegetation cannot absorb water from downpours. Floods also occur when a river outbursts its banks and the water spills onto the floodplain. Natural processes such as hurricanes, weather systems and snowmelt can cause floods. Other floods following tsunamis and coastal surges have natural causes like earthquakes in the seabed and high tides attributed to the pull of the moon [2]. There are many human-induced causes of flooding.

Urbanization has also become a major cause of flooding in cities [8], such that, a river is more likely to flood when its drainage basin is in an urban area. Inadequate drainage in some urban areas is a major cause of flooding [3], while in others, it is the lack of proper management of the drainage systems. Unplanned urban living has been identified as a significant contributor to flooding events in many developing countries. In a study into causes of flooding in Asamankese in the Eastern region of Ghana by [9] for instance, a resident succinctly summarises the problems as below;

> *[T]he main problem in Old Zongo and Abaase areas is the gutters. The gutters are not enough to carry the water when it rains heavily, and secondly, they pour*

> *so much rubbish in the gutters, so some of the gutters are also full of rubbish. So, when it rains heavily, where will the water go, it must flood the area. ....the way we build in this area too is a problem. I even think government is not hard on people so we just build anyhow in the waterways. We in this area also experience floods but it is not serious like in Old Zongo areas, that is why we are always trying to tell people here not to build in the waterway, because of what is going on in Old Zongo and Abaase.*

This summarises the major contributions of improper managed urbanisation to flooding, a phenomena which characterises many developing countries. In Ghana, for example, perhaps the most devastating flood in the history of the country occurred in its capital, Accra, on 3rd June, 2015 where 159 people lost their lives and several people rendered homeless [10]. NADMO [11], suggests that although Ghana is vulnerable to certain disasters, flooding has become the major disaster the country has suffered in recent years especially in its urban areas due to improper management of these spaces [12].

Figure 1 shows the nature of some gutters in the urban areas of Ghana. **Figure 1** shows a partially completed drainage in the flood prone zone of Asamankese, in the Eastern region of Ghana. Most of the gutters in the community, like **Figure 1**, are left open and easily gets silted by inflow of sand and other waste materials. The situation of improper management of urban spaces is worse in the major Central Business Districts in many developing countries. Plastic wastes and other debris have been left to clog urban drainage which results in flood disaster when heavy rains are experienced.

2.2 Impacts of flood disasters

What flood events share in common, is their ability to cause widespread community disruption, displacement, economic loss, property damage, deaths, injury as well as profound emotional suffering. Infrastructure and property, agricultural endeavours as well as historical and cultural sites may also be affected in flood disasters.

According to the United Nations Regional Coordinator in Dakar (October 2007) the worst flooding in 30 years that battered West Africa from July 2007 caused more than 210 death and affected more than 785,000 people [12].

The aftermaths of flood disasters in Ghana are the large-scale destruction of infrastructure, displacement of people from their dwellings, the loss of human lives, outbreak of diseases and water-borne infections, chemical exposure due to toxic pollutants being released into flood waters, huge loss of investments among other things.

Africa, which is one of the poorest continents in the world (in terms of GDP growth and income) has seen an increase in flood disasters in recent times [9]. For instance, torrential rains and flooding affected 600,000 people in 16 West African nations in September 2009 [13]. Countries with most devastating impacts were Burkina Faso, Senegal, and Niger. Another instance include the 2007 floods that displaced more than a million people in Uganda, Ethiopia, Sudan, Burkina Faso, Togo, Mali, and Nigeria, which claimed over 500 lives, and the 2008 floods in Mozambique which killed seven people and displaced tens of thousands residence [14]. Heavy seasonal rainfall starting in December 2014 also caused flooding in southern Africa [15]. As of January 2015, 135,000 people were affected by flood hazard in Malawi, Mozambique, Madagascar and Zimbabwe [15].

The impact of flooding varies both spatially and temporally. It could also be direct or indirect. Rahman [16] indicated that the direct impacts of floods are

Figure 1.
Showing a partially completed drain in the Asamankese community, Eastern region of Ghana. Source: Author.

closely related to the depth of inundation of floods water. The extent of a flood has a direct relationship for the recovery time of crops, pastures and the social and economic dislocation impact to populations. The impact of floods is considered far reaching with the aftermath effects such as flood-induced disease epidemics. Disease outbreak is common, especially in less developed countries. Malaria, Typhoid and Cholera outbreaks after floods in tropical countries are also common [17]. [9] further stated that physical damage to property is one of the major causes for tangible loss in floods. This includes the cost of damage to goods and possessions, loss of income or services in the floods aftermath and clean-up costs. Some impacts of floods, on the other hand, are intangible and are hard to place a monetary figure on. Intangible losses also include increased levels of physical, emotional and psychological health problems suffered by flood-affected people.

According to [15] the cumulative number of people affected by rains and floods in 2007 in Southern Africa was more than 194,103 persons. This included 60,995 in Malawi (Mostly damage to property and crops), 94,760 people in Mozambique (all were evacuated into resettlement camps); more than 16,680 in Zambia (1890 persons had temporary accommodation, the rest were taken in by host families); and 15,168 in Zimbabwe. An estimated additional 4000 people had been affected in Lesotho and another 2500 persons in Swaziland.

Extreme events affect both the formal and informal economies, making it difficult to assess impacts which include direct and indirect ones. Depending on how well they are constructed and the severity of the event, buildings may be partially or totally destroyed by flooding. A look at **Figure 2** will explain the partial damage that often happens to buildings as a result of flooding.

Flood destructions also hit roads and cause delays to infrastructure development initiatives and political processes [18, 19] observed that the economic impact of natural disasters shows a marked upward trend over the last decades. The hazards

Figure 2.
Depicting the impact of flood events o residents' household in Asamankese, Eastern region of Ghana. Source: Author.

tend to hit communities in developing and least developed countries more. Flood disasters have led to the loss of human life, destruction of social and economic infrastructure and degradation of already fragile ecosystems [20]. It follows therefore that social impacts include changes in people's ways of life, their culture, community, political systems, environment, health and wellbeing, their personal and property rights and their fears and aspirations.

Rahman [21], established that social impacts of floods cause significant problems for the long term functioning of specific types of households and businesses in affected communities. The type of construction influenced the extent of flood damaged (e.g. thatched homes versus concrete high rise buildings will experience different degrees of impact). It follows that vulnerability is a key element in assessing the impact of floods. Different population segments are exposed to varied relative risks because of their socioeconomic conditions of vulnerability. Because of this, disaster reduction has become increasingly associated with practices that define efforts to achieve sustainable development. The links between flood disaster and economic systems, have become another pillar of consideration for sustainable development. Floods, however, cannot be totally prevented but their devastating impacts can surely be significantly minimized if advance warning of the event is available.

3. Flood disaster risk awareness

Disaster risk awareness is the extent of common knowledge of a person or group of persons about disaster risks, the factors that lead to disasters and the actions that can be taken individually or collectively to reduce vulnerabilities to hazards. It also includes the need to build and increase the knowledge and understanding of the many issues about disaster risk reduction, to build the capacity of the people who learn and teach others about the disaster [9]. Changes in patterns of human behaviour and decision-making at all levels of government and society could, therefore, lead to a substantial reduction in disaster risk [22]. In this respect, recent experience has shown that public awareness of natural hazards and disaster risk reduction education constitutes a foundation and pre-requisite for effective catastrophic risk management

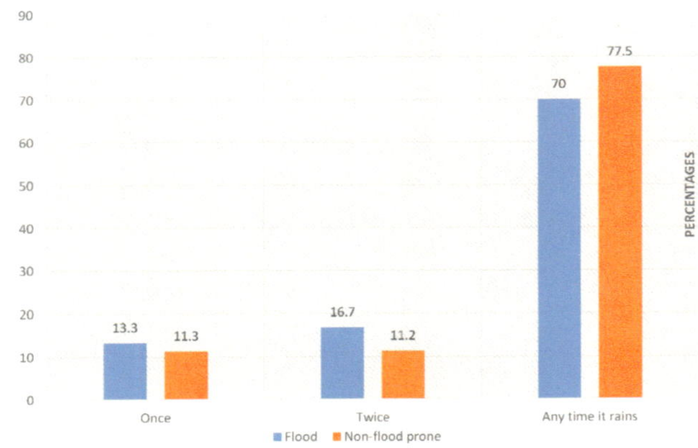

Figure 3.
Showing respondents awareness to flood disaster risk at Asamankese, Eastern region, Ghana. Source: Author.

strategies at country and regional levels. More importantly, by influencing human actions and perceptions through societal behaviour and behavioural adaptation, information and education can increase flood risk awareness and play a more effective role in reducing the costs of catastrophes associated with natural perils [22].

To proffer an understanding on the flood risk awareness of residents in the Asamankese Municipality of Ghana, [9] surveyed some residents and sought to know how likely the area was susceptible to flood through the major rainy season from March to July. **Figure 3**, which summarises the respondents result, pointed out that an overwhelming proportion of respondents (70.0% and 77.5% within the flood prone and Non-flood prone zones respectfully) indicated that flooding has become a regular phenomenon, and the community was likely to be flooded every time it rained.

Awareness is a very crucial element for a society to effectively adapt to a flood risk. As stated by [23] awareness is diminished when the provision of an appropriate information is minimal or when memories of past experiences or events are diminished. Awareness can generally be uplifted through efforts that are centred on local issues, contain the simple solution to reduce the flood risk and are repeated on a regular basis [24].

Scholars like [25] posit that worry is an important risk characteristic that serves as a normative value for awareness. Society can be aware of a flood risk, however, if it is not afraid of the risk, it will not take any action to prepare for it. A higher level of worry is more likely to result in a higher level of awareness and preparedness. There is a positive correlation between these two variables. This was reinforced by [26] with their assertion that most people become aware and prepared for disasters based on their previous personal experiences with flood disasters. Early warning information can, therefore, allow the disaster managers to be pre-informed and take steps which may significantly reduce the loss of life and damage to property.

4. Flood disaster preparedness and resilience strategies and practices

The argument now is that adequate preparation can make it possible to significantly reduce the impacts of flood disasters through a good understanding of preventive action as well as knowledge of some life-saving techniques during disasters [27]. Nowhere has the issue of floods become a developmental issue than in poor and developing countries where systemic problems and institutional constraints have increased vulnerability (social, economic and physical) to flood risk and thus, reducing resilience to flood disasters.

Disaster preparedness is defined as the state of taking measures to reduce to the minimum level possible, the loss of human lives and other damages from flood disasters through prompt and efficient actions of response and rehabilitation. That is, preparedness is to put in place the necessary measures for effective and timely response to an event. The objectives of preparedness are to ensure that appropriate mechanisms and resources are in place to assist those afflicted by the disaster and enable them to help themselves [9].

Flood disaster preparedness consists of a wide range of activities and protective measures that might be instigated from the physically or procedurally. Preparedness is very important in the disaster management process, and includes the knowledge, capacities, activities and measures carried out in advance by governments, professional response and recovery organizations, communities and individuals to effectively anticipate, respond to and recover from the impacts of imminent or current disaster situations or conditions.

The [26], conceives preparedness as a medium-term plan that involves the development and the implementation of disaster management plans. It involves the development and implementation of early warning systems, resource inventory and stockpiling of resources, coordinating of agencies and ensuring evacuation plans work. Preparedness is seen as tools for ensuring the effective coordination and enhancement of capacities to prevent, to protect, to respond to, recover from and mitigate the devastating effects of natural and man- made disasters [9].

Apart from personal experiences informing residents' awareness and preparedness towards flood disasters, external factors such as occupation, level of education of an individual, radio programs and community meetings serve as conduits for disseminating flood information. Hence, these factors may act to increase awareness and preparedness levels to flood disasters.

Studies show that residents' awareness of flood disaster is usually high. For instance, in [9]'s study of residents' level of flood disaster awareness in Asamankese, in Ghana, showed that more than 65 percent of residents in both the Flood prone and Non-flood prone settlements ranked themselves to be at a high risk of flood disasters. Furthermore, the study showed that, the awareness of human factors that exacerbate their risk to flood was also high. However, preparedness of residents' in most cases were poor, and in Asamankese, like in other developing countries, victims of flood had to usually depend on extended social networks, and government institutions for support to regain their livelihoods after being hit by flood events, a situation which results from their ill preparedness to flood disasters especially financially.

5. Strategies and technologies for improved flood disaster management

5.1 Flood disaster management strategies

Flood Disaster Management Strategies refer to a bundle of processes and activities that are aimed at reducing the overall impacts of floods on societies. Flood management needs to be considered within the overall national development planning strategy of every country and must involve strategic institutional arrangements and collaborations for a sustainable flood management.

The management of floods as problems in isolation almost necessarily results in a piecemeal, localized approach (World Meteorological Organization [28]). The flood disaster management process should also be coordinated with efforts made in closely related fields. For example, the disaster mitigation process should consider human health impacts during flooding (e.g. cholera, malaria), thereby more effectively address a health issue that arises during and after flooding [28].

The management of floods takes several approaches ranging from traditional approaches to integrated approaches. The traditional management response to a severe flood was typically an ad- hoc reaction, the quick implementation of a project that considered both the problem and its solution to be distinct and self-evident. Traditional approaches usually gives no thought to the consequences for upstream and downstream flood risks [28]. Thus, flood management practices have largely focused on reducing flooding and reducing the susceptibility to flood damage. Traditional flood management has employed structural and non-structural interventions, as well as physical and institutional interventions. These interventions have occurred before, during and after flooding, and have often overlapped.

There has been a paradigm shift in flood management. Traditionally, controlling floods has always been the main focus of flood management, with the emphasis on draining flood water as quickly as possible, or storing it temporarily, and separating the river from the population through structural measures such as dams and levees [28].

The concept of integrated flood management has led to a paradigm shift: absolute protection from floods is a myth, and focus should aim at maximizing net benefits from the use of flood plains, rather than trying to fully control floods [28].

A proactive approach towards the management of floods over a traditionally reactive approach is rapidly gaining recognition among flood managers. The proactive approach does not treat floods only as an emergency or an engineering problem, but as an issue with social, economic, environmental, legal and institutional aspects. The proactive approach is not limited to a post-event reaction but includes preparedness (including flood risk awareness) and response measures to flood management at different stakeholders' levels [28].

Recent calls in flood management are geared at taking a transboundary approach since floods do not respect borders; neither national nor regional or institutional [29]. The great advantages of transboundary cooperation are that it broadens the knowledge/information base, enlarges the set of available strategies and enables better and more cost-effective solution. Furthermore, widening the geographical area considered by basin planning enables measures to be located where they create the optimum effect [29].

5.2 Early warning systems

The term 'early warning' is used in many fields to describe the provision of information on an emerging perilous circumstance where that information can enable action in advance to reduce the risks involved. The early warning system comprise the set of capacities needed to generate and disseminate timely and meaningful warning information to enable individuals, communities and organizations that are threatened by hazards to take necessary preparedness measures and act appropriately in sufficient time to reduce the possibility of harms or losses [30].

Early warning systems exist for natural geophysical and biological hazards, complex socio-political emergencies, industrial hazards, personal health risks and many other related hazards. Studies have demonstrated that disaster prevention can pay high dividends and found that for every Euro invested in risk management, broadly 2 to 4 Euros are returned in terms of avoided or reduced impacts on life, property, the economy and the environment [31]. Early warning systems can be set up to avoid or reduce the impact of flood hazards and other disasters such as, landslides, storms and forest fires and volcanic eruptions. The significance of an effective early warning system lies in the recognition of its benefits by the members of the general public.

Early warning is a major element of flood disaster risk reduction. It saves life and reduces economic and material losses from flood disasters. To be effective, community-based early warning systems need the active involvement of the community people, a strong public education on and awareness of risks, an effective communication system ensuring a constant state of preparedness [31]. Early warning systems contribute with other Disaster Risk Reduction (DRR) interventions to protect and support sustainable economic development and early detection of undesirable situations. The society benefits from early warning systems if they are in place. Many governments have failed to take early warning into account while formulating their development and disaster risk reduction policies. Subsequently, it results in heavy losses to human lives and economic entities when disasters strike [31].

A people-centered early warning system necessarily comprises four key elements: (I) knowledge of the risks; (II) monitoring, analysis, and forecasting of the hazards; (III) communication or dissemination of alerts and warnings, and (IV) local capacities to respond to the warnings received. The expression "end-to-end warning system" emphasizes that early warning systems need to span all steps from hazard detection to community response. It is essential to link downstream communities and upstream communities for the effective operation of an early warning system [31]. There are several instances where early warning systems have helped to mitigate the impact of disasters. As an example, the Bangladesh cyclone preparedness program has successfully warned, evacuated and sheltered millions of people from cyclones since its inception in the early 1970s by the International Federation, the Bangladesh Red Crescent Society and the government of Bangladesh. In the Caribbean, during 2004's hurricane season, most countries successfully alerted their populations to approaching storms and saved many lives as a result. The key to their success was putting people, not just technology, at the centre of their warning systems. As a result of early warning systems, there were no deaths reported in La Independencia, Guatemala during the hurricane season in October 2005 [26].

The importance of early warning has been underlined in various UN General Assembly resolutions as a critical element of disaster reduction. Early warning received very high attention after the 26 December 2004 tsunami, when it became clear that a tsunami warning system and associated public education could have saved thousands of lives. The UN Secretary-General in his report; *In Larger Freedom: Towards development, security and human rights for all*, proposed that the United Nations system should take a leadership role in developing comprehensive global capacities for systematic people-centered early warning systems, which would cover all hazards for all countries and communities. Subsequently, he requested that a global survey is undertaken, with a view to advance the development of a Global Early Warning System (GEWS) for all natural hazards [32]. Thus early warning systems are very important for disaster management and should be a priority at global, national and local levels.

5.3 Flood disaster mitigation strategies

Flood Disaster Mitigation measures tend to be potentially more efficient long term sustainable solutions to water-related problems and should be enhanced, in particular, to reduce the vulnerability of human beings and goods exposed to flood risk. Flood forecasting and warning is a prerequisite for successful mitigation of flood damage. In the field of environmental engineering, flood mitigation involves the managing and control of flood water movement, such as redirecting flood run-off through the use of floodwalls and flood gates, rather than trying to prevent floods altogether. It also involves the management of people, through measures such as evacuation and dry/wet proofing properties for example. The mitigation of flooding can be done on an individual, community and at city authority or national levels.

5.4 Flood disaster adaptation strategies

Flood disaster adaptation refers to actual adjustments made that are geared towards mitigating the severity of flood disasters. Flood disaster adaptation strategies vary from before flood adaptation, during flood adaptation, to post flood adaptation strategies. It also ranges from individual, community, to citywide adaptation strategies. Discussions on flood adaptation strategies pointed out that embankments, for instance, either concrete or sandy may be constructed to prevent water from entering residential houses [33]. Adaptation options that would be effective for flood disaster in developing nations include Environmental policy reforms, changes in urban and housing design, removal of laws that can inadvertently increase flood vulnerability [34]. Capacity building is also required to integrate climate change and its impact on urban development planning, engaging local communities, raising public awareness and education on climate change and enabling wider representation at stakeholder meetings. Planting of vegetative cover to reduce runoff speed, terracing hillsides to slow flow down hills as well as control of man-made channels to divert flood water among others, serve as adaptation strategies. Generally, adaptation strategies adopted in flood disasters range from structural to non-structural [35].

5.5 Monitoring, evaluation and mainstreaming of flood disaster management

Being able to count on institutionalized capacities to mobilize and coordinate resources when and where they are needed is crucial in all phases of the disaster cycle, sometimes with very little room for delay or errors of judgment. Coordination among agencies and stakeholder groups is important for flood mitigation, in particular, the design and execution of programmes and policies to help address underlying causes of extreme vulnerability [36]. Monitoring of activities is necessary because there is often the need to link responsibilities and budgets for programmes over time. The performance of institutions and organisations responsible for disaster management should be monitored and evaluated on a regular bases. The relevance of monitoring and evaluation as a means of reducing flood disaster events cannot be overemphasised.

The capacity to monitor and evaluate flood prevention, mitigation, relief and recovery operations and institutional arrangements would create opportunities for learning and improve the accountability of authorities [36]. Monitoring is a key element in pre-flooding, flooding and post flooding stages of flood disaster management. Evaluation goes hand in hand with monitoring to assess the impact of flooding and the effects of key interventions engaged in mitigating the impact of flooding on people, infrastructure, and the environment in general. In a study on the environmental aspects of integrated flood management, [28], noted that adaptive management requires continuous monitoring of the state of the environment and evaluation at regular intervals.

The importance of monitoring has been recognized from various perspectives [28] Pre-plan monitoring of various natural processes provides the basic input for assessment of resource, risks and development options. Monitoring at a development planning level is based on actions taken in line with selected plan and factors of environmental impacts indicated in environmental assessment at the strategic level [28]. In the context of awareness of flood hazards, monitoring encapsulates awareness of causes and how these causes change over time, knowledge of interventions and how these interventions are shaping the frequency and nature of flood events in an area. Monitoring is essential in flood management from a first-hand point of providing timely and efficient early warning information. Immediate and post-implementation monitoring is important in order to assess whether the flood management measure has succeeded [28].

Mainstreaming of disaster risk reduction into development planning, policy, and implementation should be at the heart of every sustainable Development Planning agenda. Disasters, such as floods, have an enormous impact on development. There is, therefore, the need for mainstreaming disaster planning into development planning. The importance of mainstreaming is also recognized by the Hyogo Framework for Action (HFA) adopted at the World Conference for Disaster Risk Reduction (WCDRR), where integration of disaster risk reduction into the development programmes is a priority [5]. There has been increasing recognition by both governments and donors for the need to mainstream disaster risk reduction into development planning [37]. Mainstreaming disaster planning into development planning considers risks emanating from natural hazards in medium-term strategic development frameworks, in legislations and institutional structures, in sectoral strategies and policies, in budgetary processes, in the design and implementation of individual projects and in monitoring and evaluating all of the above [37].

5.6 Sustainable flood management

The concept of sustainable development is firmly rooted in all flood management. Sustainable flood management involves: ensuring quality of life by reducing flood damages but being prepared for floods, mitigating the impact of risk management measures on ecological systems at a variety of spatial and temporal scales, the wise use of resources in providing, maintaining and operating infrastructure and risk management measures, maintaining appropriate economic activity [38]. Sustainable flood management as a concept is not new, its methods have been practiced on many continents for years [39]. With increasing scrutiny of traditional engineering solutions, there is a growing realisation throughout the world that there is a huge and urgent need for pro-active and sustainable flood management solutions.

The notion of sustainability in the context of flood management is still rather ambiguous but generally embraces economic, environmental and social objectives. Sustainable flood management therefore refers to the provision of possible social and economic resilience against flooding, by protecting and working with the environment, in a way which is fair and affordable both now and in the future [40]. In practice a sustainable approach should integrate a range of flood management requirements using best practices and involving the economics of a scheme, good planning, understanding flood generation processes, protecting natural environments and working with communities [39]. Sustainable flood management is, therefore, an integrated set of procedures linked into a physical catchment.

6. Technological advancements in flood disaster management

Advanced technologies have been developed and integrated into higher institutional level decision support systems to aid the prediction, monitoring and management of flood disasters in some countries. These advanced flood decision support systems' architecture include technologies such as Geographic Information Systems, remote sensing and photogrammetry, and hydrologic models.

The Flood Decision Support System (FDSS) refers to interactive computing environment designed for specific contexts which include interlinked models/analytical tools, databases, graphical user interfaces and other systems. The FDSSs according to [41] have the potential to improve flood disaster assessment and mitigation through improved data collection and rapid dissemination of flood information to affected areas. For an effective FDSS on the technology aspect of disaster management, analysts have to ensure effective interoperability of the technologies. This will ensure that, all aspects of the technology that singularly may be responsible for data capture, storage, manipulation, analysis, retrieval or display of information, work in a smooth interwoven network and relay information to other parts of the system without technical hindrances to ensure the overall goal is achieved.

There are three main components to the Flood Disaster Support System. These include the Database component, the Modelling component, and the Display component also known as the Graphical User Interface (GUI) component. The Database component of the FDSS comprises the data used in the modelling functions. This component uses to tools to capture and store flood related data. Some data stored include historical rainfall data, geological data, soil and ecological data, population data, boundary and administrative data. Tools used in data capture for the Database varies depending on the data to be captured. For example, Remote sensing techniques are used to capture satellite data on flood zones, flood buffer zone monitoring. Sensors are also deployed to monitor flow, volume and carrying capacities of rivers while rain gauges capture precipitation volumes. These data may be complemented with census data on population and livelihoods of residents. All these various data are kept in the Database component of the DSS.

The second component of the FDSS are functions of analytics and modelling. Various analysis are carried out and the data in the database taking through several processes of manipulation. These processes of data manipulation and analysis differ in approach and are tailored to meet various goals in the decision making process. Prominent among the tools used at this stage is Geographic Information Systems (GIS) tools. Regarding flood modelling, advanced tools available to flood managers include advanced technological tools in soft computing, for instance, evolutionary computing, as well as probabilistic predictions techniques of inundation recurrence intervals [41]. These tools afford flood managers varieties of techniques that can be applied in simulation, modelling, analysis and management of flood.

The User Interface component of the FDSS provides flood decision makers an interactive graphical interface, enabling users to query the data stored in the system. It again enable users to display and visualise the models and reports from the manipulations of the data. This component of the advanced FDSS enables users to prepare and appreciate maps and animations of the hydrologic phenomena being studied.

6.1 Remote sensing and geographic information systems in disaster management

Advances in remote sensing tools and techniques over the past few years have provided disaster managers, especially flood disaster managers with powerful tools in the acquisition of flood sense data, in forecasting and monitoring of flood occurrences and in the management of watersheds, rivers and wetland areas.

Remote sensing refers to the Science of obtaining information about objects, areas or phenomena from a distance [42]. Typically, these information are collected through sensors that are planted on aircrafts or satellites. In flood disaster management, remote sensing can be applied to monitor and map events such as changes in river volume, changes in coastline, map wetlands and flood prone zones and boundaries of inundation.

A Geographic Information System(s) (GISs) refers to a framework for gathering, managing and analysing location-based data. This framework is used to analyse and organize several distinct layers of location-based information into concise visualizations through maps and 3D scenes. Ultimately, GISs present powerful capabilities that proffer deeper insights into data, which may include revelation of patterns and relationships for smarter decision making [43].

Reliable flood maps are therefore produced using GIS techniques and remotely sensed data to manage floods. GIS tools aid in the preparations to Digital Elevation Models (DEMs) for high level hydrological modelling using sensors such as The Light Detecting and Ranging (LiDAR) sensors.

With the help of data interpretation techniques of GIS, remotely-sensed imageries are interpreted to create suitable flood risk mitigation frameworks and FDSSs. Although flood disasters have increased in scale and frequency in recent years, there has been a commensurate improvement in flood data capturing and analyses techniques, that when applied in time, can significantly mitigate the risks and impacts of floods. As summarised in **Table 1**, GIS and RS are of great importance in the pre and post disaster management processes.

6.2 Internet of Things (IoT) and Big Data in flood disaster management

The Internet of Things (IoT) refers to a network of devices connected over the internet to sense, track and respond to issues. Patel and Patel [45] defines the IoT as "*a type of network to connect anything with the internet based on stipulated protocols through information sensing equipment to conduct information exchange and communications in order to achieve smart recognitions, positioning, tracing, monitoring and administration.*"

The network of physical objects are able collect data on a regular bases and in a structured form, perform high level analysis and predict changes, as well as initiate actions based on results from the analyses. IoT is hence a powerful technological tool that can provide a wealth of high level intelligence which is needed in planning and management.

There are three levels of IoT. The first is people to people interconnectivity, the second is people to machine interconnectivity and the third being machine to

Phases in Flood Disaster Management	*Relevance of GIS and RS*
Flood Prevention	Capturing imageries for hazards and risks assessment. Preparation of flood prone maps. Management of large volume of flood sense data.
Flood preparedness	As tools for planning evacuation routes. Designing centers for emergency operations. Integrating and Simulating live satellite data with other dataset to inform early warning systems.
Flood relief	Planning and execution of search and rescue operations. Planning distribution of relief items to flood victims
Flood rehabilitation	Flood impact assessment. Rehabilitation planning.

Table 1.
Showing GIS and Remote Sensing Application in flood disaster management. Source: Authors' Construct with reference to [44].

machine or things to things interconnectivity [45]. In all interconnectivity of things and people, the internet remains the main driver. This interconnectivity of Things, enables the swift transmission of meteorological, hydrological and geological data pertaining to flood events.

In flood disaster management, providing a quick feedback on the occurrence of floods can be a great step in preventing and mitigating flood disasters and their impact on livelihoods in society. Deploying IoT in flood management puts disaster managers at a position to create enhanced early warning systems that do not only measure the water levels and the speed of inundation, but early warning systems that could also send alerts to residents and flood managers through mobile phones and other personal electronic devices, and additionally, prescribe the best prevention and mitigation strategies based on data such as direction of runoff, speed of rise of water levels and the time at the disposal of residents to take necessary action.

Big Data on the other hand, refers to "*the evolution and use of technologies that provide the right user at the right time with the right information from a mass of data that has been growing exponentially for a long time in our society*" [46]. Digital data collection has not only seen growth in volume but also in variety in storage formats, hence Big Data is often described as high-volume, high-velocity and/or high variety information assets that demand cost-effective, innovative forms of information processing to enable enhanced insight, decision making and process automation [47].

Big data typically defines data that exceeds the storage, processing and computing capacity of conventional database [46]. Hence Big Data analytics typically involves automated software that assist in the collection, organisation and analysis of the data being generated to discover trends, correlations and other useful results to prompt necessary action.

Through Big data process automation, precipitation data, soil moisture data, temperature data, water content data of water bodies, data on evapotranspiration, ground water data, etc., are collected and processed in real-time without human supervision to make predictions and early warnings about flood disasters' occurrence [47].

7. Conclusion

Flood disasters have had very devastating impacts on societies and have destroyed livelihoods and investments of staggering monetary value and importance to development. However, adequate involvement of technology are leading to the creation of people-centered early warning systems that enhances residents' awareness and preparedness to flood events to significantly reduce the adverse impacts of these disasters on people. This chapter discussed various aspects of flood disaster management including early warning systems, flood mitigation and adaptation strategies, the relevance of monitoring, evaluation and mainstreaming flood disaster management into national level development planning. The chapter again discussed and encourage the integration of advanced technological tools into the frontier of flood disaster management, as these tools have the capacity to capture, analyse and disseminate real-time flood data to all stakeholders to safeguard lives and precious investments.

Author details

Frank Jerome Glago
Akatsi College for Education, Akatsi, Ghana

*Address all correspondence to: frankjerome96@gmail.com

References

[1] Ghapar, A. A., Yussof, S., & Bakar, A. A. (2018). Internet of Things (IoT) architecture for flood data management. International Journal of Future Generation Communication and Networking, *11*(1), 55-62.

[2] Adams, A.G. (2008). Perennial flooding in the Accra Metropolis: The human factor. (Unpublished Master of Science Thesis). Kwame Nkrumah University of Science and Technology, Kumasi.

[3] Ahmad, K., & Moeeni, S. (2019). Recent flood risk scenario of Bihar: a preventive strategy. NDCWWC Journal (A Half Yearly Journal of New Delhi Centre of WWC), 8(2), 3-14.

[4] Blake, E. S., Landsea, C. W., & Gibney, E. J. (2011). The deadliest, costliest, and most intense United States tropical cyclones from 1851 to 2010. NOAA Technical Memorandum NWS NHC-6, NOAA National Hurricane Center. Dept. of Commerce (Available at http://www.nhc.noaa. gov/Deadliest_Costliest. shtml).

[5] Dodman, D., Bicknell, J., & Satterthwaite, D. (2012). Unjust Waters: Climate Change, Flooding and the Urban Poor in Africa: Ian Douglas, Kurshid Alam, MaryAnne Maghenda, Yasmin McDonnell, Louise McLean and Jack Campbell. In *Adapting Cities to Climate Change* (pp. 216-238). Routledge.

[6] Abbott, P.L., (2006), Natural disasters, 5th Edition. New York: McGraw-Hill Companies Inc.

[7] USGS, (2020). What are the two types of floods? Retrieved from https://www.usgs.gov/faqs/what-are-two-types-floods?qt-news_science_products=0#qt-news_science_products. Retrieved on 27th September, 2020.

[8] Zhang, W., Villarini, G., Vecchi, G. A., & Smith, J. A. (2018). Urbanization exacerbated the rainfall and flooding caused by hurricane Harvey in Houston. Nature, 563(7731), 384-388.

[9] Glago, F. J. (2019). Household disaster awareness and preparedness: A case study of flood hazards in Asamankese in the West Akim Municipality of Ghana. Jàmbá: Journal of Disaster Risk Studies, 11(1), 1-11.

[10] Daily Graphic (2015). Ghana among the worst hit by torrential rains, Daily Graphic, 28 July 2015.

[11] NADMO, (2015). Unpublished Annual Report 2015, NADMO, Asamankese.

[12] Oppong, B. (2011). Environmental hazards in Ghanaian Cities: The incidence of annual floods along the Aboabo River in the Kumasi Metropolitan Area (KMA) of the Ashanti Region of Ghana. (Unpublished Mphil. Thesis). Kwame Nkrumah University of Science and Technology, Kumasi-Ghana.

[13] Baldassare, G., Montanari A., Lins, H., Koutsoyiannis, D., Brandimarte Luigia, and Bloschl, G. (2010). Flood fatalities in Africa: From diagnosis to mitigation. Geophysical research letters, 37(22). doi:10.1029/2010GL045467, 2010.

[14] Church World Service, (2008). CWS situation report: 2008 Mozambique floods. Retrieved January 16, 2008, from http://reliefweb.int/report/mozambique/cws-situation-report-2008-mozambique-floods.

[15] OCHA, (2015). Situation report for Southern Africa floods. Retrieved January 16, 2015, from reliefweb.int/map/malawi/southern-africa-floods-and-cyclones-update-16-Jan-2015.

[16] Rahman, S. U. (2014). Impacts of flood on the lives and livelyhoods of people in Bangladesh: A case study of a village in Manikganj district (Doctoral dissertation, BRAC University).

[17] Mwape, Y. P. (2009). An impact of floods on the socio-economic livelihoods of people: A case study of Sikaunzwe Community in Kazungula District of Zambia. University of Free State Press.

[18] Octavianti, T., & Charles, K. (2019). The evolution of Jakarta's flood policy over the past 400 years: The lock-in of infrastructural solutions. Environment and Planning C: Politics and Space, 37(6), 1102-1125.

[19] Know Risk (2005). *United Nations, Geneva, Switzerland.* Available at: https://www.unisdr.org/we/inform/publications/654.

[20] Cui, Y., Cheng, D., Choi, C. E., Jin, W., Lei, Y., & Kargel, J. S. (2019). The cost of rapid and haphazard urbanization: lessons learned from the Freetown landslide disaster. Landslides, 16(6), 1167-1176.

[21] Rahman, S. U. (2014). Impacts of flood on the lives and livelyhoods of people in Bangladesh: A case study of a village in Manikganj district (Doctoral dissertation, BRAC University).

[22] Aerts, J. C., Botzen, W. J., Clarke, K. C., Cutter, S. L., Hall, J. W., Merz, B., ... & Kunreuther, H. (2018). Integrating human behaviour dynamics into flood disaster risk assessment. Nature Climate Change, 8(3), 193-199.

[23] Shen, X. (2009). Flood risk perception and communication in different cultural contexts-a comparative case study between Wuhan, China and Cologne, Germany. (Ph.D. Dissertation), University of Bonn.

[24] Poortinga,W., Bronstering, K., and Lannon, S. (2011). Awareness and perceptions of the risks of exposure to indoor radon: A population based approach to evaluate a radon awareness and testing campaign in England and Wales. Risk Anal., 31, 1800-1812.

[25] Miceli, R., Sotgiu, I., and Settanni, M. (2008). Disaster preparedness and perception of flood risk: A study in an Alpine valley in Italy. Journal of Environmental Psychology, 28, 164– 173.

[26] International Federation of Red Cross and Red Crescent Societies (IFRC), 2007, Disaster response and contingency planning guide, IFRC, Geneva.

[27] Ibem, E. O. (2011). Challenges of disaster vulnerability reduction in Lagos Megacity Area, Nigeria, disaster prevention and management. An International Journal, 20(1), 27-40.

[28] World Meteorological Organization (2011). .Partnerships in weather, climate and water for development: Integrated flood management. Fact sheet 11, Version 1-2011.

[29] UNECE (2009). Transboundary flood risk management: Experiences from the UNECE region. New York and Geneva: United Nation Publication.

[30] UNISDR, (2009). Global assessment report on disaster risk reduction: Risk and poverty in a changing climate, Geneva.

[31] Practical Action, (2014). Poor people's energy outlook 2014. Available at: http://policy.practicalaction.org/policy-themes/energy/poor-peoples-energy outlook/poor-peoples-energy-outlook-2014.

[32] UN-Chronicle, (2007). Global early warning systems needed: Creating partnership to cope with natural disasters. The Magazine of the United Nations. 64(2), 2007.

[33] Terungwa, C. U. and Torkwase, C. I. (2013). Current issues in flood disaster: Challenges and implications for science and technology to enhance environmental education. Academic Journal of Interdisciplinary Studies, 2(6), 61-65.

[34] Kolawole, O.M, Olayemi, A.B. and Ajayi, K.T. (2011). Managing flood in Nigerian Cities: Risk analysis and adaptation options – Ilorin City as a case study. Archives of Applied Science Research, 3(1), 17-24.

[35] Sakijege, T., Sartohadi, J., Marfai, M.A., Kassenga, G.R. and Kasala, S.E., (2014). Assessment of adaptation strategies to flooding: A comparative study between informal settlements of Keko Machungwa in Dar es Salaam, Tanzania and Sangkrah in Surakarta, Indonesia', Jàmbá. Journal of Disaster Risk Studies, 6(1), 131-141.

[36] Lebel, L., Nikitina, E., Kotov, V., Manuta, J., & Birkmann, J. (2006). Assessing institutionalized capacities and practices to reduce the risks of flood disasters (pp. 359-379). United Nations University Press, Tokyo.

[37] Miyan, A. (2014). Challenge of Mainstreaming Disaster Risk Reduction in Development Initiatives with Special Reference to Bangladesh. *An input paper of the 2015 Global Assessment Report on Disaster Risk Reduction, UNISDR, United Nations, Geneva*.

[38] Gouldby, B. and Samuels, P. (2005). Language of risk -project definitions. Floodsite report T32-04-01. Available at: http://www.unisdr.org/files/657_lwr1.pdf.

[39] Johnson, R. Watson, M. and McOuat. E. (2008). The way forward for natural flood management in Scotland, Scottish Environment link: MNV/WWF/0808/1038.

[40] FIAC, (2007). Sustainable flood management report, Scotland. Available at: http://reliefweb.int/report/mozambique/cws-situation-report-2008 Mozambique-floods.

[41] Levy, J. K., Gopalakrishnan, C., & Lin, Z. (2005). Advances in decision support systems for flood disaster management: Challenges and opportunities. Water Resources Development, 21(4), 593-612.

[42] National Oceanic and Atmospheric Administration, (2020). What is remote sensing? National Ocean Service, U.S Department of Commerce. Retrieved from https://oceanservice.noaa.gov/facts/remotesensing.html Retrieved on 20th July, 2020.

[43] ESRI, (2020). What is GIS? Retrieved from https://www.esri.com/en-us/what-is-gis/overview. Retrieved on 20th July, 2020.

[44] Faisal, A., & Khan, H. (2018). Application of GIS and Remote Sensing In Disaster Management: A Critical Review of Flood Management. In *International Conference on Disaster Risk Mitigation*.

[45] Patel, K. K., & Patel, S. M. (2016). Internet of things-IOT: definition, characteristics, architecture, enabling technologies, application & future challenges. International journal of engineering science and computing, 6(5).

[46] Youssra, R., & Sara, R. (2018). Big data and big data analytics: Concepts, types and technologies. Int J Res Eng, 5(9), 524-528.

[47] Zakaria, M. M. K. N. H., & Rashid, A. (2017). Big Data Value Dimensions in Food Disaster Domain. Innovation, 11(1), 25-29.

Chapter 3

Disaster Resilient Rescue of Coastal Community on Cyclone Warning

Md. Shahin, Sanjida Akter, Prome Debnath and A.K.M. Mostafa Zaman

Abstract

Bangladesh is in the front line of battlefield of disasters due to geographical location and global warming faced over 200 natural disasters in past 40 years and most of the disasters were cyclones. People need to be evacuated and rescued before a cyclone landfall. In current practice, multipurpose cyclone shelter (MPCS) provides short-term safety for the disaster victims in Bangladesh, where people are rescued after disasters which cannot ensure survival of lives. This study aims to develop a method for efficient evacuation and rescue to reduce death tolls in the events of disasters. This study used Wi-Fi scanner and smartphones to detect people. An inbuilt index that includes name, address, mobile number, photo, service set identifier (SSID), and media access control (MAC) of smartphone was developed for 90 registered participants. In this controlled experiment, few new participants turned on hotspot in every five minutes. A new index of people with MAC/SSID was developed in MPCS simulating an emergency. Missing people were detected by comparing inbuilt index and new index, and ordered them self-evacuation. This method captured 100% evacuees. Most importantly, the proposed method will reduce death tools because the people are rescued earlier to a disaster hits a specific area.

Keywords: Wi-Fi scanner, MAC/SSID, cyclone, disaster, hotspot, rescue, evacuation

1. Introduction

Bangladesh is one of the topper rank disaster prone countries due to its geographical location and global climate change. It is one of the largest delta areas and comprises three major rivers including Ganges, Brahmaputra and Meghna [1, 2]. Climate change accelerates catastrophic tropical cyclones and storm surges during the period of pre-monsoon (April–May) and post-monsoon (October–November) in Bangladesh [3]. In recent decades, devastating cyclones with storm surges, e.g., Amphan (2020), Fani (2019), Bulbul (2019) landfall almost every year that make the coastal area of Bangladesh more vulnerable [4]. During the period of 1969–1990 on an average annually thirteen depressions were formed in the Bay of Bengal and adjacent Indian ocean where nearly 4.6 was altered on to cyclones and almost in all the causations were resulted into disasters [3]. In 21st century, 13-cyclones have already hit Bangladesh [4]. So, cyclone is a common phenomenon in the coastal zone of Bangladesh. Furthermore, flood is a regular event in Bangladesh, which causes extensive damage to properties and seasonal crops [2]. Following the great Bhola cyclone (1970) and cyclone Gorky (1991) caused death toll of 500,000 and 138,000 people, respectively [3]. The huge death tolls in 1970 is attributed to the inefficient cyclone tracking system, improper early warning and issuance of evacuation order.

In order to increase the awareness and reduce the death tolls, GoB developed early warning system and cyclone preparedness program (CPP). Bangladesh Meteorological Department (BMD) and Storm Warning Centre (SWC) are involved in weather forecasting and warning. Warning is passed as special weather bulletin via fax, teleprinter, telephone etc. to radio, television, press media/news agency, CPP, disaster preparedness programme (DPP) for necessary action and messages to the prime minister's office, ministry of disaster management and relief (MoDMR), directorate of relief, concern ministries, airport, seaport, naval base etc. Fishermen get messages from the seaport at least one/two days advance [5]. CPP volunteers circulate warning via hand siren, megaphone (e.g., miking), signal light and flags [5], while local stakeholders claimed that siren and flag do not remain active in maximum cases. Besides, most of the poor coastal residents do not have access to online newspapers, radio, television and they do not get the warning in proper time and this miscommunication of information makes them vulnerable and discourage to evacuate [1, 3, 6].

As a part of preparedness, GoB, NGOs and development partners constructed more than 2500 MPCSs in 19 coastal districts of Bangladesh including the study area Barguna [7, 8]. Few shelters have been extended with the killas as a shelter for livestock. Approximately 200 raised earthen platforms called killa have been constructed in the cyclone prone areas and one killa can accommodate 300–400 livestock. Most of the killas have been found are full of bushes and become a habitat for snakes and harmful insects due to lack of maintenance [9]. The CPP was established in 1972 was collaborated with Bangladesh Ministry of Disaster Management and Relief (MoDMR) and the Bangladesh Red Crescent Society (BDRCS) with 49,365 trained volunteers and 16,455 of those are women [9]. Those volunteers are dedicatedly involved to disseminate cyclone warning signal, assist people to move in MPCSs, rescue distressed people and provide first aid to the injured people [1].

The capacity of MPCSs varies from 400 to 1600 people [8]. Sometimes small capacity and inadequate facilities for women discourage the people to be evacuated there. Community people are encouraged to be evacuated in MPCSs as safer places before a cyclone landfalls.

The conventional evacuation systems include early warning, notice for evacuation, miking by volunteers, and cyclone shelters preparedness. Emergency medical and food supply are ensured by the local authority with the support of central authorities. When early warning about devastating cyclones is disseminated, the community people need to be evacuated in MPCSs. Generally, rescue operation is conducted after disasters which cannot save some lives from disasters. Current system cannot ensure evacuation of all vulnerable community people because the evacuation is a volunteer/self-evacuation process. During cyclone SIDR 90% of vulnerable people were warned but only 10% people evacuated [1]. Therefore, community people should be rescued before a disaster occurs.

While evacuation order is issued, most vulnerable people should evacuate to MPCSs. Many people stay in their houses to protect their properties, e.g., domestic animals. In MPCSs, maintaining a manual register to confirm the name and number of people is very difficult during disasters. Therefore, the shelter authority cannot immediately identify the missing people in the community before the disaster. Therefore, a system should be developed to ensure rescue of all people before a disaster hits a specific area in the era of information and technology. Recently,

Wi-Fi scanner, a cheap and available technology, has received much attention to detect the human mobility. The Wi-Fi scanner makes an active scan by using the probe requests. A probe request is a special frame sent by a client station requesting information from either a specific access point, specified by SSID, or all access points in the area, specified with the broadcast SSID [10].

This research aims to develop a more disaster resilient evacuation system where authority can easily identify the missing people and rescue them before a cyclone landfalls. The smart-phone and Wi-Fi scanner can contribute for the identification of the missing people. The Wi-Fi scanner can detect all the people within the shelter with turned on Wi-Fi in their smartphones. Therefore, it will be very easy to indexing the missing people to be rescued before cyclone landfalls using Wi-Fi scanners and smart-phones. This method helps to rescue the people earlier to a cyclone hits the area. In the proposed method, the people will be rescued earlier to the cyclone. As a result, all the stakeholders will be safe in the event of a cyclone.

1.1 Related works

As discussed above evacuation and rescue response depends on different factors. Manual evacuation through warning messages cannot be assured successfully. In the era of information and technology, technological interventions are noticed for disaster mitigation and evacuation responses. Global Positioning System (GPS) and Global Service for Mobile Communication (GSM) web services are together referred as Smart Life Tracking and Rescuing (SLTR) system are being used for the disaster management in India. This GPS and GSM web services are effective to identify the affected areas and possible routes to reach the location [11]. Wireless sensor networks are also used for detecting disaster, providing alert signals and completing rescue operations immediately [12, 13].

Sensor using internet of things (IoT) is being used in vehicular tunnels, and it detects the vehicle where an accident occurs [14]. A server that receives location of mobile phone user in the affected area with their environmental condition is also provided. After that automated voice inquiries store their responses and forward to emergency assistance agencies, are also displayed in electronic map and involved in rescue operation [15]. Requesting rescue by a user terminal is controlled by an informing server and a rescue centre through mobile communication network. When informing, server provides signal to rescue centre and they become able to detect the position of user terminal and evacuation can be maintained successfully [16]. Twitter can also be used for rescue operation considering some disadvantages [17].

During disaster and/or rush hour, traffic congestion can be created in the road for evacuation and this problem can be managed through contraflow evacuation method. This type of technology helps to evacuate the distress people safely without making any congestion in the road [18]. Using of Doppler Radar for cyclone wind field monitoring and identifying location of tornadoes is prescribed in some cases [19]. For effective and reliable network management in disaster prone area collaborative rescue robots are used [20].

Now-a-days, smart devices are being used for the detection of the mobility of vehicles and people by identifying the Media Access Control (MAC) of such devices, e.g., smartphones, Wi-Fi scanners. Abbott-Jard et al. [21] studied the travel time using Bluetooth scanners by detecting MAC of smartphones within its communication range and the vehicle is detected while it passes through the range of a detector. Wi-Fi Scanners are also being used for the detection of vehicles using smart devices [22–26]. Shiravi et al. [26] used Wi-Fi scanners for travel time estimation with combination of Bluetooth and Bluetooth increases the reliability of data. Similar applications are noticed in different existing studies [22, 25].

Despite the application of GPS and GSM for disaster management in different parts of the world, Wi-Fi scanners are not still being used for the detection of people in MPCSs in order to promote the early rescuing of people to reduce death tolls. From this point of view this research focused on developing disaster resilient method for successful disaster evacuation and reducing death toll by detecting human mobility in the cyclone shelters.

2. Methodology

The study proposes a disaster resilient evacuation and rescue method in case of a disaster, e.g., cyclone, flood, and storm surge. This research develops a very simple method of detecting the missing people in the catchment area of a cyclone shelter in order to rescue them before a cyclone landfalls. There are several techniques for the detection of human mobility, e.g., Wi-Fi scanner, Bluetooth scanner, global positing system (GPS) and call record data (CDR). This study proposes a method for detecting human mobility and rescuing them earlier to disasters by using Wi-Fi scanner and smartphones. Wi-Fi probe request identifies the MAC of a smartphone while Wi-Fi is turned on. Wi-Fi probe request identifies the SSID of a smartphone when the hotspot is turned on. By observing MAC/SSID, it observes the mobility of people.

A field experiment was conducted in a MPCS at Patharghata in Barguna district of Bangladesh. Total 90 people participated in the experiment who were community people. The index of the participants was made by name, address, mobile number, SSID and MAC. A MacBook Pro was used as a host computer and Wi-Fi scanner. The host computer used Wi-Fi probe request. All participant used their smartphones. They turned on the hotspot of their smartphones on request. The turned-on Wi-Fi scanner detected the preregistered participants who turned on hotspot of their smartphone in every five minutes of time interval. The index of the preregistered participants was compared with the new index during the experiment, that gave the index of missing people. The index of missing is required to be updated continuously. If some people go to different shelter other than host shelter, the inter-shelter data processing will identify the actual missing people. The proposed method includes: a) inbuilt index of people, b) indexing of people in the shelter, and c) updating of index and determining the missing people continuously. Due to the privacy reason, we cannot present all information of participant, e.g., photos, mobile number etc.

2.1 Inbuilt index of people

An index of the registered participants (resembling community people in the catchment area) of a multipurpose cyclone shelter (MPCS) was prepared earlier to the events of an emergency. The registration of people was facilitated by the authors as imaginary shelter authority. The registration included the name, address, photos, mobile phone number, and MAC and SSID of smartphone. This inbuilt index was used for the search and rescue of the people. Each MPCS is concerned of people who have already registered. **Figure 1** shows the process of developing inbuilt index.

2.2 Identification of missing people

The administrative staff of the MPCS is to prepare the control room and necessary instruments while hoisting the danger signal of cyclone or flood. The people are suggested to move to the shelters while the evacuation order is issued by the authorities of disaster management (DM). The participants were divided in some groups to facilitate the experiment. The few groups of participants in the shelter were requested to turn on the Wi-Fi/hotspot of their smartphones at a time. The inbuilt Wi-Fi scanners detected them by the MAC/SSID and made a new index of people in MPCS. The new index of people in the shelter comprises the MAC and SSID. The inbuilt index and new index were compared which gave the index of missing people. The procedure of detecting the missing people in the shelter is shown in **Figure 2**.

2.3 Updating the index of the people in the shelter

The index of the missing people is required to be updated continuously. This study considered five minutes interval for updating the index of missing people. In every five minutes, the newly arrived people were observed in the shelter. A preinstalled software is to develop that has to proceed data and develop the index of arrived people in the shelter. The inbuilt index and new index have to be cross

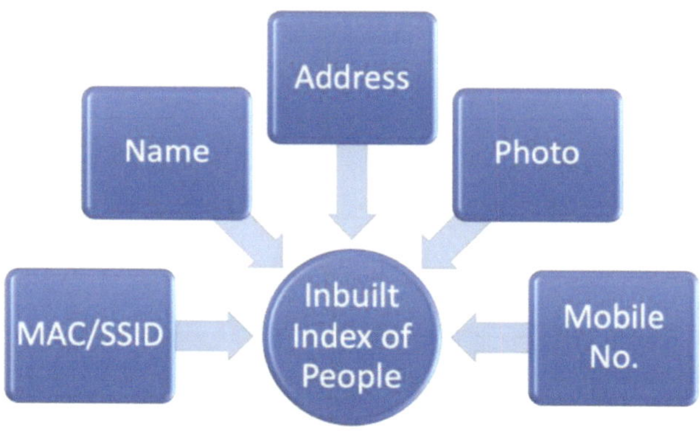

Figure 1.
Inbuilt index of the stakeholders in the catchment area of a MPCS.

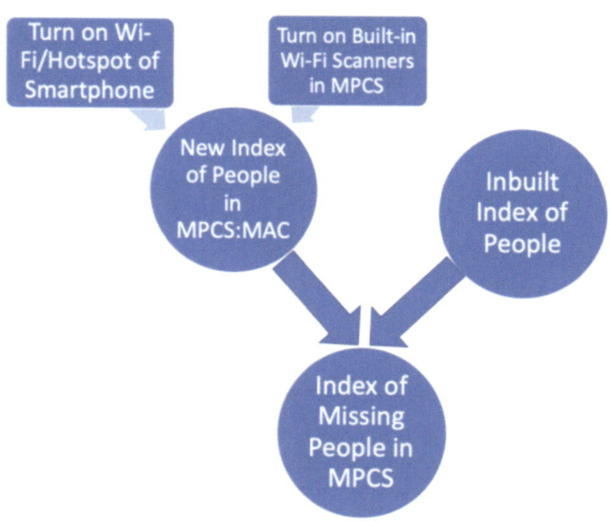

Figure 2.
Indexing of the people in the cyclone shelter and detection of missing people.

matched that may produce the index of missing people in the shelter. The updating of index notified the list of the missing people to all refuges in the shelter. **Figure 3** shows the process of updating the index of missing people.

The evacuation of the missing people needs to be confirmed by the shelter's authorities as shown in **Figure 4**. The administrative staff of MPCS will call the missing people and request to join the shelter. The local elected bodies have to contribute to bring the community people into the shelter. Mobile network operators will be given the mobile phone number and they have to send message and call for evacuation. Some people may go out and move to those shelters in emergency situations. Few people may visit their relatives and stay in the neighbouring shelters.

Figure 3.
Updating the index of the missing people in the shelter.

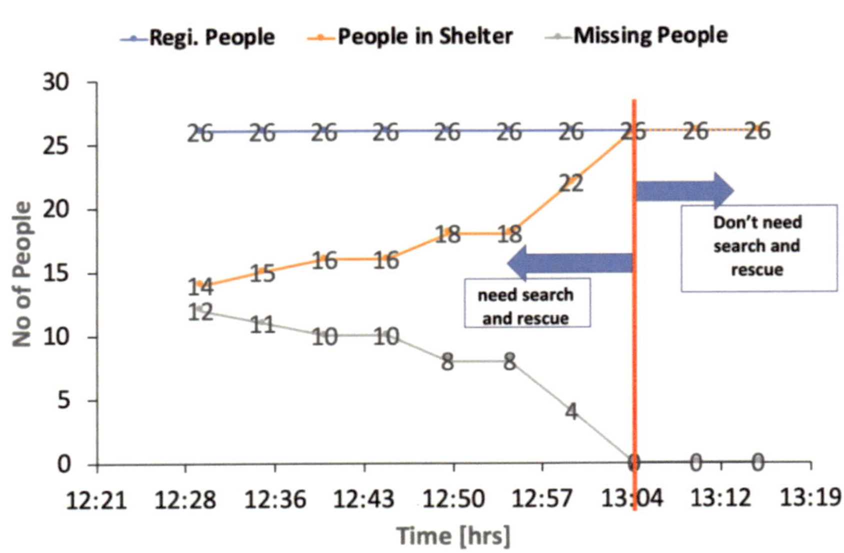

Figure 4.
Conceptual observation curve of the missing people in the shelter.

The dataset of a MPCSs will be shared with the surrounding shelters to ensure the presence of missing people in those shelters. **Figure 4** shows a conceptual graphical observation curve to facilitate the search and rescue operation.

3. Empirical analysis

The smart devices are being used for the mobility detection of the car and people, but for the rescuing of people during cyclone that had very limited application. This research proposes a very simple method of detecting the missing people in the catchment area of a MPCS by using Wi-Fi scanner in order to rescue death tolls before a cyclone landfalls. The proposed method includes indexing of the community people in the catchment area, indexing of people in the shelter, determining the missing people, and updating the index of missing people.

This study used smartphones and a Wi-Fi scanner. A MacBook Pro laptop was used as a Wi-Fi scanner. Experiment was conducted on 11/12/2020, 8/01/2020 and 9/01/2020. The data obtained on 9th January 2020 was used for this analysis. Total 90 participants joined in the experiment. The experiment was a control experiment to simulate the field conditions. All participants were requested to turn off their hotspot of smartphone. Few participants were asked to turn-on the hotspot of their smartphones and Wi-Fi scanner detected the MAC address with SSID of the participants. This process was continued till all participants were detected with a time interval of five minutes. The penetration rate was 100% for this experiment that means all participants were detected during the experiment. The MACs and SSIDs are not disclosed for the privacy reason. Only number of people detected and participated are being used for this study.

3.1 Inbuilt index

An inbuilt index was prepared for this investigation for all participants which comprised the name, address, MAC, SSID, picture and mobile number. There were total 90 registered participants. So, inbuilt index included 90 participants for this experiment. An example of the inbuilt index is shown in **Table 1**. The name was used to identify the people by which he/she is familiar in the community. MAC and SSID were be used to build-up the new index in the MPCS for the comparison and detection of missing people. The missing people could be communicated using their mobile number during the search and rescue operation.

3.2 Detection of people

Manual counting of people takes long time to identify and to rescue them. To facilitate the identification of people in MPCS, this study proposes the application of Wi-Fi scanners and smartphones for automatic detection of people. According to the Bangladesh Telecommunication Regulatory Commission (BTRC), 156 million people subscribed mobile phone at the beginning in 2019 in Bangladesh where penetration rate of smartphone is satisfactory. For detecting people in MPCS, all participants turned on Wi-Fi/hotspot of their smartphones. Wi-Fi has a total inquiry time of as little as 8 ms. This allows detection of devices every second, allowing people with using Wi-Fi who stays in the range of a detector at a much quicker rate. In this controlled experiment, the number of participants was captured for each 5-minutes. There were 90 participants who joined the experiment. At the beginning of the experiment, all participants turned-off Wi-Fi/hotspot of their smartphones. In each five minutes, few new participants

Name	Address	SSID	MAC	Mobile No.
X	House # 32, BRTC Road, Patharghata, Barguna-8720.	Y	11:3A:09:3B:8D:22	017xxxxxxxx
A	House#1, BRTC Road, Patharghata, Barguna-8720	A	7c:a1:77:1e:39:9a	017xxxxxxxx
B	House#2, BRTC Road, Patharghata, Barguna-8720	B	a4:50:46:16:51:ed	017xxxxxxxx
C	House#3, BRTC Road, Patharghata, Barguna-8720	C	40:D3:AE:73:01:88	017xxxxxxxx
D	House#4, BRTC Road, Patharghata, Barguna-8720	D	94:B1:0A:92:33:EB	017xxxxxxxx

Table 1.
An example of inbuilt index of the community people.

turned-on Wi-Fi/hotspot of their smartphones and were detected. For each time interval, the number of detected people was increased with the increase of active devices as observed. **Figure 5** shows the detection of people during the experiment.

3.3 Index of missing people

Wi-Fi scanners can identify the MAC address of the smartphones that has turned on Wi-Fi [22]. People come to the shelters while evacuation order is issued by the disaster management authority. The people who were in the shelter turned on Wi-Fi/hotspot and detected. In every five minutes, few new people were detected. A new index of people was developed with SSID/MAC. A SSID is simply a wireless network name to distinguish it from other networks in neighbourhood [27]. MAC addresses are unique identifiers defined by IEEE as a communication protocol for wireless Wi-Fi connection [26]. The inbuilt index (**Table 1**) and new index were compared that produced an index of missing people. An example of index of missing people is shown in **Table 2**.

In inbuilt index, there were 90 registered participants. While evacuation order was issued, participants started to come into the shelter and Wi-Fi scanner detected people by MAC/SSID and developed a new index. After that the inbuilt index and new index were compared that produced the index of missing people as shown in **Table 2**. The number of missing people was decreased with time as shown in **Figure 6**. At the beginning of the experiment more than 50 participants were detected as missing people. For every five minutes, few new people were detected and missing people were decreased. At a certain time, there were no missing people that meant all participants were detected and rescued. The search and rescue operation will be stopped at red line. The index of the missing people is required to be updated continuously in order to observe the newly arrived people in the shelter. The updating of index notifies the list of the missing people. The index of missing people should be displayed in a bigger screen. The family members can easily identify the missing people and contact with them. The evacuation and rescue team of the MPCS can rescue the missing people.

From the index of missing people, phone calls could be made to confirm the location of people and they would be requested to take shelter in MPCS. From phone call, we could identify the current status such as age, disability, and other burdens that prevent themselves from evacuation. Beside more information could be provided about the MPCSs to him/her for taking shelter there.

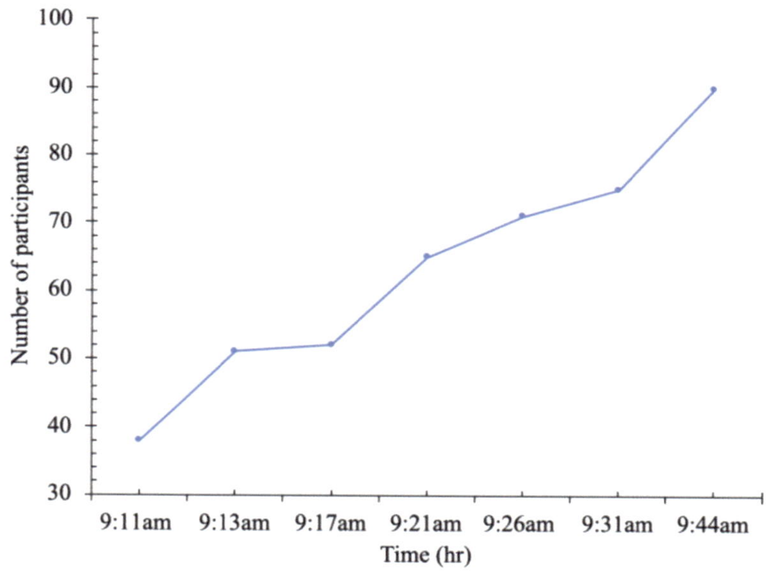

Figure 5.
Detection of people in the MPCS.

Name	Address	SSID	MAC	Mobile No.
X	House # 32, BRTC Road, Patharghata, Barguna-8720.	Y	11:3A:09:3B:8D:22	017xxxxxxxx

Table 2.
An example of index of missing people in MPCS at a certain time.

Figure 6 shows the number of missing people in different time stamp. There were 90 registered participants. In the very beginning of experiment, 52 participants were detected as missing people. Later, missing people were recorded as 39, 38, 25, 19, 15, and 0 with elapsing of time and increasing the participants. While reaching zero, all community people were rescued in the MPCS. The proposed method of detecting the missing people is an efficient method as it can detect 100% of participants. In **Figure 6**, the red line was drawn as the demarcation of search and rescue operation. This is the end point of rescue operation.

The evacuation decision and destination are influenced by many factors, e.g., distance of MPCS, facilities in MPCS, access road, weather condition, crowdedness, socioeconomic condition, psychological, physical, cultural and personal factors etc. Level of education and household income affect the access to radio, television and online newspapers. However, the ownership of radio and television, listening to cyclone warning, improper understanding of signals, late warnings, sudden change in signals, and issuance of premature evacuation order are the obstacles to successful evacuation [3, 28]. Signals are hoisted for seaports and people cannot understand the signals outside the port areas and evacuation becomes uncertain in that cases [5]. Past experience on the failure of warning known as false alarm effect, disbelief on existing warning signals, hampering income earning activity with pro-active evacuation responses, missing the target position and weaker condition of cyclone demotivated people for evacuation [5, 6, 29].

Lack of killas and public transport discourage community people to evacuate. In some cases, people avoid evacuation due to over age, perceived level of risk, fear of

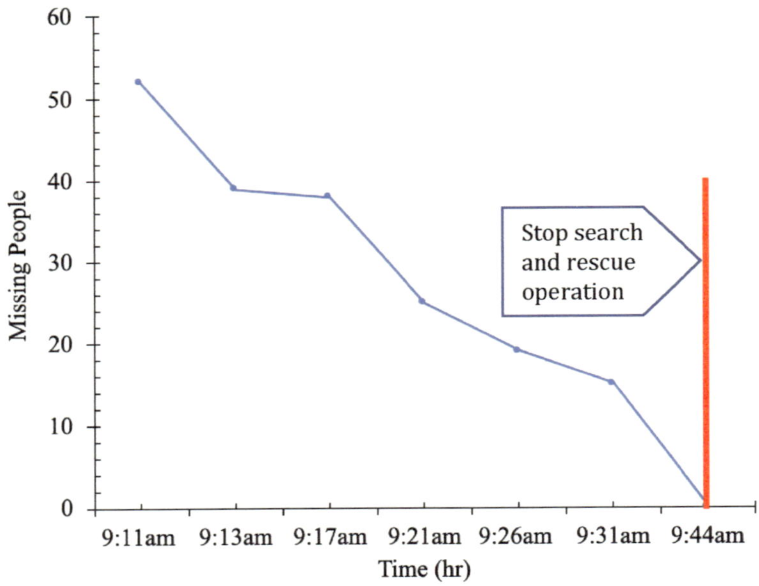

Figure 6.
Detection of missing people in the MPCS.

robbery, belief on fate, false sense of secured house and embankment. Some people believe cyclones come as punishment from god and stay home is safe by praying to god. In conservative societies women cannot leave their houses without husbands' permission, and putting men and women under the same roof affects women's status in their family and society. This fatalistic attitude makes the people in great danger. Women also desire separate facilities and sufficient water supply as they hold more dependent members like children and adults [1, 9].

Super cyclone SIDR (2007) caused 3460 death tolls [1, 9]. Strong disaster management committees, construction of MPCSs, and monitoring cells decrease death tolls in recent cyclones, e.g., eight death tolls in cyclone Fani. Reduced death tolls do not indicate successful evacuation of all vulnerable people. Despite the dynamic efforts of GoB, evacuation of people has become difficult and most of the respondents did not evacuate in some cases of cyclones, e.g., Amphan. Successful evacuation depends on evacuation preparation, order and timing, and rescue operation before cyclones landfall. If the warning become location specific and order is given within 2–3 hours before cyclone landfalls then evacuation will be effective [9]. The evacuation prior to cyclone landfall is considered as one of the best practices to minimise death tolls from a catastrophe by moving people from exposed areas to the safer places and MPCSs, temporarily.

Though there are several man-made and natural factors that discourages self-evacuation of community people, disaster management authorities are responsible to encourage and rescue them into MPCSs. This study developed an easier search and rescue operation that could be implemented in the coastal zone of disaster-prone countries like Bangladesh. This investigation emphasises on the evacuation and rescuing of vulnerable community people earlier a cyclone hits an area to reduce death tolls to zero.

4. Conclusion

This study was conducted to reduce the death tolls in disasters. The current existing practice is to rescue the local people after cyclone landfalls that cannot ensure survival of people in the affected area. To save lives, people should be identified and rescued before the occurrence of extreme events. Manual detection of people is time consuming that is not available during an extreme event but it could be easy by adopting technological intervention. The smartphone can contribute to this issue for the identification of the people in the shelter. The Wi-Fi scanner can detect all the people within the shelter with turned on Wi-Fi in their smartphones. The proposed method recommends the application of Wi-Fi scanners and smartphones for the detection of human mobility in MPCSs. The results showed that the proposed method can detect 100% people in the shelter area and this method will accelerate the evacuation and rescue operation.

To identify missing people, it had only 4 easy steps: 1) developing inbuilt index, 2) developing new index of people in MPCS, 3) producing index of missing people by comparing inbuilt index and new index, and 4) evacuation and rescuing of missing people. Therefore, it is very easy to indexing the missing people to be rescued before cyclone landfalls and efficient to reduce death tools. Thus, this proposed method maximises the safety of stakeholder and minimises life risk against disasters.

The proposed method is utmost important to reduce the death tolls because all vulnerable people will be rescued before a cyclone landfall. This study is to ensure the evacuation of all community people in the catchment area of a MPCS. Forced evacuation is suggested when we have enough information about their location and family burdens. People will be safe in the MPCS. We expect "zero death tolls" that could be a dream in case of a disaster. Though this investigation has some important implications, it requires improvement in future research. This study investigated for small number of participants. This study should be extended for 400 to 2500 people to simulate the capacity of the MPCS. The penetration rate of smartphone does not reach 100%. Several MPCSs and community people could be included in the experiment for successful implementation.

Acknowledgements

The authors are wholeheartedly thankful to the people who participated in the experiment and contributed to the success of this research. This research was funded by the Patuakhali Science and Technology University.

Conflict of interest

The authors declare no conflict of interest.

Appendices and nomenclature

Pre-monsoon	April–May
Post-monsoon	October–November
false alarm	failure of warning
MPCS	multipurpose cyclone shelter
CS	cyclone shelter
MoDMR	Ministry of Disaster Management and Relief
CPP	cyclone preparedness program
BMD	Bangladesh Meteorological Department
SWC	Storm Warning Centre

DPP	disaster preparedness programme
SSID	service set identifier
MAC	media access control
BTRC	Bangladesh Telecommunication Regulatory Commission
DM	disaster management
BDRCS	Bangladesh Red Crescent Society
GoB	Government of Bangladesh
NGO	Non-profit Organisation
GPS	global positing system
CDR	call record data
GSM	Global Service for Mobile Communication
SLTR	Smart Life Tracking and Rescuing
Participants	the people who participated in the experiments. These participants are assumed as community people

Author details

Md. Shahin*, Sanjida Akter, Prome Debnath and A.K.M. Mostafa Zaman
Patuakhali Science and Technology University, Patuakhali, Bangladesh

*Address all correspondence to: shahin_dre@pstu.ac.bd

References

[1] Parvin GA, Sakamoto M, Shaw R, Nakagawa H, Sadik MS. Evacuation scenarios of cyclone Aila in Bangladesh: Investigating the factors influencing evacuation decision and destination. Prog Disaster Sci. 2019 Jul;2:100032 (1-13). Available from: https://doi.org/10.1016/j.pdisas.2019.100032

[2] Choudhury AM. Managing natural disasters in Bangladesh, Sustainable Development in Bangladesh. Achievements, opportunities and challenges. In: Rio+10, Dhaka, 16-18 March, 2002. 2002

[3] Paul S. Determinants of evacuation response to cyclone warning in coastal areas of Bangladesh: A comparative study. Orient Geogr. 2014;**55**(1):57-58

[4] Wikipedia. List of Bangladesh tropical cyclones. In: The Free Encyclopedia (accessed on 25th Aug 2020). 2020. Available from: https://en.wikipedia.org/wiki/List_of_Bangladesh_tropical_cyclones

[5] Disaster Management AMH. Cyclone warning system in Bangladesh. In: Early Warning Systems for Natural Disaster Reduction. Berlin, Heidelberg: Springer Berlin Heidelberg; 2003. pp. 49-64. Available from: http://link.springer.com/10.1007/978-3-642-55903-7_8

[6] Ahsan MN, Takeuchi K, Vink K, Warner J. Factors affecting the evacuation decisions of coastal households during Cyclone Aila in Bangladesh. Environ Hazards. 2016 Jan 2;15(1):16-42. Available from: http://www.tandfonline.com/doi/full/10.1080/17477891.2015.1114912

[7] Faruk M, Ashraf SA, Ferdaus M. An analysis of inclusiveness and accessibility of Cyclone Shelters, Bangladesh. Procedia Eng. 2018;212:1099-106. Available from: https://doi.org/10.1016/j.proeng.2018.01.142

[8] Shahin M, Billah M, Islam MM, Parvez A, Zaman AKMM. Cyclone shelters need sustainable development. Int J Disaster Resil Built Environ. 2020 May 15;**11**(5);659-678. Available from: https://www.emerald.com/insight/content/doi/10.1108/IJDRBE-12-2015-0061/full/html

[9] Paul BK. Factors affecting evacuation behavior: The case of 2007 cyclone Sidr. Bangladesh. Prof Geogr. 2012;**64**(3):401-414

[10] Hak5Darren. Haktip 23 – WiFi 101: Probe Requests and Responses. Hak5 LLC. 2020 [Accessed on 18 November 2020]. Available from: https://www.hak5.org/episodes/haktip-23#top

[11] Nagashree C, KavyaRao B, Lobo MJ, HarshithaB S. Smart life tracking and rescuing disaster management system. International Journal of Computers and Applications. 2012;**45**(23):10-17

[12] Aziz NA, Aziz K. Managing Disaster with Wireless Sensor Networks. 2011.

[13] Kaur H, Sawhney RS, Komal N. Wireless sensor networks for disaster management. International Journal of Advanced Research in Computer Engineering and Technology. 2012;**1**(5):129-134

[14] Kim Y-D, Son G-J, Kim H, Song C, Lee J-H. Smart Disaster Response in Vehicular Tunnels: Technologies for Search and Rescue Applications. Sustainability. 2018 Jul 18;10(7):2509. Available from: http://www.mdpi.com/2071-1050/10/7/2509

[15] Pfleging GW, Wilkin GP. Automated search and rescue call generation to mobile phones in a defined geographic disaster area. US; US. 2007;2007/0202927 A1

[16] Chung T-S. Method for requesting rescue by receiving disaster alarm signal

and apparatus thereof. United States; US 2009/0061891 A1, 2009.

[17] Mihunov V V., Lam NSN, Zou L, Wang Z, Wang K. Use of Twitter in disaster rescue: lessons learned from Hurricane Harvey. Int J Digit Earth. 2020 Feb 19;1-13. Available from: https://doi.org/10.1080/17538947.2020.1729879

[18] Clark AE, Hagelman RR, Dixon RW. Modeling a contraflow evacuation method for tropical cyclone evacuations in Nueces County, Texas. Natural Hazards. 2020 Sep 12;103(3):2757-86. Available from: https://doi.org/10.1007/s11069-020-04101-w

[19] Joseph PV. Tropical cyclone hazards and warning and disaster mitigation systems in India. Sadhana. 1994 Aug;**19**(4):551-566. Available from: http://link.springer.com/10.1007/BF02835639

[20] Sugiyama H, Tsujioka T, Murata M. Collaborative Movement of Rescue Robots for Reliable and Effective Networking in Disaster Area. In: 2005 International Conference on Collaborative Computing: Networking, Applications and Worksharing. IEEE; 2005. p. 1-7. Available from: http://ieeexplore.ieee.org/document/1651217/

[21] Abbott-Jard M, Shah H, Bhaskar A. Empirical evaluation of Bluetooth and Wifi scanning for road transport. In: Australasian Transport Research Forum (October, 2013), Brisbane, Australia. 2013. p. 1-14.

[22] Shahin M, Kusakabe T, Oguchi T. Cleaning of Wi-Fi Probe Request Data to Estimate Travel Time Reliability in Mixed Traffic. In: The 6th International Symposium of Transport Simulation (ISTS'18) and the 5th International Workshop on Traffic Data Collection and its Standardization (IWTDCS'18), 4-6 August 2018, Ehime University, Matsuyama, Japan. 2018. Available from: http://www.cee.ehime-u.ac.jp/~keikaku/ists18/pdf/ISTS_IWTDCS_2018_paper_36.pdf

[23] Browning RC, Baker EA, Herron JA, Kram R. Effects of obesity and sex on the energetic cost and preferred speed of walking. Journal of Applied Physiology. 2006;**100**(2):390-398. Available from: http://www.physiology.org/doi/10.1152/japplphysiol.00767.2005

[24] Mai VH, Kusakabe T, Suga Y, Oguchi T. Travel time estimation in mixed traffic using Wi-fi detector based data. In: ISCTSC 11th International Conference on Transport Survey Methods. 2015. pp. 1-14

[25] Purser KM. Exploring Travel Time Reliability Using Bluetooth Data Collection: Case Study In San Luis Obispo. California Polytechnic State University: California. California Polytechnic State University; 2016

[26] Shiravi S, Hossain K, Fu L, Ghods A. Evaluation of using Wifi signals to estimate intersection travel time. In: Proceedings, Annual Conference - Canadian Society for Civil Engineering (2016) 4. 2016

[27] Bell Canada. What do SSID and WPA2 meaning?. 2020 [Accessed on 21 Aug 2020]. Available from: https://support.bell.ca/internet/connection-help/what-do-ssid-and-wpa2-mean

[28] Islam M, Paul A. Community response to broadcast Media for Cyclone Warning and Disaster Mitigation: A perception study of coastal people with special reference to Meghna estuary in Bangladesh. Asian J Water, Environ Pollut. 2004;**1**(1):55-64

[29] Paul S. Determinants of evacuation response to cyclone warning in coastal areas of Bangladesh : a comparative study. 2017;(July).

Chapter 4

Medical Provision of the Population within an Outbreak of a Traumatic Defeat an Earthquake: A Fundamental Tool of the Staged Health Risk Management

Diana Dimitrova

Abstract

The most destructive and unpredictable disasters around the world are determined earthquakes. Various consequences are reported as possible negative effects and therewithal health-related of them. The identification and classification of the different types of health risk factors is an initial goal in an uncomplicated earthquake setting and a fundamental tool to a good understanding and effective organization of the health care system (HCS) in case of complicated medical situation. The health care system works at high tension with considerable difficulties due to the calamity of a large magnitude outbreak of a traumatic defeat such as an earthquake. In conditions of the worst-case earthquake scenario with the subsequent provoked multi-secondary disasters and with multi-secondary risk factors possibilities to take accurate solutions is a real challenge for the health risk manager. They are available critically low resource constraints. Two main critical points are formed. On the one hand the description of a structure of mass victim and achievement high quality medical triage in complicated setting due to earthquakes is a conceptual medical stage of health risk management. On the other hand it is a main step of medical provision of the population and a step of risk reduction strategy.

Keywords: earthquake, mass victim structure, medical triage, medical provision of the population, health risk management, risk reduction, prevention measures, emergency and disaster medicine, good medical practices

1. Introduction

The medical provision of the injured population (MPIP) is a key task in the activity plan of the health system (HS) in case of emergencies and disastrous situations (EDS) [1–3]. This is regulated and specified for each level of government and for each type of structural unit within the HS of the country [2]. The timely and correct updating of the action plan of EDS by the healthcare leaders and health risk managers is a consistent, up-to-date annual task in order to maintain a highly optimal readiness for rapid response [2–4]. According to the type of EDS for each individual critical care potion the availability of action protocols guarantees not only confident firmness but also readiness for effective operational response and good medical practices. The productive operational activities at each stage of the MPIP with responsible and expert risk health management in a dynamically changing environment at the EDS require mobile flexibility and maneuverability in the development of the plan and its implementation [2, 5–7]. The provision with sufficient resources and their maneuver contextually according to the specific EDS is the basis for a good organizational culture [7, 8] Essential step for rescuing and providing successful medical care to the affected population is a practical interaction with other emergency systems and organization [8, 9]. Coordination of the EDS activities on a large scale is a solid fundament principle for the correct direction of the actions [7–9]. In case of the destructive EDS with a serious territorial scope and severely affected

Region	**Probability (low, moderate, high)**	**Power (low, moderate, high)**	**Expected losses (low, moderate, high)**		
			People	**Finance**	**Ecology**
Blagoevgrad	High	High	Moderate	Moderate	High
Burgas	Low	Low	Low	Moderate	Moderate
Varna	Moderate	Moderate	Moderate	High	High
Veliko Tarnovo	High	High	Moderate	Moderate	Moderate
Vidin	**Low**	**Low**	**Low**	**Low**	**Low**
Vratsa	Low	Moderate	Low	Moderate	Moderate
Gabrovo	Moderate	Moderate	Low	Moderate	Low
Dobrich	High	High	Moderate	Moderate	Moderate
Kardzhali	Moderate	Low	Low	Moderate	Moderate
Kyustendil	Moderate	Moderate	Low	Low	Moderate
Lovech	Low	Moderate	Low	Low	Low
Montana	**Low**	**Low**	**Low**	**Low**	**Low**
Pazardzhik	High	High	Moderate	Moderate	Moderate
Pernik	Moderate	Moderate	Moderate	High	High
Pleven	Moderate	Moderate	Moderate	Moderate	High
Plovdiv	High	High	Moderate	High	High
Razgrad	Moderate	Moderate	Low	Low	Low
Ruse	High	High	Moderate	High	High
Silistra	High	High	Moderate	Low	Moderate
Sliven	Moderate	Moderate	Moderate	Moderate	Moderate
Smolyan	Moderate	Moderate	Low	Low	Low
Sofia-city	**High**	**High**	**High**	**High**	**High**
Sofia	Moderate	Moderate	Low	Moderate	Low
Stara Zagora	High	High	Moderate	High	Moderate
Targovishte	High	Moderate	Low	Low	Low
Haskovo	High	High	Moderate	High	High
Shumen	Moderate	Moderate	Low	Moderate	Moderate
Yambol	High	Moderate	Moderate	Moderate	Low

Table 1.
The earthquake risk assessment in Bulgaria by regions (sources: BAS).

available resources the participation of international organizations, rescue teams and means of medical specialist is a possible option [10].

As a particular type of EDS, large magnitude earthquakes are characterized not only with mass casualty but also with the diversity of damage among the population of any natural community [2, 7–9].

According to the World Health Organization (WHO) evidence-based databases, around the world more than million earthshakings annually are occured, of which about 100,000 have a magnitude of 3–8 Richter and are felt by humans. Some of the strongest earthquakes in the world are: the Assam (June 12, 1897) in Northeast India; The Japanese (September 1923), in which the cities of Tokyo and Yokohama were destroyed; Gobi - Altai (December 4, 1957); Chilean (May 29, 1960); the Alaska Earthquake (March 28, 1964); the Armenian Earthquake (December 7, 1988); the earthquake of December 26, 2004 after which almost 230,000 missing and presumed dead [11]; the earthquake in Haiti (January 12, 2010) that killed more than 230,000 people and another 300,000 were injured [12]; the earthquake in Japan (March 11, 2011), which killed more than 15,800 people, injured more than 6000 and disappeared more than 2500 people [13]. The earthquake risk assessment in Bulgaria is made by regions. The number of regions in Bulgaria is 28 in total (**Table 1**).

The data indicate that the probability is low in only five regions in Bulgaria (almost 18%). In contrast, the probability is high in 12 regions (almost 43%) and moderate level in 11 regions (around 39%). It is interesting to note that in 12 regions the risk of a strong earthquake is rated as high and only in 4 is indicated as low. According to prognostic data, human losses are assessed as high only in Sofia-city region. On the one hand only in Sofia-city region is reported high values prognosis for each of the indicators. On the other hand in two of them (Montana and Vidin) a low risk assessment is given.

2. Medical provision of the population - goals, principles and tasks

The medical provision of the population (MPP) is an element of activity of the health system especially due to EDS. In case of emergencies it is based as much as possible on the existing health care system and only with an organizational approach moves to a new mode of work with available staff and sources. MPP according to real practice, results of an epidemiological survey and documentary research is defined as a complex of interconnected organizational, medical and hygienic-anti-epidemic measures [7–10, 14].

It seems that the Aim of MPP is organized in a few groups of actions:

1. *Preservation of the health* and strengthening of the physical condition and working capacity of the population.

2. *Saving the lives* of the affected people and reducing mortality and disability and the fastest recovery of health and work ability.

3. *Prevention of long-term* and infection diseases.

The occurrence of different types of traumatic defeat due to huge EDS, which are characterized by diverse in nature injuries and mass effect determine the main principles of MPP [1, 2, 7–10, 14, 15].

The main Principles of MPP due to contemporary scientific knowledge are collected in some main key points [1, 2, 7–10, 14–16]:

1. *Universality* of using of medical resources.

2. *Maximum allowable economy* of using of available resources.

3. *Implementation of the medical evacuation* into a MPP activity.

4. To *use correctly* Unified rescue system with unified emergency number for providing MPP.

5. To use *unified doctrine* for rendering medical aid and treatment from the epicenter of defeat to the ends in the multi-profile and specialized medical establishments until the final outcome.

For the correct understanding and optimization of the activities connected with EDS **the tasks** of the MPP are divided into three groups depending on the time for their implementation – before, during and after disaster strike [1, 2, 7–10, 14–18]:

1. Before a disaster occurs:

    - Study of the devastating effect of various factors in EDS and the means for prevention, diagnosis and treatment.

    - Planning the activities of the health system for MPP.

    - Construction and maintenance of formations and resources of the MPP.

    - Creation of the medical and sanitary property.

    - Training, drills, workshops for preparing medical staff to react as better as possible.

    - Constant and regular current hygienic control.

2. In case of threatening and an ongoing EDS:

    - Deployment of the medical formations according to the operational plan, keeping them ready for work and protection.

    - Bringing the health risk management system into readiness.

    - Strengthening the medical provision (MP) system and teams and start MP of the evacuated population from epicenter, during transport to emergency room.

    - Strengthening and targeting the operational and special preparation of medical staff to starting and working into epicenter.

    - Strengthening the epidemiological surveillance of the territory.

3. After a disaster occurs:

    - Medical reconnaissance.

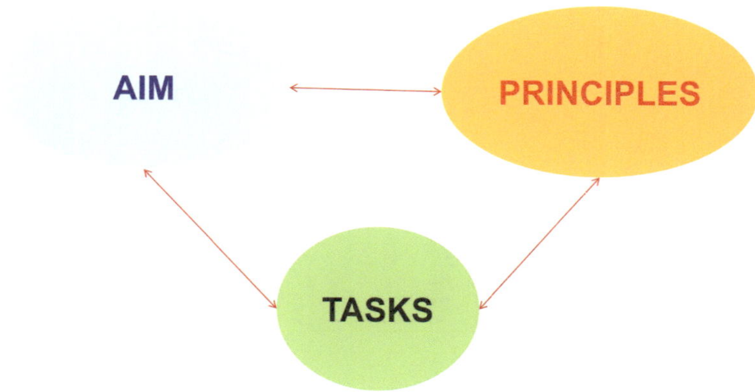

Figure 1.
Main interconnected components of successful medical provision of the population.

- Introduction of the formations in the center of defeat.
- Organization and rendering of timely first medical aid in the center of defeat.
- Removal and evacuation of the victims for appropriate medical care and subsequent treatment.
- Provision of complex of hygienic and anti-epidemic measures.
- Continuous management and maneuver for the most appropriate use of the medical forces and means.
- Conducting a forensic medical examination of the victims and their identification.

Proper performance of these tasks is a prerequisite for providing in the shortest possible time the optimal amount of medical care to the largest possible number of victims [2, 7–11, 14, 16–18]. The good success of medical provision of the effected population is guaranteed with a clear goal, streamlined tasks and by following the basic principles (**Figure 1**).

3. Earthquakes as an outbreak of traumatic damage: risks and consequences

3.1 Mass destruction after earthquakes

Affecting large areas with mass destruction and mass loss is a typical effect after hug magnitude earthquakes. A typical example of possible consequences in large-magnitude earthquakes is presented in **Figure 2** as number of victims and financial losses due to some of the most destructive earthquakes in XX century (**Figure 2**) [2–6]. Industrial sites, homes, hospitals, public buildings, utilities, underground and aboveground technical facilities, transport hubs, etc., are damaged or destroyed [2, 7–10, 13, 19–22]. This significantly impedes rescue operations, effective enough immediately after the earthquake occurs.

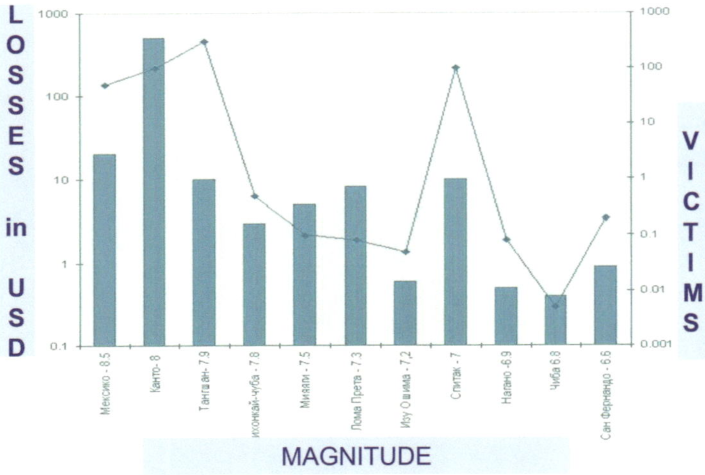

Figure 2.
Victims and financial losses (USD) due to some of the most destructive earthquakes in XX century (resources: Primary data from USGS, 2005 г.)

3.2 Secondary disasters due to earthquakes

Analysis of primary and secondary statistical data shows that occurrence of secondary defeats after an earthquake is a common effect. This increases number of risks factors and character of injuries among the affected population. According to analysis some of possible secondary disasters after earthquakes are [2, 7–10, 13, 19–21, 23–26]:

1. *An outbreak of chemical contamination* by industrial poisons. Observed in the area of the incident due to the destruction of chemical enterprises, warehouses with agricultural or industrial poisons. The chemicals produce not only acute intoxication but also long-term heath effect and cancer diseases as well [7, 10, 19].

2. Occurrence *of explosions and fires* due to destruction of flammable or explosive objects as petrol station after large magnitude earthquakes. They can develop dangerous additional consequences as fire burn, chemical burn, traumas, toxicological injuries etc [7, 10].

3. Destruction of buildings and severe *air pollution* with fine dust particles in large enough areas because of an earthquake leading to casualties, poisoning and suffocation [7, 10, 13, 19, 23–26].

4. *An outbreak of nuclear damage* such as Fukushima Daiichi nuclear accident, 2011 due to the large earthquake and tsunami is a real possible consequences. Radiation as danger risk factor produce some particular health problems, which can be acute, or chronic, somatic or genetic, stochastic and non-stochastic diseases, and long-term effects [7, 10, 13, 23–26].

5. *A tsunami* is generated when undersea earthquakes occurs. Fukushima 2011 and Haiti 2010 are significant examples and it is shows that many health problems influence not only primary effected county, but also much more people around the world for many years [1, 7–13].

6. Occurrence of *catastrophic floods* [1, 7–9, 11–13] after earthquakes can increase number of traumas due to new risk factors with leading place and role for the speed and depth of the water flow. On the other hand can cause hypothermia [7, 10, 27] especially for infants, adults, for the chronically ill and for people with special needs like people with compromised vision or limbs. Can be formed a new defeat of waterborne disease such as Cholera as well (like Haiti, 2010) [1, 7–12].

7. *An outbreak of biological contamination* and infectious diseases can be created after a destructive earthquake within or without a tsunami wave (Like Haiti, 2010; like Nepal, 2015) [1, 7–12, 21, 22].

8. Occurrence *of landslides and avalanches* (like Nepal, 2015) [1, 7–12, 21, 22].

9. *An ecological disaster* and even local environment can be changed totally (like Kresna, Bulgaria, April, 1904, with 7,8 M) [7–9].

10. *A volcano* generated by an earthquake. It is believed that mixed of danger toxic substances after that can affect society due to discarded volcanic ash into the environment [7–9].

11. *Traffic accidents* due to earthquakes [7–10].

12. *Social related and financial crisis* due to earthquakes [1, 7–10, 13, 17].

3.3 Victims: main structure and frequency of traumatic health related effects

It seems that after huge earthquakes, a large number of victims and injuries are occurred. According to experts, many numbers of victims need emergency medical care at the same time [7–10]. Nature and structure of the injuries can be very diverse [7, 13]. Injuries to the musculoskeletal system, extensive burns, prolonged compression syndrome due to prolonged compression of individual limbs or parts of the body, injuries to large blood vessels predominate [7–10]. Significant percentage of victims may be in a state of shock [7, 10, 27], with acute respiratory and cardiovascular insufficiency, in need of urgent respiratory and cardiovascular resuscitation [7, 10], neuropsychiatric disorder due to the experienced mental stress and others. According to scientific research [7, 10], by severity, medical losses are divided into the following groups: lightly injured (40%), moderately severely and severely injured (60%), of which 20% need specialized medical care.

3.4 Pollution from different origin as health risk factor

Creating a severe hygienic-epidemiological situation on the territory of the outbreak of traumatic defeat is defined as a big health risk factor. Prerequisite for this is the pollution of the territory from the destruction of water supply and sewerage systems, difficulties in finding the corpses of dead people and animals, the appearance of rodents, insects etc. The appearance of diseases of infectious and non-infectious origin is possible, as well as the appearance of epidemics - typhoid fever, paratyphoid fever A and B, salmonellosis, hepatitis, cholera etc [7–10].

4. Health risk management: essential principles

4.1 Role and place of the medical forces and resources for MPIP

4.1.1 Emergency care system (ECS)

Usually the signals for victims of an accident or disaster are received by the medical director (manager) of the emergency room and emergency hospitals as medical institution from ECS. Firstly, emergency medical care center (EMCC) as a front line of health system is informed by Unified rescue and emergency number 112. The teams of EMCC are the first to go to the place and take place of the scene of the EDS as event [7–10, 14, 16–18, 27].

In large-scale disasters, the EMCC teams are not enough to provide the necessary amount of medical care to those in need. This requires in advance formation, preparation and equipment additional teams of staff of medical institutions for their inclusion in the provision of medical care. In addition to these medical teams for the population, especially in earthquake-prone areas or areas with chemical sites in anticipation of numerous medical losses, emergency hospital medical teams and emergency military teams can be used. These are medical formations built on a functional principle with opportunities to provide emergency qualified therapeutic and surgical care for vital indications [7, 14, 16–18, 27].

4.1.2 Hospitals

The main tasks of the hospitals in EDS are the provision of medical care and treatment of the victims and hygienic and anti-epidemic provision of the affected regions, and by order of the chairman of the respective commission (district, municipal) and neighboring regions [7].

The network of health facilities and their infrastructure must be ready to provide timely emergency and specialized medical care to the population in emergency and disaster situations [2, 7–9, 16–18, 27].

Knowledge of the factors that can lead to damage to health or endanger the lives of people in EDS allows them to predict the medical consequences, to clarify ways to combat them, to take the necessary preventive measures to limit the medical consequences, to organizing emergency medical measures and eliminating the consequences of emergencies [7, 8].

The hospitals provide the necessary human and material-technical resources, create an effective organization and keep in constant readiness the forces for immediate action in EDS [2, 8].

Before the occurrence of the disaster, the head of the medical institution – hospital must make a comprehensive assessment of the condition and the ability of the health institution to work in such a situation. During this period, the action plan for the EDS and the work of the medical institution for the medical provision of the population of the respective territorial unit must be developed, in accordance with the plan of the Ministry of Health [2, 8, 16–18, 27].

The plan is developed in different variants depending on the expected nature and severity of the medical losses and includes the following [7, 8]:

1. Creation and maintenance of a system for notification of the employees of the hospital.

2. Calculation of the medical losses and of the necessary forces and means for rendering medical aid to the victims.

3. Necessary medical teams and formations for rendering emergency and urgent medical aid, as well as inpatient medical care.

4. Creation of an appropriate structure (restructuring) of the bed base of the medical institution if necessary.

5. Organizational scheme for providing medical care at the site of the lesion, during transportation and in the medical institution.

4.1.3 Regional health inspectorates (RHI)

The Regional Health Inspectorates (RHI) are developing a work plan for the Hygiene and Epidemiological Provision and Inspectorate in the disaster area. It must be in accordance with the plan for conducting rescue and other urgent works of the regional and municipal commission [7, 8].

In case of disasters, the director of RHI clarifies the place and nature of the event, then organizes and conducts research and control of environmental hygiene parameters in the affected areas, in industrial and other sites in terms of toxic substances, dust, noise, vibration, microclimate, radiation and other harmful factors. This activity is carried out by pre-formed and trained anti-epidemic teams of RHI on the territory of the disaster [7, 16–18, 27].

The main task of the RHI is the organization and implementation of disinfection, disinsection, deratization and control of the degassing and decontamination activities in the affected areas after the normalization of the situation [8].

The number and nature of foodstuffs affected by the disaster, the type and quantity and the nature of the damage must be clarified, and enhanced sanitary control must be organized over all foodstuffs in the disaster area. This requires organizing and conducting intensive laboratory control over the affected catering establishments and food industry establishments.

Based on the conclusions of the analysis, the RHI prescribes measures for compliance with hygiene standards and requirements for all factors of the working environment. After conducting a control for hygienic efficiency of the conducted measures, a conclusion is given for safe working conditions with a view to resuming regular operation of the affected sites [7].

4.2 Organization of MPIP

4.2.1 Surgical and trauma care

The relevant clinics from the Multidisciplinary Hospital for Active Treatment (MPHAT) and Hospital for Emergency Medicine, as well as the clinical departments of surgery, orthopedics and traumatology, resuscitation and anesthesiology in the district, regional and municipal hospitals are used as a base. If necessary, the bed capacity of the same hospitals is used. In some cases, staff and facilities from other surgical units (ophthalmology, maxillofacial surgery, etc.) of hospitals can be used. This allows in EDS for a short time and without significant difficulties to be included in the organizational scheme of medical care. About 60% of the inpatients can be discharged and a specialized bed stock can be released for the needs of the victims. For work in a trauma center, if necessary, medical teams (trauma, surgical, etc.) are formed on a functional principle, without seriously violating the readiness of these wards for admission and treatment of victims. These teams must arrive at the scene no later than one hour after the emergency medical teams. At this time, the medical situation, the scope of work and the possible number of required specialized surgical teams should be clarified. If necessary, they can be strengthened with teams from medical institutions in neighboring regions [7, 8].

4.2.2 Radiological care

In case of accidents at the NPP, in case of incidents with sources of ionizing radiation, in case of cross-border transfer of radioactive substances in the therapeutic wards of the MPHAT, an opportunity must be created to provide radiological assistance to the victims. All therapeutic wards of the medical establishments, in the vicinity of the NPP, must be ready for possible admission of radiation patients and those with combined radiation injuries. For this purpose, it is necessary for physicians-therapists to have radiobiological training and in case of radiation conditions to organize the work of the ward in radiological terms and to conduct radio protective measures in the medical institution. The existing departments of radiotherapy and isotope diagnostics, based most often in oncology dispensaries, oncology hospitals, etc., have a corresponding place in this functional radiological system. The medical staff from these radiology departments are involved in providing radiological assistance to the victims [7, 20–26].

The duty and responsibilities according to the International Atomic Energy Agency (IAEA) require doctors to have the relevant knowledge of radiation protection, which enables them to initiate preliminary treatment and provide assistance to specialized units in the event of a radiation accident. Another task of health care in the section of radiological care is the control of the radiation parameters of the working and living environment, which directly affect the person [22]. The radiation control department must organize and conduct the necessary radiation-hygienic measures on the given territory.

The organization of the radiological assistance is related to the plan for radiation protection of the country in case of an accident at the NPP, which ensures the implementation of the plan in its medical section [7, 20–26].

4.2.3 Toxicological care

The organization of toxicological care uses a mixed approach, including the establishment of staff and functional units. The expanding chemical pathology necessitated the establishment of full-time clinics and toxicology departments in the settlements with large sites of the chemical industry. These units, in addition to providing toxicological assistance to the population, also serve to train medical personnel in this field. In the other hospitals, the therapeutic wards are re-profiled into toxico-therapeutic ones for admission of toxicologically ill patients for emergency toxicological care. Good interaction should be ensured with the intensive care unit of the hospital [7, 10].

On the basis of the staff clinics and toxicology departments, specialized medical teams are established, provided with medical and sanitary equipment and transport. These teams must be constantly prepared to work in a chemical outbreak or to strengthen the therapeutic wards of neighboring hospitals where toxicologically ill patients are hospitalized [7–10].

4.2.4 In outbreak of biological contamination (OBC)

When creating an outbreak of biological infection all types of medical care (first medical, first medical, qualified and specialized) are within the area of the outbreak. For this purpose, the medical and prophylactic establishments on the territory in the OBC mainly are used [7–10].

4.2.5 In the outbreak of combined defeat

The first medical and qualified medical care for vital indications is provided at the medical center according to various schemes. Most often, two groups of medical forces are created for rescue operations in the center of a combined defeat: in the biologically infected area and outside it [2, 7–10].

5. Medical triage in an OTD

Medical sorting is performed in the OTD. According to the severity of the OTD, the distribution of the victims by sorting groups allows for homogeneous treatment and prevention measures [7–10, 16–18, 27].

Depending on the severity of the injuries the victims are sorted into two main groups in Bulgaria [7, 10]:

a. Slightly injured (40%). This group includes victims of soft tissue injuries who do not need hospital treatment;

b. Moderately and severely injured (60%). These are victims who need urgent medical attention and inpatient treatment. This group can be divided into four subgroups (**Figure 3**):

- Group T4. Dying and agonizing;
- Group T3. Injured with some slightly graded cranio-cerebral injuries or some insignificant spinal cord injuries and others slightly injured (20–40%). These are persons with injuries of the small bones of the frontal part of the skull (mandible, nose), medium and small soft tissue injuries, etc.;
- Group T 2. Persons to whom medical care can be postponed for 6–8 hours (20%). These victims have an advantage of transportation, but do not need extreme treatment. These include victims with some surface thoracic or

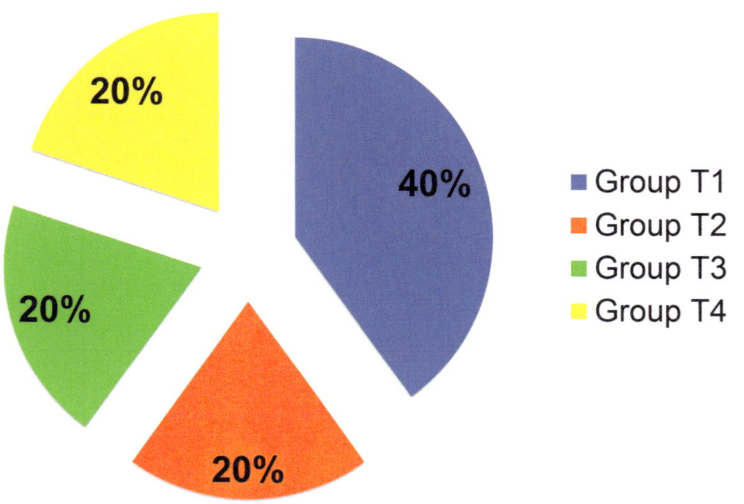

Figure 3.
Triage groups in an OTD. (the worst prognostic option).

abdominal injuries, or some not penetrating injuries to the uro-genital tract, or some negligible blood vessels rupture, burns less than 20% of the body surface in people of active age, but without other injuries;

- Group T 1. Persons whose trauma is defined by immediate vital disorders (20–40%). This group includes victims with respiratory failure, cardiac arrest, ventricular fibrillation, huge bleeding, shock, increased intracranial pressure, burns of the face and respiratory tract, or extensive burns occupying more than 20% of the body surface; poly-trauma, etc. The victims of this group receive emergency first aid in order to stabilize the basic vital functions and have the highest priority in treatment.

Particular attention in medical sorting should be paid to groups dangerous to others and in need of urgent medical attention. Dangerous for others are those infected with poisonous substances, radioactive substances, bacterial agents and patients with particularly dangerous infections (PDI), acutely unlocked or exacerbated chronic psychiatric diseases, some acute mental disorders etc. This danger imposes the need for sanitary treatment of the infected and isolation of patients with PDI, mental disorders as well.

6. Preventive measures for primary risk reduction

Prevention of the population in favor of public health and the future of the nation is fundament for stabilization of the health system in case of EDS [7–10, 16–18, 27]. Prophylaxis measures (PM) used for primary risk reduction is only first group of tasks and primary step for government and national health system. By **Figure 4** is presented risk reduction actions before, during and after earthquakes (**Figure 4**). Some of the most important groups of methods for prevention of public health are:

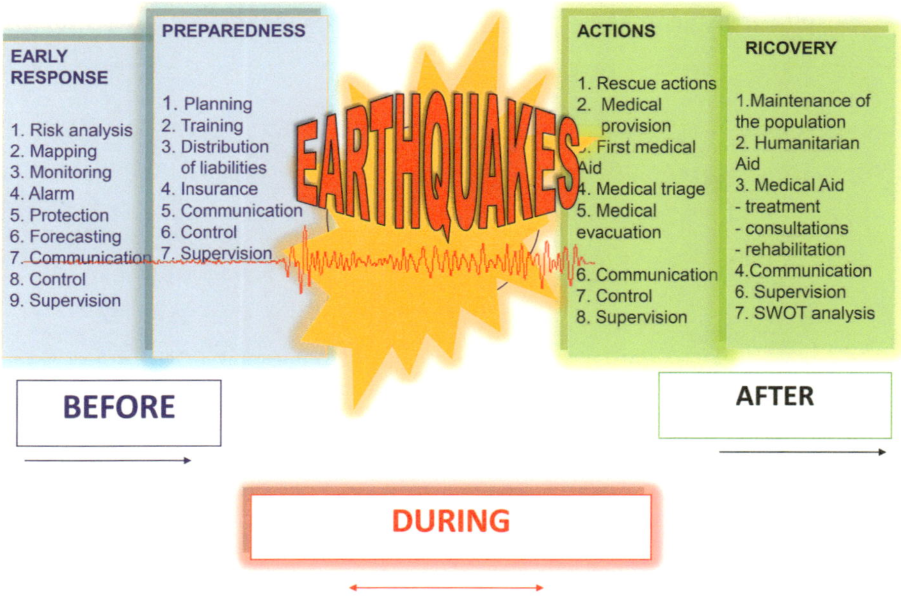

Figure 4.
Risk reduction actions before, during and after earthquakes.

1. Earthquake prognosis.

2. Establishment of well-equipped and prepared seismic stations and a notification system.

3. Scientifically based anti-seismic construction.

4. Compilation and timely updating of a map of OTD threats.

5. The approaches to the hospitals should be known and organized in case the entrances are covered with destructions. For each hospital, which is located in an area with high seismicity, a helipad should be provided.

6. Systematic training of medical staff.

7. Systematic preparation of the population to react and provide main first medical aid steps.

7. Conclusion

Unlike other disasters, earthquakes by sudden onset and rapid flow are characterized. Trough detailed analysis of earthquakes a number of specific features are described. First instance place, unlike other natural risks, in earthquakes the response time is practically a very limited resource. Good preparation and collaboration of various institutions with the healthcare system in case of the OTD with the multi-factorial nature of the risks due to an earthquake and the possible consequences is definitely connected. All levels of health risk management of the health care system for medical provision of the affected population are included. On the EMCC is based the field work on the medical provision of the injured people. A real staged process, but not a condition is proper medical triage and the provision of medical care in OTD. Two-stage system with evacuation by appointment as method for MPIP in case of traumatic defeat of EDS is used. A guarantee for the adequate provision of good medical practices in case of EDS is defined with good enough developed health risk management system, together with the necessary number and structure of resource capacity.

Conflict of interest

The author declares no conflict of interest.

Author details

Diana Dimitrova
Emergency CC, Department of EM&MCS, Stanke Dimitrov, Bulgaria

*Address all correspondence to: d.dimitrova.phd.md@abv.bg

References

[1] The UN. Natural Hazards, Un Natural Disasters. The Economics of Effective Prevention 24-30 [Internet]. Available from:http://documents.worldbank.org/curated/en/620631468181478543/pdf/578600PUB0epi2101public10BOX353782B.pdf

[2] EC.JRC.IPSC. European Commission. Joint Research Centre. Institute for the Protection and Security of the Citizen. Current status and best practices for disaster los data recording in EU member states. 30 [Internet]. Available from: http://pprdeast2.eu/wp-content/uploads/2015/11/JRC-SOTA-Best-Practices-Loss-Report.pdf

[3] WHO. Earthquakes [Internet]. Available from: https://www.who.int/health-topics/earthquakes#tab=tab_1

[4] BGS. Earthquake database search [Internet]. Available from: http://www.earthquakes.bgs.ac.uk/earthquakes/dataSearch.html

[5] NOAA. The Significant Earthquake Database [Internet]. Available from: https://www.ngdc.noaa.gov/nndc/struts/form?t=101650&s=1&d=1

[6] CGS. Earthquake Data and Reports [Internet]. Available from: https://www.conservation.ca.gov/cgs/earthquake-data

[7] Dimitrova D. Medical provision of the population during earthquakes - readiness of Emergency medical care center in Blagoevgrad region. [PhD thesis].Sofia: MU; 2015.

[8] Dimitrova D., Management of medical provision of the population during earthquakes in Bulgaria. Current state and development perspectives. 1st ed. Propeller: Sofia; 2016.

[9] Dimitrova D.: Challenges in emergency medical care center in Bulgaria in case of mass traumatism during disaster situations – Epidemiological survey, 2018; 130:e139. DOI:10.1016/j.resuscitation.2018.07.299

[10] Dimitrova D., Medicine of the disastrous situations. Educational exercises and seminars - from theory to practice. V1. 1st ed. Propeller: Sofia; 2016

[11] NOAA.Sumatra, Indonesia Earthquake and Tsunami, 26 December 2004 [Internet]. Available from: https://ngdc.noaa.gov/hazard/26dec2004.html

[12] The balance. Haiti Earthquake Facts, Its Damage, and Effects on the Economy. The 2010 Earthquake Caused Lasting Damage [Internet]. Available from: https://www.thebalance.com/haiti-earthquake-facts-damage-effects-on-economy-3305660

[13] NPAJ. Damage Situation and Police Countermeasures.8 March 2019. National Police Agency of Japan. Retrieved 13 March 2019

[14] Dimitrova D., Evidence based medical practice. Aspects of emergency and disaster medicine. 1st ed. Propeller: Sofia; 2015.

[15] WHO. Mass casualty management systems: strategies and guidelines for building health sector capacity [Internet]. Available from: https://www.who.int/publications/i/item/mass-casualty-management-systems-strategies-and-guidelines-for-building-health-sector-capacity

[16] Dimitrova D., Role and place of emergency medical care center in Republic of Bulgaria for providing prehospital care during disaster situations. PDM, vol 28, supplement 1, Publ. CUP, 2013, ps97

[17] Dimitrova D., Psychiatric aid in earthquakes.-In: Medical and

psychiatric provision in missions and operations abroad - contemporary challenges. S., 2012, p13

[18] Dimitrova D., Effective management of the working shift in EMCC [thesis]. Blagoevgrad, Atera: SF-SWU; 2001.

[19] WHO. Chemical releases associated with earthquakes [Internet]. Available from:https://www.who.int/publications/i/item/chemical-releases-associated-with-earthquakes

[20] Data world. US DOE/NNSA Response to 2011 Fukushima Incident: March 2011 Aerial Data [Internet]. Available from: https://data.world/us-doe-gov/46ef4d73-da7c-4f15-9648-154344708268

[21] Data world. Nepal - Causalities caused by earthquake, 2015 [Internet]. Available from: https://data.world/opennepal/d63ec0b5-2e99-4991-b2c2-97b41eecec6c

[22] WHO: The response to the 2015 Nepal earthquakes: the value of preparedness [Internet]. Available from: https://youtu.be/P29RknVelNM

[23] WNA. World Nuclear Association. Chernobyl Accident 1986. [Internet]. Available from: https://www.world-nuclear.org/information-library/safety-and-security/safety-of-plants/chernobyl-accident.aspx

[24] Health Effects of the Chernobyl Accident and Special Health Care Programmes, Report of the UN Chernobyl Forum, Expert Group "Health", World Health Organization, 2006 (ISBN: 9789241594172)

[25] WNA. World Nuclear Association. Fukushima Daiichi Accident [Internet]. Available from:https://www.world-nuclear.org/information-library/safety-and-security/safety-of-plants/fukushima-daiichi-accident.aspx

[26] IAEA. Fukushima Nuclear Accident [Internet]. Available from: https://www.iaea.org/resources/safety-standards

[27] Dimitrova D., Emergency conditions in disastrous situations. Shock.-In: Medicine of the disastrous situations. 1st ed. ARSO, S.; 2011, 374-383

Chapter 5

Question of Livelihood in the Light of Disaster: With Special Reference to Flood of Bahraich, India

Keyoor Pathak and Chittaranjan Subudhi

Abstract

Floods in India is a repetitive one due to natural reasons like excessive rain and man-made mistakes like encroachments of water bodies, heavy rain-falls and so on. The chapter is based on a field study of Bahraich, a district of eastern Uttar Pradesh, India. Interestingly, the district shares its boundary to neighboring country Nepal that influences the occurrence of floods in the region. The district is also prosperous in water resources such as the great Ghaghara river and many small and big ponds and lakes are in the district that becomes a cause of sorrow in the rainy seasons. The key concern of the chapter is to understand the challenges of livelihood of the rural communities which is annually threatened by devastative floods.

Keywords: capital, flood, disaster, livelihood, vulnerability

1. Introduction

> *We do not see our hand in what happens, so we call certain events melancholy accidents when they are the inevitabilities of our projects (Stanley Cavell)*[1].

India, a country of rivers and is a big centre of the flood. It is a regular and the most destructive disaster in India. The earliest evidence of devastative flood dates back to the flood of the Indus Valley Civilization that sabotaged the great civilization. In independent India, the first major flood was recorded in 1953, and following the flood's impact first national policy on the disaster was set up in the year of 1954. Notably, India accounts for 1/5th of the global lives' loss, and around 30 million people are evacuated every year. The area vulnerable to flood is more than 40 million hectares, and the average area affected is 8 million hectares[2]. Whatever the disastrous flood has done vandalisation should not be measured only in economic terms; however, from 2011 to 2016, INR 144665.79 crore has been estimated as the total damage in the country. In 1953, total damage around INR 52.40 crore had been reported,

---

[1] Ted [1]. *Acts of God: The Unnatural History of Natural Disaster in America*. Oxford University Press, New York.

[2] *Flood: Trends & social Impacts in Indian Context*. https://www.ssvk.org/koshi/analytical_articles/joshi.pdf Accessed: 19.09.2020

while in the year of 2016 it reached to INR 57291.098 crore. The most vulnerable states to flood are Uttar Pradesh, Bihar, Assam, West Bengal, Gujrat, Odisha, Andhra Pradesh, Maharashtra, Punjab and Jammu & Kashmir. Uttar Pradesh, Bihar, West Bengal, and Assam account for 17 percent of India's geographical area but disproportionately account for 43–52 percent of all flood-prone areas of the country [2].

Iwasaki [3] has reported that, these floods or cyclones severely squeezed people's livelihood. It affects the traditional economic base of the family that changes the family's main source of income [4]. Sina et al. [5] mentioned that, displacement due to flood is a major challenge of restoring the livelihood which needs timely assessment in building resilience in livelihoods. The aid packages are usually short-term fulfill of the needs of the victims but rarely focus the long-term revival of their livelihoods. The current institutional mechanisms are lagging behind the large-scale post-disaster reconstruction [6]. These frequent occurrence of floods have caused a blow to the livelihood resilience of the poor and marginalized sections [7]. The severity of the disaster can be reduced with the help of local and national commitments [3] along with the intervention of livelihood diversification programs in the flood affected area [3]. Along with this, social cohesion is playing a pivotal role for restoring the livelihood of disaster-displaced people [8].

Bahraich (part of eastern Uttar Pradesh) is one of the districts which have been under the regular influence of flood for decades, but in recent years, an alarming increase in the devastation can be visited in the region. It has serious repercussions, such as displacement, migration, poverty, hunger, unemployment, diseases, and many more issues. People are compelled to accept the tragedy of their life and trying to compromise themselves with the unfortunate socio-economic condition.

2. Reviewing livelihood and disaster

Livelihood: the term 'livelihood' as a systematic study developed from basically the rural studies, since then it has been used in various dimensions in academia. The conventional meaning of the term is an economic activity, while in fact, apart from the economic aspect it comprises a wide range of issues such as ecological, political, cultural, environmental and so forth. The formal analysis of the term can be given as "a livelihood comprises the capabilities, assets (both natural & social) and activities required for a means of living; a livelihood is sustainable which can cope with the recovery from and shocks, maintain or enhance its capabilities and assets, both now and in the future, while not undermining the natural resource base" [9]. This analysis primarily consists of three concepts: capability, equity and sustainability; and these are interconnected and interdependent in its proper implication. Scholars emphasize the relationship as "each is also both and means.... linked together, capabilities, equity and sustainability present a framework or paradigm for development thinking which is both normative and practical" [9]. Here, through separate descriptions, the core meaning of the terms can be understood.

Capability: in the developmental studies this was firstly used and systematically studied by Amartya Sen as a theoretical foundation to understand human development, he maintains that income, utilities, resources, and wealth act as means towards an end not an end in themselves, further he states that development should be understood as the removal of major barriers to our freedom like poverty, illiteracy, poor economic opportunities and so on [10].

Equality: the traditional method of measuring the equity has been concerned with only relative distribution of income, while it keeps a vast meaning like property, opportunity and capability; and avoids discrimination on any ground [9].

Sustainability: there is a prosperous series of intellectual criticism on the developmental model in which production, employment, and poverty reduction are a key concern. It claims that this is mainly made for the welfare of the capitalism and market, and lesser responsible towards the people of the rural and marginalized communities. In recent decades, the developmental model moved to the diverse framework of the society and inclined towards the social, cultural and political field of the communities [9].

Hence, the term 'livelihood' does not merely denote the issue of employment or source of income, but the entire construction of a community viz., polity, society, culture, economy, wherein people live for centuries, comes under the periphery of 'livelihood'.

Disaster: The disaster is a natural phenomenon or just a deleterious consequence of human's unnecessary intervention in the natural world, which is a controversial debate across the world. Difficult to conclude that this is an 'act of God' or the human's self-made tragedy, but cannot be denied the fact that modern technological advancement has multiplied the intensity and consistent occurrence of the disaster. The term 'disaster' is derived from astrological science and has been understudies for more than one century, but in terms of its specific academic studies, it has expanded since the 1950s [11]. To formulate the term, the World Health Organization on Emergency and Humanitarian Action (EHA) states that a disaster is an occurrence a level of suffering that exceeds the capacity of adjustment of the affected community[3]. The Centre on the Epidemiology of Disaster, Brussels holds the view as "a disaster is a situation or event which overwhelms local capacity, necessitating a request to a national or international level for external assistance, an unforeseen and often sudden event that causes great damages, destruction and human suffering" [11]. Similarly, "disaster is seen as a process leading to an event that involves a combination of a potentially destructive agent from the natural or technological sphere and a population in a socially produced condition of vulnerability" [12]. There is no major dispute on the character of disaster as 'massive human suffering', but has big dissension over it is either 'man-made' or the 'revenge of god'. Of course, nature naturally generates a number of disasters like cyclones, earthquakes, floods, famine and many more, but the role of modern science and technology should not be overlooked that has escalated the pace of disaster [1]. For instance, the flood of Uttarakhand in 2013, and the Kosi flood of Bihar in 2008 are basically the consequences of unnecessary human encroachment in the domain of nature [13].

3. Glimpses of the tragedy

The history of flooding in the region goes back in the deep past. In the known history the region has been gravely affected by flooding from 1922 to 1925. 1946, 1954, 1955, 1960, 1961 and 1963 were also the year of flooding, but the flooding of 1969 was much more devastative than previous years. 1978, 1980, 1981, 1982 and 1983 were also the year of tragedy. In the year of 1983, the sudden release of water from Girija Bairaj of Nepal caused severe damage to the region, while this was not the year of heavy rainfall. Hundreds of people died due to the sudden release of water without any prior information. 1986, 87 and 1988 were not untouched by the flood. In the year 1990 flood-affected a few pockets of the region. In the year 1991, there was no flood, but the Ghaghara river induced soil erosion. Again, in the year 1992 floods made a catastrophe. In the year 1993, it broke the previous records. The

[3] *World Health Organization, 2002*. World Disaster Report.

year 1998 was also the year of flood due to the sudden release of water by Nepal. The year of 2000 multiplied the pain of the region. Gopiya Bairaj of Nepal released water and hundreds of people died and thousands of displaced. In the last few decades, dozens of villages have been physically disappeared and the population is displaced by the flood and many more are under threat of disaster. A few of the examples of the villages that are already collapsed in recent decades or under the process are given below:

Collapsed village	Year of displacement	Number of displaced people
Khargapur	1968	10,000
Gangapur	1994	4000
Panchadupur	1998	4000
Baharpur	1998	5000
Silauta	2001	7000
Umaraiya	2001	3000
Bhauri Sipahiya	2002	8000
Munsari	2003	4000
Maikapurva	2003	6000
Sansari	2004	2500
Kapraul	2004	4000
Golaganj	2006	8000
Magraul	2007	7000
Pipri	2007	4000
Bansgadhi	2011–2012	6000
Tarapurva	2012	3000
Jarwal	Till the date	4000
Kayampur	Till the date	3000
Baundi	Til the date	3000
Jogapurva	Till the date	3000

Source: Super Idea, Sept.2012[4].

4. Nature's revenge or human's mistake?

Tracing the reasons for the flood in the region, we find a few major human activities that induce disaster. If we blame nature for this disaster it would be an injustice, not only to nature but also to the people of the area. A sudden release of water by the government of Nepal through dams like Gopiya & Girija is one of the main reasons. The people claim if river water comes slowly in a natural way, it is lesser devastative compared to sudden releases from dams. The rapid flow of water comes only after the collapse of dams or highways or such kinds of big constructions, not by the natural processes of the flood. These unplanned and unmapped developmental projects have made hindrances in the ways of floodwater, therefore the water stuck up to 20–30 feet for many days. Apart from these, sand mining is also a major cause of soil erosion and changes in the direction of

[4] *Super Idea, Sept.2012*. It is a local magazine of the district that covers regional news usually.

the flow of water. Illegal sand mining in the basin of the Ghaghara is rampant. And astonishingly, the nexus of the local public representatives, contractors and bureaucrats are involved in this activity as villagers claimed and were also deeply observed during the field study.

5. Loss of capital & vulnerability of livelihood

The loss of the community of the region can be best understood through the theoretical lens of Piero Bourdieu's 'capital' that he describes as "accumulated, human labor, which can potentially produce different forms of profit" [14]. Further, he classifies it in three sections as cultural, social and economic capital.

Social capital: "Social capital is the aggregate of the actual or potential resources which are linked to possession of a durable network of more or less institutionalized relationships of mutual acquaintance and recognition or in other words, to membership in a group" [14]. The displacement caused by flood displaces and damages not only the physical body of a community but also crushes the social institution and relationships like kinship, lineage that has been a product of centuries-old social interactions and behaviors. One villager Kamlesh Kashyap (Baundi, 35 years old man) asked "earlier we used to marry in our own communities, but now it is not possible, since most of the people of our community have migrated towards other states for their survival. We are left here and staying on the dams or roadside, so how is it possible to get a suitable bride or groom?". The structure of kinship and marriage were such kinds of capital they could face and challenge the problems of daily life through.

Cultural capital: cultural capital can exist in three forms as 'embodied state' (mind & body), 'objectified state' such as books, instrument, etc., and 'institutionalized state' like academic qualification etc. [14]. An around seventy years old man Shankar Kashyap of Jogapurva showed me the pity of the loss of cultural capital. "Till ten years ago we used to make our house by the wild bushes, unfortunately, it is not flourishing due to the annual occurrence of flood, so now we purchase it from other areas to build the houses. Similarly, there were several herbal plants that were useful in curing many diseases, and we also used to make domestic products like toys, pots etc. New generations have neither the knowledge of wild herbs-shrubs, plants and bushes nor have the knowledge of making all that thing". Here, we see flood excludes the community along with their traditional knowledge.

Economic capital: "economic capital is at the root of all the other types of capital and that these transformed, disguised forms of economic capital, never entirely reducible to that definition, produce their most specific effects only to the extent that they conceal the fact that economic capital is at their root" [14]. Lakshmi Devi (Jogapurva, 50 years old woman) told the pain of her life how she lost her ornaments while flood water entered into the house. "I had some gold & silver made ornaments that I used to keep within the soil of the room so that no one could steal. I had purchased it by savings from my rigorously earned income. It was kept for the days of crisis or ceremony like the daughter's marriage or disease of the family members, but the spate of water entirely wiped out my house along with my ornaments. When floodwater came in, I was in my farmland that is why I could not save all that".

The vulnerability of livelihood: the term vulnerability is widely used but is basically a vague term and its meaning varies across the disciplines. However, it may be considered as "vulnerability is an individual or group's reduced capacity to cope with, resist, and recover from the impacts of a natural or human-made hazard"[5]. From the lens of the given definition we just look at the story of Sunil (Baundi, 40 years old man) who asked "I had ten bigha farmland in which I used to cultivate crops affluently. Now every year water is stuck in the land so now we are neither able to cultivate in that nor sell since purchasers will give a very little amount of money. Even we are in confusion we should start a business or migrate towards the cities or should continue in risky agriculture. Going to other cities is difficult since I have four small children. Due to the dearth of money, I am not able to provide quality education. We are just trying to alive ourselves...nothing more". The present and the future of the communities of the region is in dark. Several farmers have changed their production of crops. Sukhan (Jogapurva, 45 years old man) told "earlier we used to produce wheat and rice like commercial crops, that was helpful in fulfilling my livelihood issues, but now I am compelled to change my agricultural production, in which there is no benefit and cannot properly fulfill the need of my livelihood. Flood has converted my fertile land into desert, so now I just plant watermelon in my farmland".

Losing all forms of *capital* and displacing from its native places, the communities of the area are in the process of just being a gathering of the people or crowds in different cities or nearby places. They are uprooting from its age-old livelihood sources that were sustainable in its nature. Their vulnerability arises grave questions to the policymakers and academicians. What would be the future of such communities in terms of health, education, employment and etc.? What can we expect from such communities in the domain of creativity and productivity, which are the core value of human civilization?

6. State & its machineries

A public representative of the district, Rajesh Tripathi (Bahraich, 50 years old, man) told: "There is no strong long-term planning of administration, they just work during the flood, and once the flood is over, they get rid of their responsibilities". Here, the response and action of the local administration can be understood as in the three periods, before the flood, during the flood and after the flood. There is a serious dearth of long-term planning and preparedness before the state, so cannot be denied the fact that it is the very reason for the ravage of the flood. The eyes of the administration open just after the coming of the disaster. During the disaster, local administration works only as an agency of relief distribution like biscuits, plastics, rice, gram, matchbox and etc., unfortunately, discrimination in terms of caste, class, gender and so on, in this allocation is also clearly visible. Astonishingly, the role of the administration in the post-flood period is much more deleterious than the flood. Corruption in rehabilitation processes, social-conflict induced by the official's work-culture, unemployment, diseases and so many socio-economic problems rapidly emerge in this period. For instance, massive soil erosion takes place during the disaster, which erases the demarcations of farmlands of the people. They go to the district's land department offices for the re-demarcation of the farmlands, but officials demand a heavy amount in bribe, which is very difficult for the people, who had recently been ruined by the flood. Therefore, they try to manage it on its own community-based understating, but several times it turns into social mayhem. Sluggish and irresponsible practices of the administration can be noticed

[5] Nani Maiya Sujakhu, Sailesh Ranjitkar, Rabin Raj Niraula, Muhammad Asad Salim, Arjumand Nizami, Dietrich Schmidt-Vogt & Jianchu Xu. 2018. *Determinants of livelihood vulnerability in farming communities in two sites in the Asian Highlands, Water International*. https://www.tandfonline.com/doi/full/10.1080/02508060.2017.1416445 Accessed: 18.09.2020

in the settlement of sources of survival, like fishing, agriculture and so forth. Apart from the rampant corruption and irregularities in the rehabilitation and relief process, another important feature is, which is totally absent from the public policies; government is ignorant of the loss of social and cultural capital, their central policies move around merely economic capital.

7. Conclusion

We should not hide our face from the fact that our unmapped and uneven developmental policies have multiplied the pace and intensity of natural disasters that subsequently sabotage the rural livelihood setup, which is sustainable in its form from the time immemorial. The need of the hour is to come out from such public policies that overlook the interest of the agrarian communities since India's around seventy percent population live in the rural areas only. Apart from policy-making, one of the major problems is in the implementation of existing policies, since the bureaucracy is indulged in extreme corruption and money-making through illegal sources that finally hampers the livelihood issues of the communities.

Acknowledgements

We would like to acknowledge the contribution of Harry Raj, who helped us in shaping the chapter.

Note

The researcher has changed the original name of the respondents to protect their identity and to maintain confidentiality.

Author details

Keyoor Pathak[1] and Chittaranjan Subudhi[2*]

1 Council for Social Development (An ICSSR Research Institute, Hyderabad Regional Centre), Telangana, India

2 Department of Social Work, Central University of Tamil Nadu, Thiruvarur, India

*Address all correspondence to: chittaranjan@cutn.ac.in; chittaranjan.subudhi@gmail.com

References

[1] Ted Steinberg, 2000. *Acts of God: The Unnatural History of Natural Disaster in America*. Oxford University Press, New York

[2] Keyoor, Meenakshi R. J., & Subudhi C. 2019. Disaster, Development and Livelihood with Special Reference to Floods in India. *Disaster Advances* 12(5).

[3] Iwasaki, S. (2016). Linking disaster management to livelihood security against tropical cyclones: A case study on Odisha state in India. *International Journal of Disaster Risk Reduction*, *19*, 57-63.

[4] Naithani, S., & Saha, A. K. (2020). Social capital and livelihood strategies in response after 2013 Kedarnath disaster (India). *Disaster Prevention and Management: An International Journal*.

[5] Sina, D., Chang-Richards, A. Y., Wilkinson, S., & Potangaroa, R. (2019a). A conceptual framework for measuring livelihood resilience: Relocation experience from Aceh, Indonesia. *World Development*, *117*, 253-265.

[6] Daly, P., Mahdi, S., McCaughey, J., Mundzir, I., Agus Halim Nizamuddin, A., & Srimulyani, E. (2020). Rethinking relief, reconstruction and development: Evaluating the effectiveness and sustainability of post-disaster livelihood aid. *International Journal of Disaster Risk Reduction*, 101650.

[7] Patel, S. K., Mathew, B., Nanda, A., Pati, S., & Nayak, H. (2019). A review on extreme weather events and livelihood in Odisha, India. *Mausam*, *70*(3), 551-560.

[8] Sina, D., Chang-Richards, A. Y., Wilkinson, S., & Potangaroa, R. (2019b). What does the future hold for relocated communities post-disaster? Factors affecting livelihood resilience. *International journal of disaster risk reduction*, *34*, 173-183.

[9] Chambers, R. & Conway, G. 1992. *Sustainable Rural Livelihood: Practical Concepts for the 21st Century.* Discussion paper No. 296. Brighton: Institute for Development Studies.

[10] Kuhumba, S. 2018. Amartya Sen's capabilities Approach as Theoretical Foundation of Human Development. *Journal of Sociology and Development. Vol.1(1)*

[11] Reddy, S. 2013. *Clash of Waves*. Indos Books, New Delhi.

[12] Smith, O. Hoffman, A. Susanna, M. 1999. *The Angry Earth* (edited). Routledge, New York.

[13] Bidwai, P. June, 2013. India Floods: A man Made Disaster. *The Guardian*.

[14] Richardson, J., 1986. *Handbook of Theory and Research for the Sociology of Education.*, Westport, CT: Greenwood

Chapter 6

Simple Housing Solution Project: (Re) Building in Critical Situations

Leandro Di Gregorio, Gustavo Guimarães, Fernando Danziger, Graziella Jannuzzi, Eduardo Qualharini, Aline Veról, Assed Haddad, Marcelo Miguez, Virgílio Cruz and Daki Bianchessi

Abstract

One of the significant challenges of recovery in critical situations (post-disaster, post-conflict, refugee settlement, among others) is the prompt and adequate housing (re)construction with scarce resources, and the affected population's involvement. The Simple Housing Solution (SHS) project consists of a proposal for a methodology for (re)construction of homes and other small buildings (schools, health clinics), using low-cost construction technologies and community labour (mutual help system). The SHS project's body of knowledge was organised in the form of a course with video lessons on YouTube and a website translated into different languages. The idea is to provide material that may help affected populations to work towards their recovery, with the support of qualified professionals (engineers and architects). The purpose of this chapter is to present the SHS Methodology and its main outputs.

Keywords: disaster recovery, post-conflict recovery, humanitarian aid, housing recovery, livelihood recovery, community recovery, reconstruction, risk management

1. Introduction

In many disasters, housing loss is the second major concern, coming soon after casualties. Simultaneously, climate change is a global reality [1] and requires adaptations of the built-up environment to adverse climate scenarios, such as hurricanes and tornadoes.

In the disaster recovery phase, rescue and relief activities are conducted relatively quickly. Nevertheless, the pace is much slower in the recovery phase, and problems drag on for years after the media have withdrawn [2].

It must be highlighted that, from an economic viewpoint, the loss of a home may represent more than family savings over a generation time. The loss of housing can also significantly increase the degree of vulnerability of those affected, with consequences for the loss of livelihoods, deteriorating physical and mental health conditions, unsafe environment for study and education, and family breakdown.

According to United Nations Development Programme (UNDP) and International Recovery Platform (IRP) [3], each house built represents an individual project, and grouping together hundreds, thousands, and even millions of homes constitutes much broader reconstruction programmes. From this perspective, it should be considered that recovery needs to be approached from two aspects: collective solutions and individual solutions that reflect each family's needs and provide specific recovery routes that take into account the peculiarities of each nucleus.

Housing provision should be understood as a process (and not merely providing a product) that should involve the people stricken by the disaster and the communities directly or indirectly affected by the situation [4]. According to UNDP and IRP [3], victims who can immediately begin their reconstruction effort will want to do so as soon as possible. Although the impact that a speedy start to recovery has on morale, those responsible for planning must ensure that the earlier vulnerabilities are not repeated. The authors also claim that in areas where immediate work is possible, there is less dependence on temporary housing, and victims feel that recovery is progressing. On the other hand, Leykin et al. [5] pinpoint three aspects that contribute directly to the resilience of a community in emergencies: preparation, leadership and collective effectiveness. Barakat [6] indicates that joint reconstruction must be carefully organised and managed, requiring managerial and technical expertise by the agencies involved in the implementation.

Concerning technical aspects, Marcial Blondet comments that in developing countries most people live in non-engineered low-rise constructions made of inferior materials, thus making them more vulnerable [7]. Considering the issues related to post-disaster reconstruction, Yi and Yang [8] mention that research efforts in developing countries in Asia and South America are far behind those in developed countries, and Africa is rarely addressed.

This chapter introduces Simple Housing Solution (SHS), a methodology designed to facilitate the reconstruction process in critical situations, and necessary for recovery with few resources (i.e., post-disaster, post-conflict, relocation from risk areas, refugee settlements). It was conceived with the philosophy of gathering basic knowledge that can help build housing units and essential collective equipment (schools, health clinics), in a joint effort (community labour system), by adopting low-cost constructive technologies. The idea is to help local governments, support agencies and, above all, vulnerable communities to better organise the process for the recovery of tens, hundreds or thousands of families with the guidance of skilled technical assistants.

2. The simple housing solution project

The SHS project started in 2009 (phase 1), and from 2017 on the project underwent improvements at the Federal University of Rio de Janeiro (UFRJ), when it was transformed into a course (phase 2). The most direct users are members of technical staff (engineers, architects, building technicians, social technicians) who have the option of using the project's content free of charge in support of their work with the affected communities. It presently relies on a website [9] and a YouTube channel [10], with about 30 video lessons. The project was one of the finalists for the 2019 Sasakawa Awards, a United Nations (UN) award in the area of disasters, with impacts on Sustainable Development Goals 1, 3, 4, 5, 6, 8, 9, 10, 11, 12, 13, 16 and 17.

2.1 Strategic SHS directives

The SHS project is based on the strategic tripod Simple Professional Design, Low-Cost Construction Technologies and Community Labour System (mutual help or joint working system), targeting the fundamental principles to a sustainable housing recovery stated by United Nations Development Programme (UNDP) and International Recovery Platform (IRP) [3]: environmental, technical, financial and socio-organisational sustainability [11].

The idea is to allow aesthetically attractive and ergonomically appropriate projects in rectangular plants that can be easily built and expanded by the community itself over time, according to needs and possibilities. Gable roofs were chosen for their simple execution and aesthetic appeal in conjunction with rustic-looking brickwork. Engineering projects were developed based on technical standards, theoretical–experimental research and laboratory analyses.

Nowadays, the technology chosen is reinforced masonry with Compacted Earth Blocks (CEBs), which allows block production using manual presses and local materials. It is also considered easy to build, and reduces greenhouse gas emissions, since the blocks are not kiln-burnt but use small ratios of 6:1 or 8:1 of soil and cement as a binder. Compared to the traditional Latin American building system (non-structural brick masonry associated with reinforced concrete structure using pillars, beams, and slabs), the proposed technology can save up to 30%. This is because it does not need wood moulds for concrete or wall cladding (except for wall façades exposed to moisture). When combined with a community labour system, the savings may reach 50% [12] of direct costs.

The interest of the vulnerable population in actively participating in the recovery work is widely documented in the literature and has been confirmed in at least three field activities (interviews):

- Reconstruction in the municipality of São José do Vale do Rio Preto, Rio de Janeiro State, Brazil, after the 2011 disaster, with 90% acceptance among the homeless [13];
- After the 2010 earthquake and Hurricane Matthew in 2016, with 100% acceptance in Don de l'Amitié, Haiti [14];
- Relocation of residents from risk areas in the municipality of Barra Mansa, Rio de Janeiro State, Brazil, with 85% acceptance among vulnerable people [15].

The adaptable enterprises of the SHS project were designed for three scenarios: 20, 50, and 120 homes, aiming to complete the work within 18 months depending on the availability of the land and resources. For the construction of more units, it is possible to replicate various 120-home developments in parallel (within the same deadline) or carry out several projects in series (adding together their deadlines).

2.2 SHS framework

The SHS methodology content has been organised into three broad thematic axes (Design, Communal Labour and Materials) representing the three strategic project directives.

The project activities were organised around the thematic axes and production of slides, spreadsheets, and video-classes, which comprise the didactic material of the SHS course and are available on the project website as they are finalised. Multidisciplinarity and interdisciplinarity are outstanding elements of the contents covered, as illustrated in **Figure 1**.

The Design macro-theme combines knowledge of architectural and engineering building designs (structural masonry, foundations, electrical, hydro sanitary, rainwater facilities) and options for the urbanisation.

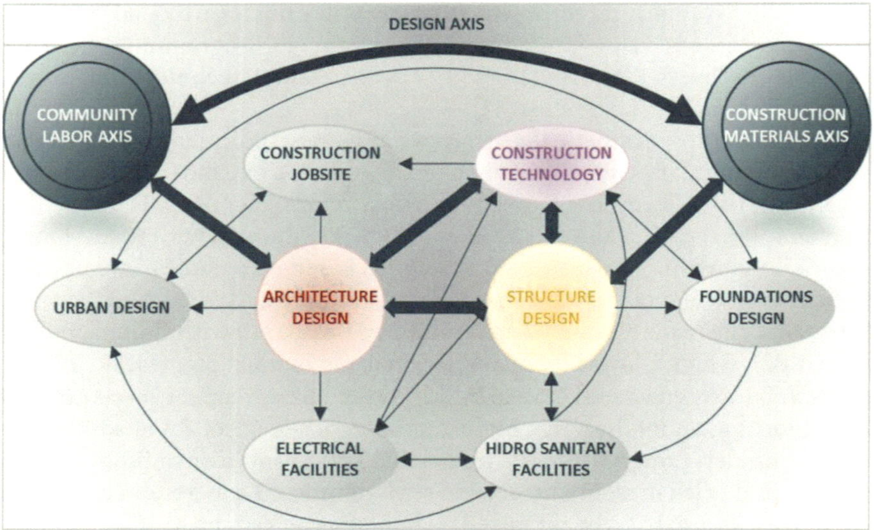

Figure 1.
Representation of multidisciplinarity and interdisciplinarity between the three axes of the SHS project. Source: [16].

WGs/Teams (Leaders)	Mission
General Coord. (1st author)	To conceive the SHS project, plan the work and manage the groups and the project as a whole.
DESIGN WG	To develop or improve basic architecture and engineering projects, compatible with the construction technology, and also develop didactic materials for the SHS course (slides and videos) on those topics.
Architecture (10th author)	To develop a basic architecture design for schools, homes, health clinics, with didactic material.
Structures (1st author)	To research and develop a basic structural design for schools, homes, health clinics, and didactic material.
Foundations (3rd and 4th authors)	To research and develop a basic foundation design for schools, homes, health clinics, with didactic material.
Electrical facilities (7th author)	To develop the preliminary design of electrical facilities for schools, homes, health clinics, with didactic material.
Plumbing facilities (6th author)	To develop a basic design for plumbing facilities (water, sewage and rain water) for schools, homes, health clinics, with didactic material.
Urban design (8th and 9th authors)	To develop basic urban planning in blocks with schools, homes, health clinics, with didactic material.
COMMUNITY LABOUR WG	Developing and or improving methods and tools to organise the project's collective labour and administration address building construction techniques, and provide didactic materials for the SHS course (slides and videos).
Construction techniques (1st author)	To address techniques and details for the construction of low buildings, with didactic material.
Construction jobsite (5th author)	To develop construction jobsite facilities for different project scenarios with didactic material.
Community labour management (1st author)	To research and develop methods and tools for community labour management, with corresponding didactic material.
Construction works admin. (1st author)	To research and develop planning and to control contents for construction work scenarios with didactic material.

WGs/Teams (Leaders)	Mission
MATERIALS WG (1st and 2nd authors)	To carry out research and experimental tests on the materials and technologies used in the project and develop didactic material on prefabrication tests, component manufacturing, and post-fabrication tests.
MEDIA WG (1st author)	To record the teams' work process and the SHS course, format documents, prepare videos, 3D animations, and develop the project's website and visual identity.
TRANSLATION WG	To translate the course material into different languages.
English*	To translate the project content into English
French*	To translate the content of the project into French.
Spanish*	To translate the content of the project into Spanish.
Creole (1st author)	To translate the content of the project into Haitian Creole.

**See Acknowledgements section.*

Table 1.
Distribution of teams and responsibilities in the SHS project.

The Community Labour macro-theme involves knowledge about community mobilisation and organisation, the construction site's organisation, the assembly of the brick factory, and the administration of construction works. Information about the construction techniques is also covered in this group.

The macro theme Materials contains the knowledge necessary to produce and test critical materials and technologies in construction technology. Currently, the project efforts are focused on structural CEB reinforced masonry technology. Furthermore, studies began on applying structural masonry with recycled aggregate concrete blocks, especially where construction and demolition waste are readily available.

The 2nd phase of the SHS project was organised in five multidisciplinary working groups (WGs), involving 16 teams, with the distribution of responsibilities according to **Table 1** and the participation of five UFRJ units: Polytechnic School, Coppe, Faculty of Architecture and Urbanism, School of Communication, Faculty of Letters.

3. Results and discussion

3.1 Design options

3.1.1 Architecture

To meet different scenarios depending on each eventual situation, three families of home projects were considered based on easy module design, environmentally friendly materials, and urgency in construction:

- Family A: Residential modules with minimal sizes, intended for conventional homes in the short/medium term, for light horizontal loads and simplified module design. It does not apply to events such as earthquakes and hurricanes.

- Family B: Residential modules comply with the *Minha Casa Minha Vida* Programme (Brazilian programme providing affordable housing for low-income populations). They are designed for conventional situations (light horizontal loads) of continuous permanence. These projects have an approximately 10% larger built-up area compared to Family A and provide greater essential comfort and expansion.

- Family C: This single storey residential module 2C was designed for moderate horizontal loads, aiming at the context of earthquakes and hurricanes. This family module will not be included in this chapter, but is addressed in detail in [11, 17, 18]. Liquefaction mitigation was analysed using earthquake drains as a low-cost solution [19].

Based on light horizontal load scenarios, two health clinic modules, six school modules, and four home modules were planned based on the SHS project directives:

- Health modules (**Figure 2**; see Video 1, Video 1 can be viewed at https://youtu.be/PGMr2FmfTxQ): Each health module was designed to function as an autonomous unit with an area of about 92m^2, but in conjunction with other units. The community can annex different health modules and thus build adaptable health clinics:

  - o Module 1 contains patients' reception/waiting room area, pharmacy, washrooms, three medical offices and a dental surgery;

  - o Module 2 has reception, restrooms, child emergency care area, area for dressings, sutures, and nebulisation, as well as two medical offices.

- School modules (**Figure 3**; see Video 2, Video 2 can be viewed at https://youtu.be/Q5gZLSOF4lE): Each school module was designed to function as an autonomous unit around 92m^2 in area, but in conjunction with other units so that the community can combine school modules and thus build adaptable schools:

  - o Module 1 contains restrooms and two classrooms with capacity for 42 students;

  - o A library, restrooms, kitchen, cafeteria and teachers' room can be found in Module 2;

  - o Module 3 has restrooms, administration and computer areas;

  - o A nursery area, bathrooms with shower and nursery can be found in Module 4;

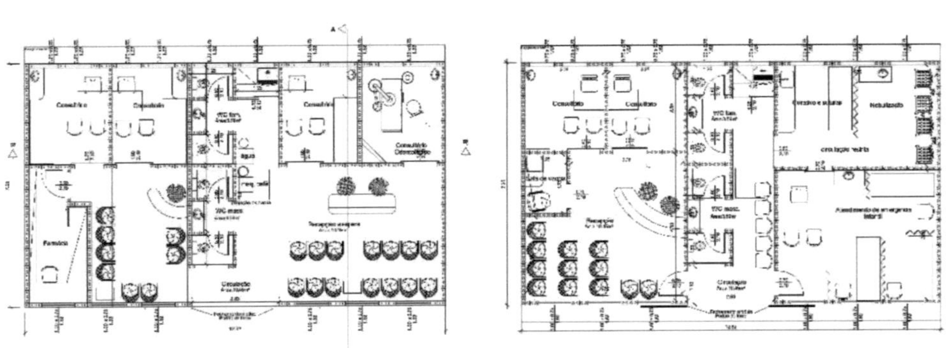

(a) (b)

Figure 2.
(a) Health module 1. (b) Health module 2. Source: [20].

- Module 5 contains an auditorium with 42-seating capacity, restrooms, kitchen and covered patio area for recreation;

- Restrooms and two classrooms for group work, each with a capacity for 32 students, can be found in Module 6.

- Residential modules (**Figure 4**; see Videos 3–6, Video 3 can be viewed at https://youtu.be/BgfojYVUTT0; Video 4 can be viewed at https://youtu.be/51-diapmCwE; Video 5 can be viewed at https://youtu.be/ZMXMYRVCb_4; Video 6 can be viewed at https://youtu.be/od1HyQq8vC0): It begins with an elementary module (embryo 1, with living room, kitchen, bathroom, laundry area) with the capacity to temporarily house a family of four or less in projects with substantial cost restrictions. Embryo 2 has a horizontal expansion of embryo 1

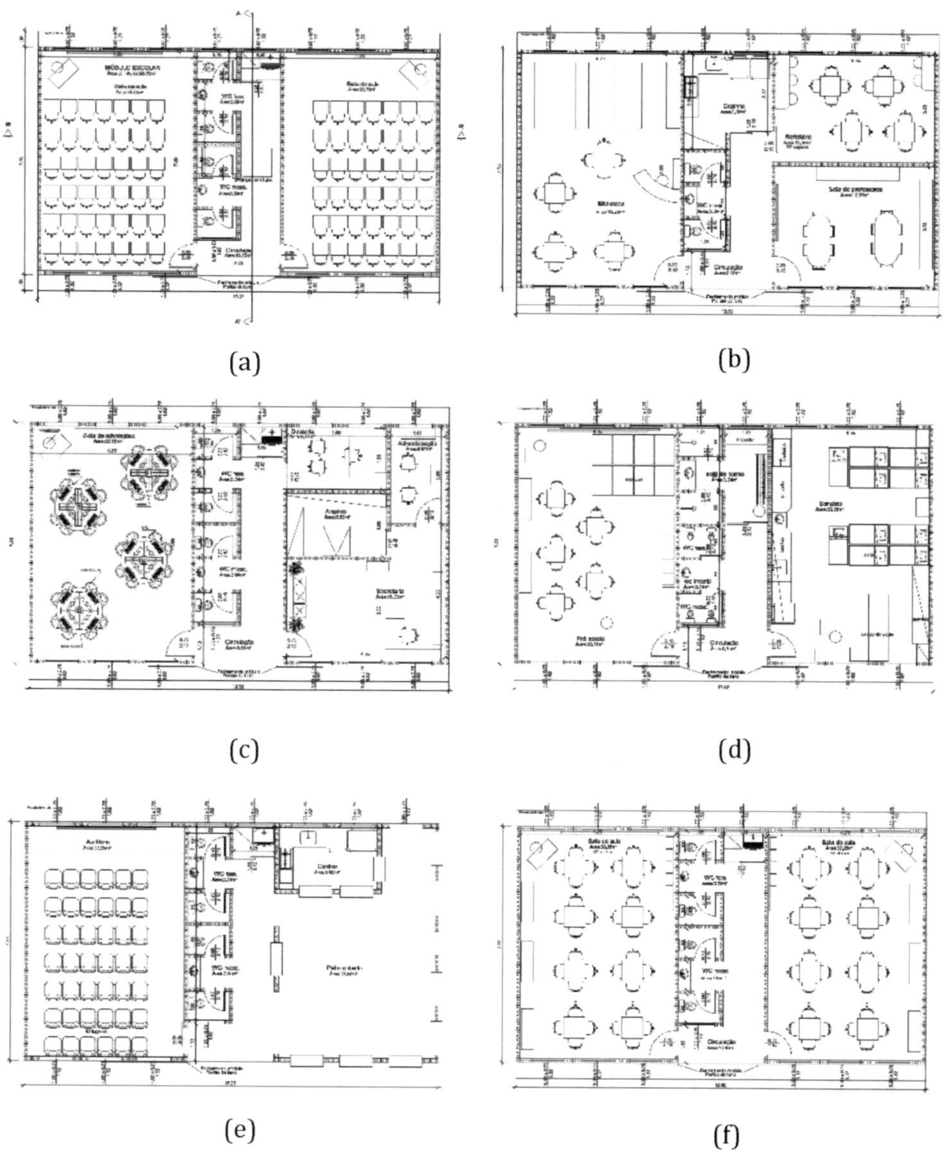

Figure 3.
(a) to (f) School modules 1 to 6, respectively. Source: [20].

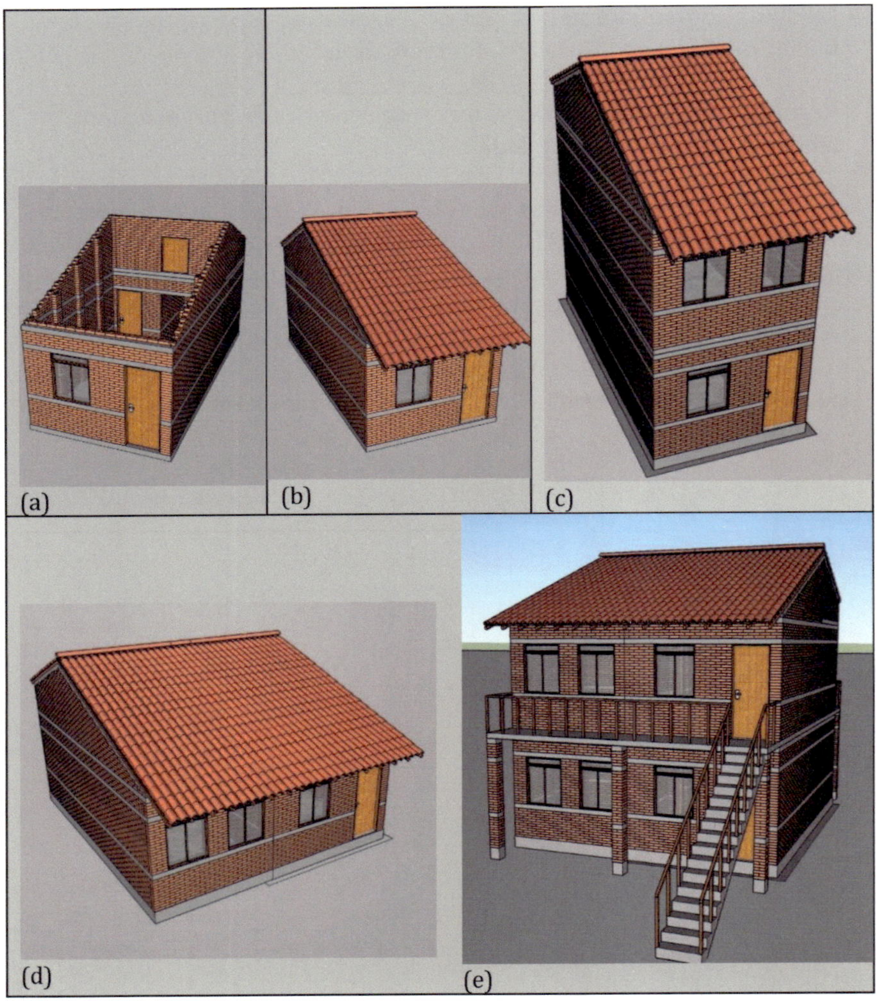

Figure 4.
Expanding possibilities for residential modules from embryo 1. (a) Internal view of embryo 1 (see Video 3). (b) External view of embryo 1. (c) Vertical expansion of embryo 1 to embryo 3 (see Video 4). (d) Lateral expansion of embryo 1 to embryo 2 (see Video 5). (e) Vertical enlargement of embryo 2 to embryo 4 (see Video 6). Source: [21].

(plus two bedrooms) and can house a family of six or less. Embryo 3 consists of vertical expansion of embryo 2 (embryo 1 plus two bedrooms) in situations with space restrictions on the building site. Embryo 4 consists of a vertical enlargement of embryo 2, in order to accommodate two families.

3.1.2 Structures and foundations

The structural system option adopted in the SHS project is that of structural masonry, noting that so far priority has been given to the reinforced masonry technology with two-hole compacted earth blocks (CEBs). However, studies on the application of masonry with concrete blocks have also been carried out [22], with favourable results.

In the CEB reinforced masonry system, the blocks are laid with mortar, and consist of masonry panels with a structural function, and stiffeners where necessary (especially at the side of doorways and windows). At points predetermined by the designer, reinforcements are positioned in the holes of the blocks (and embedded in the foundations), which are subsequently filled with fluid concrete (grout). As part of the CEB masonry system, there are reinforced concrete bond beams on five levels (below and above window openings and doorways, at the slab level, at the gable half-height, and perimeter of the roof gables). The structural system also has a set of frames that integrate the masonry with the purlins using the stiffener connections at the level of the contour concrete bond beams) (**Figure 5**).

Considering that there are still no technical standards that address the design of structural masonry in CEB, the calculations were based on the adaptation of the former British standard BS 5628–2 (2005), which was considered adequate due to its history of success and addresses different materials. The minimum strength of the brick established by the NBR 8491 (2012) standard is 2 MPa and several tests were carried out on the materials used in the project to determine the parameters used in the structural calculation, as mentioned in Section 3.3. Wind calculations were performed according to NBR 6123 (1988). The combination of loads was made according to NBR 8681 (2004).

Strip footings (with light reinforcement) under the walls were used as foundations. In order to reduce the excavation and backfilling costs, the footing installation was designed for a very shallow depth, 0.40 m. An allowable stress of only 50 kN/m^2 was adopted to allow the same foundation dimensions in almost all possible soil types. Even with this conservative value, soil investigations are recommended. Since almost all types of soil investigations require some sort of costly equipment and trained crew, the DPL (dynamic probe light, ISSMFE, 1989, ISO 22476-2: 2005) was chosen for this purpose. The DPL is a simple test that employs light inexpensive equipment, and can be carried out by local personnel with simple training. The test consists of continuous driving a standard cone (with a nominal base area of 10 cm^2) into the soil using a 10 kg hammer with a 500 mm drop height (**Figure 6**). The number of blows to drive the cone 10 cm (N_{10}) must be logged. In general, the test can reach a maximum depth of 8 m [24]. The N_{10} value can be used as an indirect measurement of soil resistance. However, there are no reliable correlations between N_{10} and allowable stresses for shallow footings, as in the case of the standard penetration test (SPT). Therefore, the use of minimum values of N_{10} = 6 to a depth of 5 m was a conservative suggestion in the present application.

Figure 5.
Elements of the structural system of the embryo 2. Source: [23].

Figure 6.
DPL (a) equipment; (b) testing. Source: [25].

3.1.3 Building facilities

The electrical facilities for the SHS project buildings were designed according to standard NBR 5410 for two types of power supply: two-phase and three-phase lines with a 220v phase-to-phase voltage, with grounded neutral and electrical apparatus.

When the brick holes are not grouted, they act as conduits for the wires, which do not require pipes on route to the points of use (switches and sockets). When the pipes are embedded in concrete holes, they must be inserted into preferably flexible conduits. The light fixtures are attached to the roof rafters. The switchboard is located outside the masonry (overlapping), close to the bathroom, in order to reduce wire consumption in the kitchen and electric shower circuits.

The apparent vertical plumbing system was designed for easy maintenance and to avoid holes in the walls, since they have a structural function. The water supply and distribution system were designed considering the use of apparent piping, preferably PVC, considering the cost–benefit ratio of this material and its easy procurement (**Figure 7a**). The system is directly connected to the public water supply network and does not require a pumping system. It was proposed to use a 1000-litre elevated water tank for smaller prototypes, and two 1000-litre units for the larger.

A standard internal solution was proposed for the water distribution in standpipes, branches and sub-branches to enable future expansion. This solution is identical in embryos 1, 2, and 4 and slightly different in embryo 3, because of its particular architecture. In this case, the pipe is built into the bedroom walls.

The sewage system was designed in PVC, using ceiling cladding to separate the pipes from the domestic environment. The waste and vent pipes were placed in the corners of the rooms and inbuilt with plaster. Considering that vulnerable communities would not have a formal sewage system, the proposed solution would be

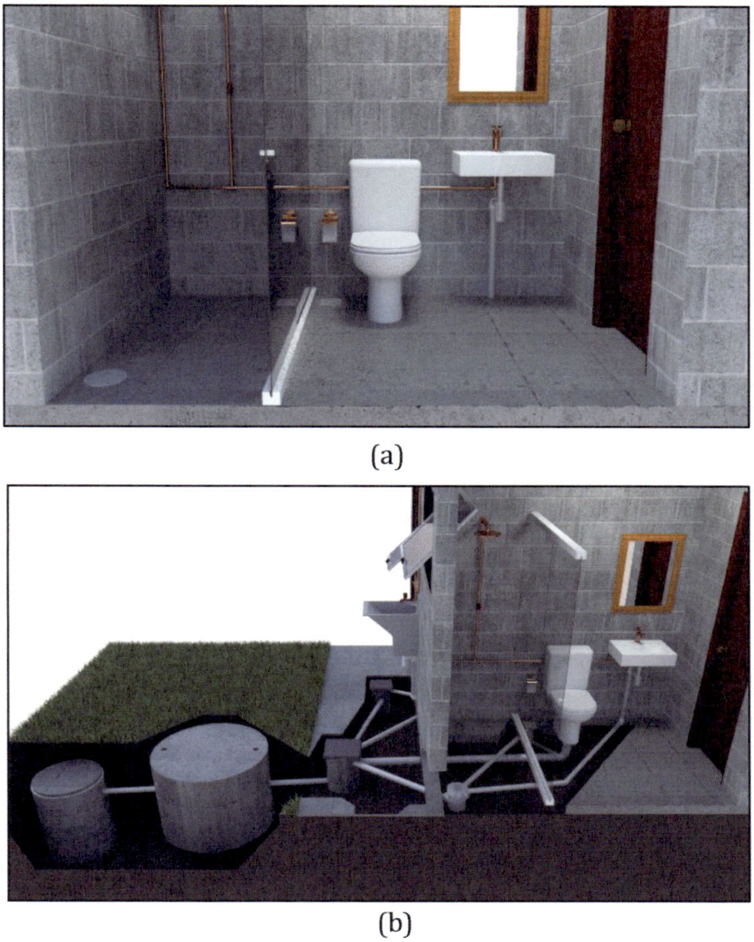

Figure 7.
(a) Internal distribution system - bathroom with apparent pipes. (b) a waste collection system in the bathroom and laundry area with a primary treatment system. Source: [26].

an individual sewage treatment system using a combined septic tank and sinkhole (**Figure 7b**).

Concerning the rainwater drainage systems, a simplified solution was proposed. The rainwater flows from the roof to catch basins and then to the public storm water system, whenever applicable. As an alternative, a rainwater harvesting system can be installed using rain barrels (**Figure 8**). In this case, it is necessary to install gutters, roof drains and downspouts connected with a rain diverter to the rain barrel. These barrels would accumulate the rainwater, reserving it for less exciting use, such as washing floors.

3.1.4 Urban design

Different typologies of the urban subdivision were conceived for a pilot project considering at least 100 homes in embryo 2, including the necessary community facilities (schools, health centres and public squares) to serve the expected population. For each 100-home model, simple replication logic was applied to extend the project in case of more significant needs, for example, when an entire neighbourhood has to be built (**Figure 9**).

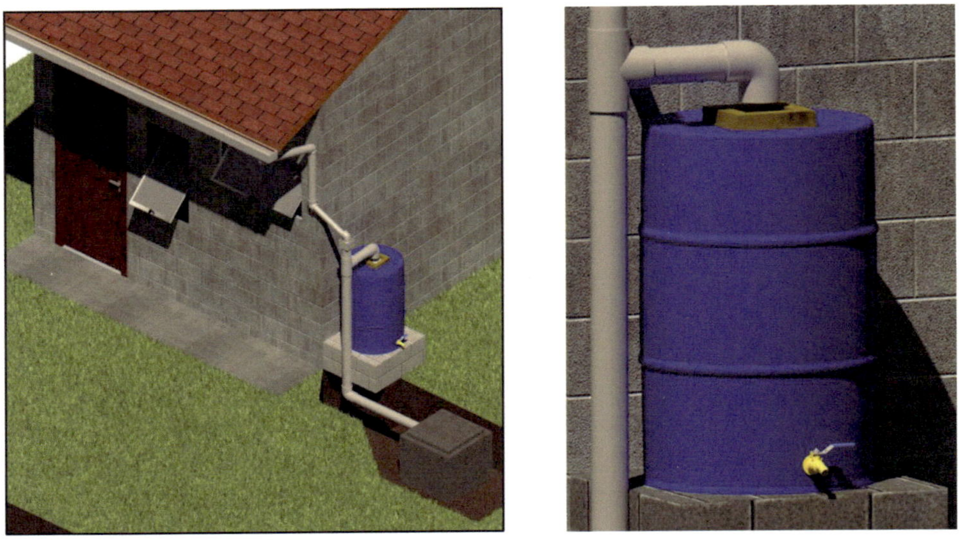

(a) (b)

Figure 8.
(a) Rainwater harvesting system; (b) rain barrel. Source: [26].

Figure 9.
Strategy for replicating residential and hybrid blocks (containing community facilities designed for the local population), based on typology 2 ("village" type). Source: [27].

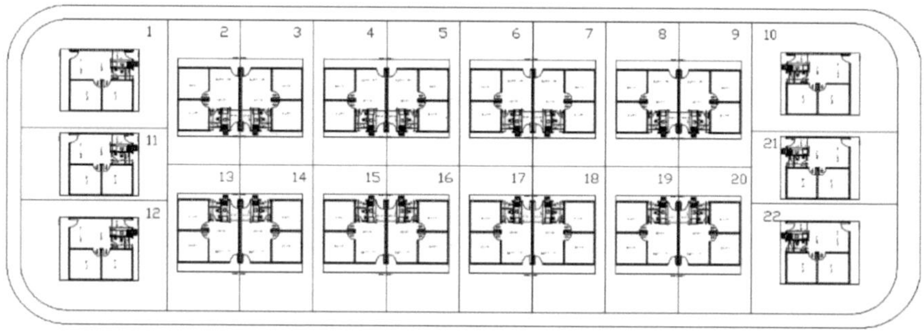

Figure 10.
Lot distribution on a typical residential block with semi-detached houses. Source: [27].

Figure 11.
Single lot with multiple houses in the "village" typology. Source: [27].

The following urban typologies were considered in detail (mainly facing a smooth topography):

- Typology 1 (**Figure 10**): Grid – single-family home lots distributed in rectangular blocks shaped by orthogonal streets. Two different designs were used: the first considered mirror image semi-detached houses to save urban space and define smaller lots; the second considered traditional homes centred in the lot area.

- Typology 2 (**Figure 11**): Grid – "village" type consists of multiple houses in a shared lot (each owner has a fraction of the site, in a condominium arrangement), with more significant block areas.

- Typology 3: Radial – based on a single-centre model that can be replicated with several interconnected centres – used for uneven and hilly ground and not fully detailed in the project's first phase.

Both typologies 1 and 2, which are suggested for flat or sloping ground, were detailed in the design process. Typology 3, more adaptable to rugged and uneven ground, was only developed conceptually, considering the inherent implementation problems, lacking the facility for community labour system.

Table 2 shows a comparative analysis for a hypothetical case of 200 houses, showing that the "village" setup saves urban space, with a considerably higher number of dwellings per area, thus consuming less area per home throughout the development.

Detailed comparative analyses were made for the proposed urban typologies planned to relocate the population from flood-prone areas, in the municipalities of

Typology	Description	Area per house (m^2)
Individual lots, with standard centred houses	193 lots +6 schools +4 nurseries +2 local health centres +1 public square	286
Individual lots with semi-detached houses	Idem	233
Shared lots of "village" type	200 lots +6 schools +4 nurseries +2 local health centres +1 public square	156

Table 2.
Average urban area implementation per house for different developing typologies.

Mesquita [28] and Barra Mansa [15], in Rio de Janeiro State, Brazil. These analyses confirmed that urban typology 2 (community villages) is more efficient both in terms of space and implementation cost of the development.

3.2 Community labour system

3.2.1 Building construction technique

Building construction techniques consider all the construction processes related to the technologies adopted (currently, CEB reinforced masonry).

The CEB reinforced masonry process starts with batter boards, excavation, and execution of the shallow foundations. The foundations are usually concrete stripped footings, with at least 20 MPa of compressive strength. They should be built over a thin layer of lean concrete. It is crucial to pay attention to the mix grade, especially the water/cement ratio, which should be as low as possible. The vertical rebar of the grouted holes in the CEB masonry has to be implanted in the footings (see Video 3).

See the "Structure and foundations" section to build the CEB masonry, the grouted holes, bond beams, and portico system. It should be mentioned that the electrical conduits could be installed inside the CEB masonry holes. Simultaneously, plumbing installations must be external (see the section of "Building facilities", which also provides rainwater and sewer drainage).

The places most exposed to water or moisture (masonry base, bathrooms – namely showers, kitchen, laundry room, service area, and base of external walls) should be waterproofed. An acrylic texture coat or any waterproofing resin should be applied to the outside of house façades. CEB masonry should be preserved as much as possible against direct action of moisture. It must be stressed that CEB masonry may become less durable when exposed to water in the long term. A layer of compacted gravel around 3–5 cm in thickness should be inserted between the compacted soil and the ground level concrete slab to protect against moisture.

The type of roof tile adopted will depend on the gable roof, so if the choice is a ceramic/clay tile, either a higher slope or extra wooden structure will be required.

3.2.2 Construction jobsite

The construction jobsite was designed to: reduce the area, increase efficiency and safety of the construction process, reduce travel distances of both construction material and people, and also to avoid any barrier to plant movement.

A couple of layouts of jobsites were designed for different developments (construction of 20, 50, and 120 homes). All arrangements at the jobsite include elements related to production (carpentry and concrete/mortar mixer), production support (warehouse, cement and lime deposits, and concrete aggregate storage bays), technical and administrative support (security cabin, site office, medical facility) and living areas (canteen, changing rooms and restrooms).

A layout design was also considered, including a CEB brick production plant and a control technological laboratory for the materials used in CEB masonry technology. This layout was designed to create a housing development with smaller areas [29].

For example, in the case of a 120-home development (embryo 2A), the average cost of the CEB masonry plant per home is above half the average cost per house of a 20-home development. Therefore, in the case of building a CEB plant, a 120-home development proves to be far more efficient.

3.2.3 Administration of community labour and construction works

In addition to technical knowledge and people management skills, the organisation and administration of developments using community labour address challenging issues and require a certain degree of sensitivity, especially considering the vulnerability of the population affected.

This section is adapted from [30] and includes processes and activities for the organisation and management of community labour, namely: viability and security of the project, types of management, creation of an association for the people affected, technical assistance, setting up committees and teams, organising tasks, legal and financing issues, providing designs and planning construction works, distribution and receipt of finished houses, creation of the social factory, registration of families and workers, task force training, and issues related to withdrawals, warnings, and exclusions.

It is worth explaining that two types of administration must work in parallel, having a high degree of interdependence between them: the administration of community labour and administration of the construction works. While the former will take care of the community, the latter will address construction planning (time, cost and resources), control of time, cost, and stock control.

Construction planning is aligned to the "lean construction" philosophy, adapted from the "lean production" concepts. Therefore, a maximum of 120 households per project has to be respected (if there are more households, then several projects should be implemented in parallel or series).

Dynamic tools (spreadsheets) for construction planning and control were developed, based on an algorithm designed for SHS [31]. These tools allow users to configure their development's specific input parameters, such as the number of homes, embryo types, number of workers per household, expected task force efficiency, payroll and automatically obtain the development's essential planning and control structure.

Five task groups were created with different duties: Foundations, Masonry, Roof, Facilities and Finishes, each with a manager in charge (see Video 7, Video 7 can be viewed at https://youtu.be/Hg0TezfNWqc). Volunteer workers would comprise these groups (in more significant numbers, from the affected community and its relationship network, preferably skilled) and contract workers (in smaller numbers, to meet qualified professionals' specific needs).

First of all, it is necessary to have time dedicated to training the task force, preferably using some hands-on activity to familiarise them with construction of the buildings. It is recommended that this "preparation time" of the project should take at least two months and no longer than six months. During this time, various measures can be taken, such as design adaptation, legal measures to start the project, and implementing the jobsite.

Another fundamental point addressed in the SHS project is how to calculate the hours for recording each household's working hours (hour bank) and the criteria adopted to prioritise the delivery of the homes. To support these tasks, a time control tool was also developed, which provides the structure for recording the hours worked (with the possible transfer between households) and an automatic list for prioritising the distribution of homes at different stages of the development.

3.3 Construction materials

Two deposits of unsaturated tropical soil (Soil 1 and Soil 2), located in Macaé, Rio de Janeiro, Brazil, were investigated, and a geotechnical characterisation was carried out in the laboratory.

Soil samples were prepared and tested according to international standards. Soil preparation, grain-size distribution, specific gravity and Atterberg limits were obtained according to ISO 23909 (2008), ISO 17892-4 (2016), ISO 17892-3 (2015), and ISO 17892-12 (2018) respectively [32, 33]. An empirical test was also carried out to check soil shrinkage during the natural drying process.

CEBs were manufactured using a mechanical hand press, from a mixture of 50% Soil 1 and 50% Soil 2, in volume. The final mixes used cement as a binder with 2% of lime and water content (CEB water absorption) next to 20% (**Figure 12a-c**). It

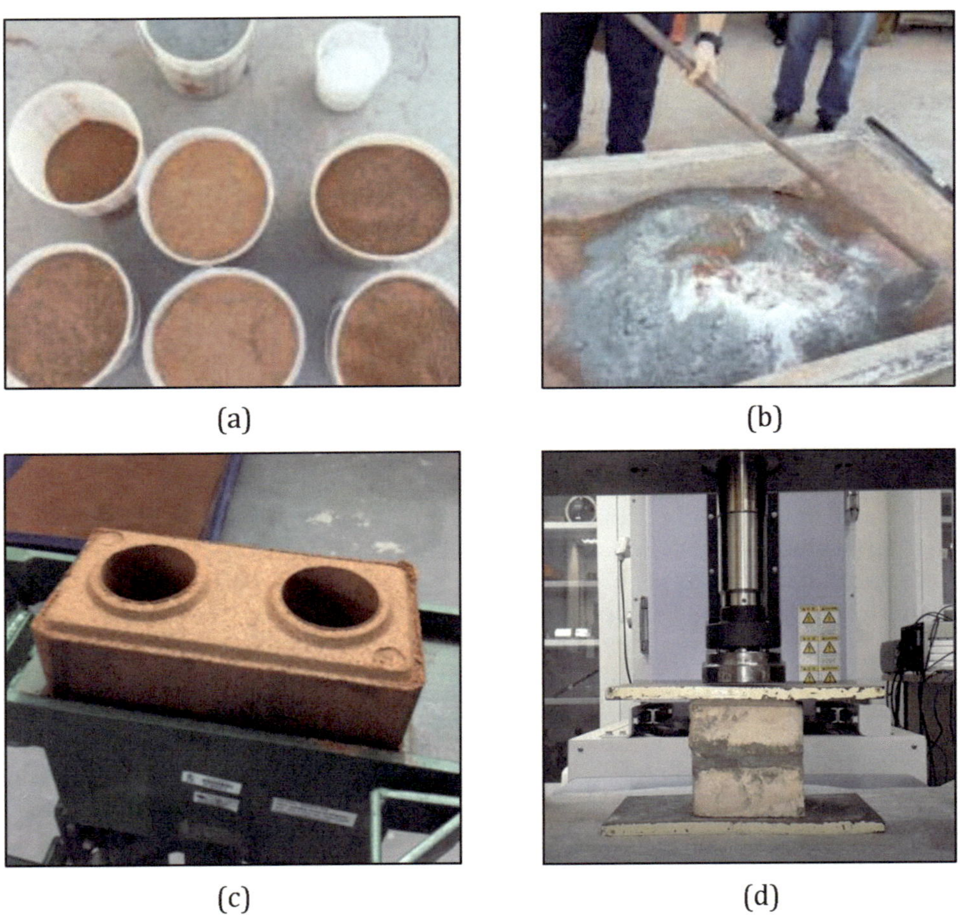

Figure 12.
(a) Soil preparation, (b) Material's mixture, (c) CEB's production (d) CEB compressive strength test. Source: [34].

should be pointed out that water content could significantly vary within different types of soils. CEB water absorption and compressive strength were verified according to NBR 8492 (2012) (**Figure 12d**) [32, 33, 35]. Tests on wallets equivalent to two grouted columns per metre and fully grouted columns were carried out to obtain mechanical properties and wallet-block efficiency [33, 17]. The main results and extensive discussion can be found at [11].

Horizontal load capacity tests were used to obtain inputs for computational models used to investigate the proposed residential model under seismic conditions, performed according to NBR 15421 (2006). The test consisted of applying a horizontal load on a wallet one metre in length (equivalent to 4 blocks per tier), a clear width of 10.5 cm, and a height one metre (equivalent to the elevation of 13 tiers of blocks and a joist at the top). All the specimens had their blocks laid manually in a ~ 1 cm thick mortar joint. The reinforced bond beams had the same width as the blocks and reflected the panel's upper confinement. Vertically, the wall had its lateral holes reinforced and grouted. In addition to these, two more intermediate holes were symmetrically reinforced and grouted, representing panels with two reinforced holes per metre. The load was manually increased and applied by a hydraulic jack placed in the middle of the joist section. The displacements were measured using linear variable differential transformer (LVDT). A load cell was used to monitor the load. Displacement and load were acquired at the same time using a data acquisition system [11, 17].

4. Conclusion

Housing recovery in situations where urgent construction is necessary but resources are scarce (such as post-disasters, post-conflicts, refugee settlements, and so on) is a challenging task requiring technical skills, management, and focus on the affected population. The Simple Housing Solution (SHS) methodology was developed to contribute to empowering governments, support agencies and, above all, vulnerable communities. It is presented in the course format and based on the tripod Simple Professional Design, Low-Cost Construction Technologies, and Community Labour System.

The first class on the SHS course was held in October 2018 for an audience of 30 people: 15 Haitians living in Brazil and intending to return to their country to participate in the reconstruction process; three members of the non-governmental organisation (NGO) Engineers Without Borders Brazil; three members of the NGO Teto; three community members affected by a relocation process near UFRJ; two employees from the municipality of Niterói (Rio de Janeiro State, Brazil); two employees from the municipality of São Lourenço (Minas Gerais State, Brazil); one graduate student from the Federal University of Santa Catarina, and one refugee from the conflict in Syria. Since then, two other reduced versions of the course were offered in 2019: one at the Federal Institute of Maranhão, Brazil, on the manufacture of soil-cement bricks, during the 5th Maranhão Symposium on Civil Engineering, and another at the Federal University of Pará, Brazil, on SHS Methodology, during the 3rd Brazilian Congress on Disaster Risk Reduction.

It should be noted that the educational material developed under the SHS project is intended to be a starting point and should be evaluated and adapted to each implementation reality by the local technical assistance team. However, the material provided can greatly expedite reconstruction work, mostly because of the project's simplicity, pre-planned work, accessible construction technology, and tools developed to administer the community labour and construction works.

Aiming to expand the project's scope, it currently seeks to improve the embryo 2C, the SHS model for multi-risk situations involving more aggressive horizontal loads in reinforced masonry technology in CEB [11]. For this reason, new tests are being proposed for SHS project phase 3, when it is necessary to perform cyclic tests in porticos and dynamic tests in full scale on shake tables.

Acknowledgements

Thank you for the collaboration of the SHS team; Sergio Santos (support in seismic analysis); Silvio Lima (support in seismic analysis); Michèle Pfeil (support in wind load analysis); Luis Otavio Araújo (support in drawings for masonry production), Claudia Morgado (Polytechnic School Director and support in safety and occupational health issues); Wilson Silva (support in lab tests); Bianca Pinheiro (head of French-language team); Thiago Aragão (head of English language team); Marcos Gallo (head of Spanish language team); Romildo Toledo and Numats team (support in lab tests); Edilberto Strauss (Assistant Director of Education and Extension); and PR5-UFRJ team. The SHS project was supported by funds from the Rio de Janeiro State Carlos Chagas Filho Foundation for Research Support (Faperj) (grant number E-26/190.212/2009), and from the company Interpro Gerência de Projetos Ltda, during Phase 1 (2009–2012).

Author details

Leandro Di Gregorio[1*], Gustavo Guimarães[2], Fernando Danziger[3], Graziella Jannuzzi[4], Eduardo Qualharini[1], Aline Veról[5], Assed Haddad[6], Marcelo Miguez[7], Virgílio Cruz[4] and Daki Bianchessi[8]

1 Environmental Engineering Program/Urban Engineering Program/ Polytechnic School, Federal University of Rio de Janeiro, Rio de Janeiro, Brazil

2 Civil Engineering Department, Federal University of Rio de Janeiro, Macaé, Brazil

3 Civil Engineering Program Alberto Luiz Coimbra Institute of Graduate Studies and Engineering Research - Coppe, Federal University of Rio de Janeiro, Rio de Janeiro, Brazil

4 Polytechnic School, Federal University of Rio de Janeiro, Rio de Janeiro, Brazil

5 Graduate Program in Architecture – PROARQ, Urbanism and Architecture Faculty, Federal University of Rio de Janeiro, Rio de Janeiro, Brazil

6 Environmental Engineering Program/Polytechnic School, Federal University of Rio de Janeiro, Rio de Janeiro, Brazil

7 Environmental Engineering Program/Polytechnic School, Civil Engineering Program Alberto Luiz Coimbra Institute of Graduate Studies and Engineering Research – Coppe, Federal University of Rio de Janeiro, Rio de Janeiro, Brazil

8 Infanti Engenharia, Nova Iguaçu, Brazil

*Address all correspondence to: leandro.torres@poli.ufrj.br

References

[1] IPCC, 2018: Summary for Policymakers. In: Global Warming of 1.5°C. An IPCC Special Report on the impacts of global warming of 1.5°C above pre-industrial levels and related global greenhouse gas emission pathways, in the context of strengthening the global response to the threat of climate change, sustainable development, and efforts to eradicate poverty [Masson-Delmotte, V., P. Zhai, H.-O. Pörtner, D. Roberts, J. Skea, P.R. Shukla, A. Pirani, W. Moufouma-Okia, C. Péan, R. Pidcock, S. Connors, J.B.R. Matthews, Y. Chen, X. Zhou, M.I. Gomis, E. Lonnoy, T. Maycock, M. Tignor, and T. Waterfield (eds.)]. In Press.

[2] Nakagawa Y., Shaw R. Social Capital: A Missing Link to Disaster Recovery. In: International Journal of Mass Emergencies and Disasters, 2004, v. 22, n. 1, pp. 5-34. Available from: http://ijmed.org/articles/235/download/ [Accessed: 2020-10-24].

[3] United Nations Development Programme, International Recovery Platform. Guidance Note Recovery: Shelter. [Internet]. 2010. Available from: https://www.undrr.org/publication/guidance-note-recovery-shelter [Accessed: 2020-10-24].

[4] International Federation of Red Cross, Red Crescent Societies. Owner-Driven Housing Reconstruction. Geneva [Internet]. 2010. Available from: https://www.sheltercluster.org/sites/default/files/docs/Owner-Driven%20Housing%20Reconstruction%20Guidelines.pdf [Accessed: 2020-10-24].

[5] Leykin D., Lahad M., Cohen R., Goldberg A., Aharonson-Daniel L. The dynamics of community resilience between routine and emergency situations. International Journal of Disaster Risk Reduction, 2016, v.15, 125-131. DOI: 10.1016/j.ijdrr.2016.01.008

[6] Barakat S. Housing Reconstruction After Conflict and Disaster. Humanitarian Practice Network [Internet]. 2003. Available from: https://odihpn.org/resources/housing-reconstruction-after-conflict-and-disaster/ [Accessed: 2020-10-24].

[7] Le Maoult A., Politopoulos I., Atanasiu G.M., Casarotti C., Pavese A., Dorka U., Nguyen V. T., Marazzi F., Molina Ruiz F. J., Pegon P. EFAST project (Design Study of a European Facility for Advanced Seismic Testing). 1st-year EFAST annual report. Commissariat à l'Énergie Atomique, Technical University "Gheorghe Asachi" of Iasi, European Centre for Training and Research in Earthquake Engineering, University of Kassel, European Union, 2010.

[8] Yi H., Yang, J. Research Trends of Post Disaster Reconstruction: The Past and The Future. Habitat International, 2014, v.42, 21-29. DOI: 10.1016/j.habitatint.2013.10.005

[9] SHS Solução Habitacional Simples [Internet]. 2018a. Available from: https://www.shs.poli.ufrj.br/ [Accessed: 2020-10-24].

[10] SHS Solução Habitacional Simples – Aulas em Português BR [Internet]. 2018b. Available from: https://www.youtube.com/watch?v=QEY0ZaOz8AY&list=PL515-FzRssVNGAsgIrQNgU3LCld3DGtrY [Accessed: 2020-10-24].

[11] Di Gregorio L., Guimarães G., Tenório M., Lima D., Haddad A., Danziger F., Jannuzzi G., Santos S., Lima S. (2020). The Potential of CEB Reinforced Masonry Technology for (Re)construction in the Context of

Disasters. Materials. 13(17). 10.3390/ma13173861.

[12] Santos R.L.R. Orçamento, planejamento e gerenciamento de obras de residências de baixo custo em regime de mutirão: caso de estudo do Projeto Solução Habitacional Simples [undergraduate monograph]. Rio de Janeiro: Federal University; 2019.

[13] Di Gregorio L.T. Proposta de ferramentas para gestão da recuperação habitacional pós-desastre no Brasil com foco na população atingida [doctoral thesis]. Niterói: Federal Fluminense University; 2013.

[14] Alerte J. Proposta de (re)construção de casas populares em regime de mutirão como alternativa ao déficit habitacional do Haiti [undergraduate monograph]. Rio de Janeiro: Federal University of Rio de Janeiro; 2017.

[15] Silva J. M. Análise de viabilidade da metodologia "Solução Habitacional Simples" para realocação de áreas de risco de inundações do município de Barra Mansa/RJ [undergraduate monograph]. Rio de Janeiro: Federal University of Rio de Janeiro; 2019.

[16] Di Gregorio L.T. Projeto SHS: (Re) construção em situações críticas com tecnologias de baixo custo. In: IV Simpósio Maranhense de Engenharia Civil – SIMEC; 12-16 agosto 2019; São Luís, Brazil. Oral Presentation.

[17] Tenório, M.C.U. Análise da viabilidade técnica da alvenaria estrutural em tijolos de solo-cimento para situações com cargas sísmicas: práticas construtivas e análise estrutural do projeto de solução habitacional simples [Undergraduate monograph]. Rio de Janeiro: Federal University of Rio de Janeiro; 2019.

[18] Gonçalves, F.S. Aspectos construtivos para residências de baixo custo sob a ação de ventos fortes [undergraduate monograph]. Rio de Janeiro: Federal University of Rio de Janeiro; 2018.

[19] Benvenuti Junior M. P. Análise de Sistemas de Fundações Visando à Mitigação de Danos Causados por Sismos: Caso de Estudo Projeto Solução Habitacional Simples (SHS) [Undergraduate monograph]. Rio de Janeiro: Federal University of Rio de Janeiro; 2020.

[20] Bianchessi D. I. P., Di Gregorio L. T. Projetos de arquitetura. In: Di Gregorio L. T. org. Projeto SHS – Material Didático. 2012.

[21] Bianchessi D. I. P., Di Gregorio L. T. Projetos de arquitetura. In: Di Gregorio L. T. org. Projeto SHS – Material Didático. 2018.

[22] Freitas J.P.B. Aspectos estruturais, construtivos e orçamentários para construção de casa popular em blocos de concreto em situações críticas [Undergraduate monograph]. Rio de Janeiro: Federal University of Rio de Janeiro; 2019.

[23] Di Gregorio L. T. Projetos de estruturas. In: Di Gregorio L. T. org. Projeto SHS – Material Didático. 2019.

[24] ISSMFE - International Society for Soil Mechanics and Foundation Engineering. Report of the ISSMFE - Technical Committee on Penetration Testing of Soils - TC 16. Reference to Test Procedures CPT-SPT-DP-WST. Suécia: Swedish Geotechnical Institute Information, n. 7, 1989.

[25] FURG Universidade Federal do Rio Grande. Pictures of DPL Equipment and Testing Source. 2017.

[26] Veról A. P., Di Gregorio L. T. Projetos de instalações hidrosanitárias e

águas pluviais. In: Di Gregorio L. T. org. Projeto SHS – Material Didático. 2018.

[27] Miguez M. G., Cruz V. N. R., Di Gregorio L. T. Projetos de urbanização. In: Di Gregorio L. T. org. Projeto SHS – Material Didático. 2018.

[28] Duarte C.C.R. Urbanização de interesse social: um estudo da organização do espaço urbano para realocação da população em área de risco de inundação no município de Mesquita (RJ) [undergraduate monograph]. Rio de Janeiro: Federal University of Rio de Janeiro; 2018.

[29] Resende D.M. Projeto e estudo da logística do canteiro de obras com uma fábrica de tijolos de solo-cimento integrada: uma análise prática do projeto solução habitacional simples [undergraduate monograph] Rio de Janeiro: Federal University of Rio de Janeiro; 2019.

[30] Di Gregorio L. T. Manual para fabricação de tijolos de solo-cimento. Rio de Janeiro. 2012.

[31] Di Gregorio L. T. Algorithm for Communitary Labour Construction System. 2018.

[32] Sousa J.F. Fabricação e análise de tijolos de solo-cimento. Undergraduate Monograph in Civil Engineering, Federal University of Rio de Janeiro, Rio de Janeiro, 2018.

[33] Lima, D.A.S.M.F. Avaliação do tijolo modular de solo-cimento fabricado em prensa manual como material para construção de baixo custo com mão de obra comunitária. Undergraduate Monograph in Civil Engineering, Federal University of Rio de Janeiro, Rio de Janeiro, September 2018.

[34] Di Gregorio L. T., Guimarães G. V. M. Component testing for soil-cement masonry. In: Di Gregorio L. T. org. Projeto SHS – Material Didático. 2018.

[35] Lobo, G.M. Efeito da umidade de fabricação na resistência à compressão de tijolos solo-cimento. Undergraduate Monograph in Civil Engineering, Federal University of Rio de Janeiro, Rio de Janeiro, 2019.

Chapter 7

Disaster Management Process Approach: Case Study by BOSS for Disaster Response under COVID-19

Arisa Yasui, Muneyoshi Numada and Chaitanya Krishna

Abstract

Comprehensive disaster response processes need to be managed and progress communicated to avoid ineffective management such as duplication with stakeholders, amendments as a result of leaders' incomplete instructions, and waiting without instruction from the EOC (Emergency Operation Center). As there is existing research on standardization and systematization of disaster response processes, a pure paper-based SOP (Standard Operation Procedure) is challenging to use in actual and practical situations concerning the standard workflow based on the SOP. For effective disaster management, this study developed a Business Operation Support System (BOSS). The BOSS characteristics have the standard workflow chart based on the related documents and experiences, such as the SOP, concerning manuals/documents, past experiences, and knowledge. The overview, checkpoints, necessary documents, related information systems linked to the disaster management plan, and document formats are defined in every workflow. Even for the young or non-experienced individuals, the BOSS can support the responders through the processes for necessary actions during disasters. This research aims to compare the effect of responses to the 2019 massive rain disaster in Kawasaki city, with or without the BOSS. First,a comprehensive workflow focusing on shelter management under the Coronavirus Disease 2019 (COVID-19) workshop with Kawasaki city staff and community people in the BOSS was created. Second, experiments (with or without the BOSS) were carried out to analyze the differences and the BOSS effect. "With the BOSS" means that the responders can follow the workflow in the BOSS for shelter management. "Without the BOSS" means that the conventional paper-based manuals are used for the operations. Two types of manuals in Kawasaki city were used; one guides the expected shelter management points, and the other contains the explanation about COVID-19. Members of both teams comprise one leader and two staff. As a result of the experiments, the big difference between the two teams is the leader's behavior. Because the BOSS team leader instructed the different staff works following the BOSS workflow, the BOSS team responded to more kinds of works compared to the manual team. The role of all members of the BOSS team was evident. On the other hand, the manual team responded to one work by all members, including the leader, without the leader's instruction. Due to no instruction from the leader, a period of waiting was observed in the next work manual. This research obtained that the leader's instructions' effect caused the effective responses by quantitative analysis of the demonstrative experiment. For future research, the leader's behavior and decision-making should be analyzed for BOSS's effective operation and team-building.

Keywords: BOSS, process management, disaster management, disaster response, COVID-19

1. Introduction

COVID-19 is prevalent all over the world. Different countries applied a large-scale lockdown [1], and although the lockdown is currently relaxed and economic activities have started again, the expected second wave of infection has occurred. Re-expansion occurred [2], and it has continued.

In Japan, a disaster-prone country, there is concern about the COVID-19 as it spreads. Particularly in recent years, disasters such as the 2011 Tohoku-Pacific Ocean Earthquake-tsunami disaster, the 2016 Kumamoto Earthquake disaster, and the 2018 massive rain disaster in western Japan have become more severe and frequent.

It is necessary to manage the entire disaster response processes in such a situation. Also, it is essential to manage the disaster response work process and effectively communicate the disaster response work so that there is no duplication, rework, or waiting by concerning human resources in an organization.

Regarding the operation of shelters among various disasters, schools and public halls become shelters for many displaced residents [3, 4], and various operations may coincide with the potential high risk of infection among the crowded residents. In the current COVID-19, outbreaks of infectious diseases at shelters can occur. There is currently no mention of the Basic Act on Disaster Countermeasures in Japan requiring improved living conditions in evacuation centers, such as distributing food, clothing, medicines, and healthcare services (Article 86–6). Therefore, there is an urgent need to establish an evacuation center management system that incorporates measures against infectious diseases.

As there is a concern that the COVID-19 outbreak will result in a lack of human and physical resources, the following five components are essential for sufficient disaster response work: (1) Standard workflow for disaster response, (2) Information sharing and distribution to understand the situation of dispersed evacuation centers, (3) Effective allocation and management of human resources for various kinds of situations, (4) Information management and distribution of materials and equipment, (5) Continuous follow-up on measures/policies against infectious diseases.

Regarding the standard workflow for disaster response, research on standardization and systematization of disaster response has been conducted to realize effective disaster response. The author developed a Business Operation Support System (BOSS) for effective disaster management. BOSS is a workflow system with a database that summarizes disaster response works in a workflow chart.

In this study, the BOSS was used to create a comprehensive workflow focus on shelter management for COVID-19, and the created workflow was verified in the situation of the scenario disaster. In particular, we compared the responses for shelter management with the BOSS or without the BOSS (with a conventional paper manual) and analyzed the effect of using BOSS. This study verified shelter management operations in Kawasaki City, Kanagawa Prefecture.

2. Literature review

2.1 Research on infectious diseases during disasters

Infectious diseases are often prevalent after the occurrence of natural disasters [5]. There are many cases of infectious diseases and psychological stress caused by a massive change in the living environment due to the way of life in evacuation centers [6], food shortages, and unsanitary environments increase the risk of infectious diseases [5]. Also, an unspecified number of people live together in evacuation centers, but the area per person is small, and the rate of infections spread is high in dense evacuation centers [7]. Medical workers get tired during disasters, and they may be affected, and insufficient medical resources provision compared with typical phases contributes to the spread of infection [6].

There were many acute respiratory symptoms and acute gastrointestinal symptoms in the 2011 Great East Japan Earthquake disaster [8]. During the 2016 Kumamoto earthquake disaster, infectious gastroenteritis due to Norovirus was prevalent [9]. After natural disasters, although not limited to Japan, the outbreak of infectious diseases such as aspiration pneumonia, skin, and wound infections occurred during the 2004 tsunami disaster in Indonesia [10], and the hurricane Katrina in 2005. At that time, Norovirus infection spread in evacuation centers [11].

2.2 Research on disaster response work

Local governments often create manuals as a prevention measure. Since many disaster response works are performed in parallel and complicated activities, omissions occur in manuals [12]. Due to detailed descriptions [13], many studies have been conducted to clarify effective manual creation methods. Also, [14, 15] clarified the effects of the new system replacing manuals in the form of empirical experiments. From the operational perspective, according to [16, 17], the actual disaster response work is quantitatively analyzed.

However, although all of these analyzes actual disaster response, the effects of documents such as manuals created in advance are not mentioned.

Therefore, in this study, using an evacuation shelter case as an example, shelter operation using the BOSS under COVID-19 was conducted to quantitatively clarify the effect of the BOSS by comparing the evacuation shelter operation manual (with or without the BOSS).

3. Disaster response workflow model under COVID-19

3.1 Overview of disaster response process management system BOSS

The disaster response process management system BOSS was developed to support a disaster response process. The BOSS has a database of disaster response procedures not only during disasters but also the preparedness. It visualizes the whole image of disaster response workflow by processing knowledge related to all disaster countermeasures. By constructing a checklist of workflow, the responders can realize a comprehensive disaster response processes. Besides, although there are various disaster prevention plans in local governments, by associating the workflow with the disaster prevention plan, the relevant part of the regional disaster prevention plan can be linked and the contents of the plan can be quickly grasped. Every work has a working detail sheet that describes the implementation method (who, when, what, how), and the contents can be easily understood. Also, since it is possible to share the work details sheet by associating manuals, guidelines, past issues, and lessons, etc. which are distributed to each work, it is easy to take over the points for responses even if the person in charge of disaster prevention was changed. Since in the disaster mode of the BOSS, it is possible to mutually understand all operations' progress at the disaster response headquarters and on-site. By making the disaster response work into a flowchart style, the point of decision-making becomes clear simultaneously, and the decision-making process can be shared among concerning stakeholders.

A process consists of three layers: large process, middle process, and small process (specific procedure). A large process is a process expressed and mainly comprises "organizations" such as departments and groups. The middle process refers to a medium-scale process expressed and mainly represents "resources" such as humans and machines and "things" such as materials. A small process is described as a small-scale process expressed in units, and mainly represents one decision-making by a single resource or an action that ends with one work [18].

3.2 How to build evacuation center operations

This research created a shelter operation workflow on the BOSS based on the Kawasaki city shelter operation manual. As a result, 244 kinds of intermediate processes were prepared (**Figure 1**).

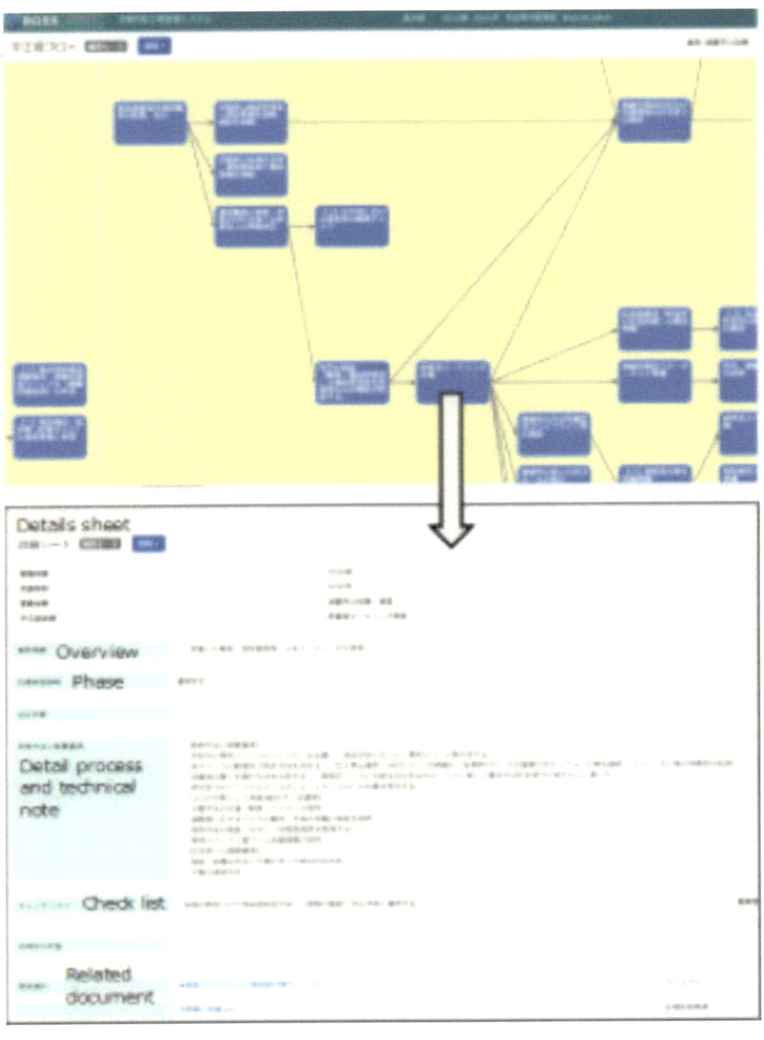

Figure 1.
Evacuation center management workflow built-in BOSS. (upper row: Workflow, lower row: Detail sheet).

As a procedure for creating the evacuation shelter, first, the shelter operation manual of Kawasaki City and the Cabinet Office's shelter operation guideline was used to list the conventional shelter operation work for each phase.

Necessary activities were extracted to construct a flow chart on the BOSS according to the workflow by the workshop with Kawasaki city members.

The following operations have been added as specific operations related to COVID-19 [19, 20].

a. Zoning inside the evacuation center. Depending on the conditions of the evacuees, zoning for non-concentrated space, vulnerable people shall be performed, and each space shall be set up for special attention, such as the position of toilets.

b. Response at reception. At the reception, it is necessary to check the physical condition of the evacuees. Temperature checks will be carried out, and self-reported physical conditions will be assigned to each space.

c. Hygiene equipment. Masks, goggles, gloves, and protective clothing are required to maintain responders' health who come into contact with an unspecified number of evacuees. If it is difficult to obtain them, alternative options should be used.

d. Securing social distance. Evacuees should be kept at a distance of more than 1 meter between each space—furthermore, the waiting line at the reception to prevent the spread of the infection.

e. Regular disinfection and ventilation. Hand sanitizers are necessary to prevent contact infection at various places in the evacuation shelter, such as at the reception, each living space, and around toilets. Regularly disinfect handrails and doorknobs. It is necessary to change the air by deciding the interval and time, such as performing ventilation for 10 minutes.

Regarding the created flowchart in the BOSS, a detailed sheet explaining the work details was also created (**Figure 1**). In this detailed sheet, each work's know-how was designed by describing the work outline, response phase, detailed procedures and notes, checklist, and related documents. The BOSS includes various manuals such as Operation manual for emergency evacuation in case of different hazards such as COVID-19 countermeasures in evacuation center operation (plan) in Kawasaki City.

4. Outline of the experiment

This study compared and verified evacuation shelters' operations with or without the BOSS in the case of flood disaster in Kawasaki City, Kanagawa Prefecture under COVID-19. Kawasaki City managed a large-scale disaster for the first time due to the 2019 East Japan Typhoon, subject to the Disaster Relief Act. During the typhoon, Kawasaki city analyzed the weather information and used a disaster warning system two days before the typhoon was closest. The city prepared and responded with the attitude of "thinking the worst-case and doing everything that should be done." However, as a result, various kinds of issues became apparent, such as the fact that information on the site was not reported to the headquarters (EOC, emergency operation center).

Based on this experience, the BOSS can be used to improve disaster response and verify how to change the operations.

Verification experiments will be conducted twice, and the differences will be discussed.

4.1 First experiment: June 24, 2019 (Wednesday)

4.1.1 Overview

On June 24 (Wednesday), 2020, at the Kawasaki City Higashi-Kokura Elementary School Gymnasium, the experiment about operations of the evacuation center with the BOSS or without BOSS (with conventional paper manual) were performed.

In the experiment, at the scenario setting, the number of people who tested positive for COVID-19 was increasing in Kawasaki City for several days, and heavy rain fell due to the linear rainbands around noon on that day. Level 3 evacuation preparation was organized at 14:10. The evacuation was issued for vulnerable people such as the elderly. It was set to prepare for the establishment of an evacuation center.

4.1.2 Verification category

In the experiment, the team was divided into two groups: the manual team (without the BOSS) and the BOSS team (with the BOSS), and both teams operated the shelter simultaneously. Each team consisted of three members. The manual team was staffed by the Saiwai Ward staff of Kawasaki city, and the Kawasaki City Crisis Management Office staffed the BOSS team. The structure of the three-members teams consists of one leader and two staff.

Also, the gymnasium of Higashi-Kokura Elementary School was divided into two spaces, the left side of **Figure 2** was the manual team, and the right side was the BOSS team.

Two manuals were used by the manual team as the shelter operation manual for wind and flood damage (Manual 1) and COVID-19 countermeasures in evacuation center management (Manual 2). The former is an evacuation shelter operation manual in the event of a typhoon disaster. However, as the manual does not include measures for COVID-19, the latter manual was also used to supplement the points of infectious disease. **Tables 1** and **2** show the principal works of each manual.

4.1.3 Characteristics of each team

The characteristics of each team member are described here. The manual team leader is a staff member in his second month, who joined the Kawasaki city office as a new graduate this year and is in charge of the disaster management department. The other members are staff members in their fifth and fourth years and contribute to promoting the city development department at the Saiwai Ward government office of Kawasaki city office. All of them are new graduates and have no experience as business/private company members. They have received training on storm and flood disasters this year but have no experience in operating shelters during real disasters.

On the other hand, the BOSS team leader is a staff member who has been working for the Kawasaki city office for five years. After working at the ward office for three years, the leader was transferred to the Kawasaki City disaster management department and has short-term experience in operating evacuation centers during typhoons in the 2019 East Japan typhoon disaster. Members consisted of the

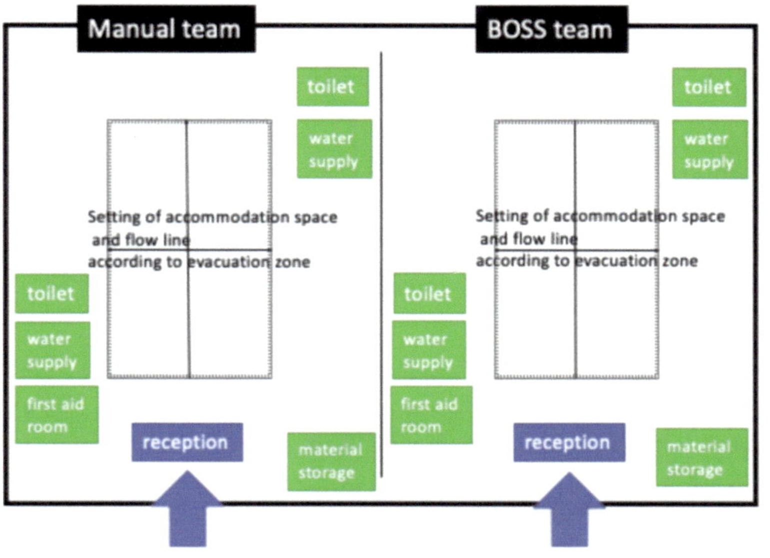

Figure 2.
Gymnasium classification.

Work Description	Specific Content
Meeting	Confirmation of work to be performed on a time series basis
	Division of roles
Reception	Reception card
Improvement of acceptance environment	Pet, animal space
	Simple air mat in space for people who need consideration
	Toilet paper
	LED lantern
Confirmation of communication	Contact to ward headquarters

Table 1.
Contents of the emergency evacuation center operation manual in case of wind and flood damage (manual 1).

Work Description	Specific Contents
Securing evacuation space exclusively for levers	separating evacuation space
	Separating toilets and flow lines
Confirmation of goods	Thermometer
	disinfection liquid
	Mask
	Goggles
	Tape for social distancing

Table 2.
Contents of measures for COVID-19 in evacuation centers (manual 2).

mid-career members. One member has passed two years at the disaster management department after twelve years of working with a private company. The other has passed seven years working for the Kawasaki city office. All members had training on HUG (Hinanzyo Unei Game, shelter management imagination game) education several times.

Besides, the BOSS installed a personal computer at the evacuation center's reception desk and arranged to check the BOSS flowchart to operate the works (**Figure 3**).

Both teams confirmed the response of a young leader with relatively little experience in real disasters.

4.2 Second experiment: Tuesday, august 4, 2020

4.2.1 Overview

Under COVID-19, both teams tried to operate evacuation shelters at Kawasaki City Nakanoshima Elementary School with shelter management committee members, school staff, and ward office staff held at Nakanoshima Elementary School in Kawasaki city on Tuesday, August 4, 2020.

Table 3 summarizes the positioning of each experiment. In the first experiment, although the BOSS team leader has limited experience in actual shelter management, the BOSS team is more advantageous. Therefore, in the second experiment, the leader of the BOSS team should be a new employee who has no experience of actual shelter operation and how to utilize BOSS, and the members should be senior staff rather than the leader. Also, verify that young leaders can effectively direct older members. On the other hand, the leader is a staff member who has experience operating shelters for the manual team.

In the first experiment, the BOSS team leader was a young staff member with experience, but the leader can be instructed by the BOSS with past knowledge.

4.2.2 Verification category

As in the first session, participants were divided into a manual team and a BOSS team, and each team carried out evacuation center operations at the same time. The number of people in each team consisted of 5 people, with 1 leader and 4 members.

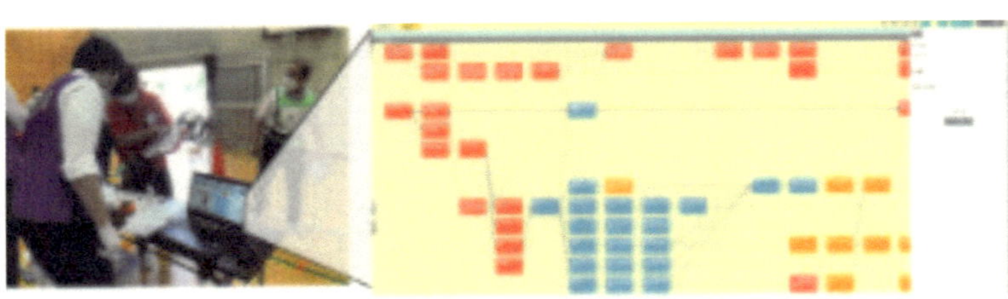

Figure 3.
How the BOSS team uses BOSS.

Experiment eases	Team type	Leader	Member	Preparation
First experiment	Manual team	Inexperienced young	Inexperienced young	None
	BOSS team	Experienced young	Inexperienced young	None
Second experiment	Manual team	Experienced mid-level	Experienced mid-level	Prepared before the experiment
	BOSS team	Inexperienced young	Inexperienced mid-level	none

Table 3.
The positioning of each experiment.

The manual team is a staff member of the Tama Ward, the evacuation center staff member at Nakanoshima Elementary School. The BOSS team is a staff member of the Kawasaki City Crisis Management Office, but a staff member different from the first member was assigned. The manual team also set up a reception area near the staff entrance on the first floor, and the BOSS team set up a reception at the adjoining doorway for children.

4.2.3 Characteristics of each team

The manual team leader is instructed by a mid-level staff who has experience in disaster response and gives them instructions. On the other hand, the BOSS team leader is in charge of the young staff member who had no disaster response experience, and the member is a middle-ranked staff member. This research confirmed whether a young leader without know-how could instruct members using BOSS.

4.2.4 Preparation

In the second experiment, the manual team conducted similar training a day before the experiment to confirm the flow of the experiment's day. On the other hand, the BOSS team had no experience establishing shelters and only checked the manual and BOSS in advance.

The manual used by the manual team was the shelter operation manual at the time of storm and flood damage (**Figure 4(a)**) in the first instance, and the content was mainly text. However, in the second instance, the storm and flood damage kit (As shown in **Figure 4** (b) was used as a mission to understand the work even by paper-based documents.

4.3 Contents of verification experiment and verification method

In the first experiment, both teams carried out preparations to open an evacuation center, accept evacuees, and cooperate with the evacuation center and the disaster response headquarters. Regarding cooperation with the disaster response headquarters, it was requested to contact the ward headquarter while the shelter

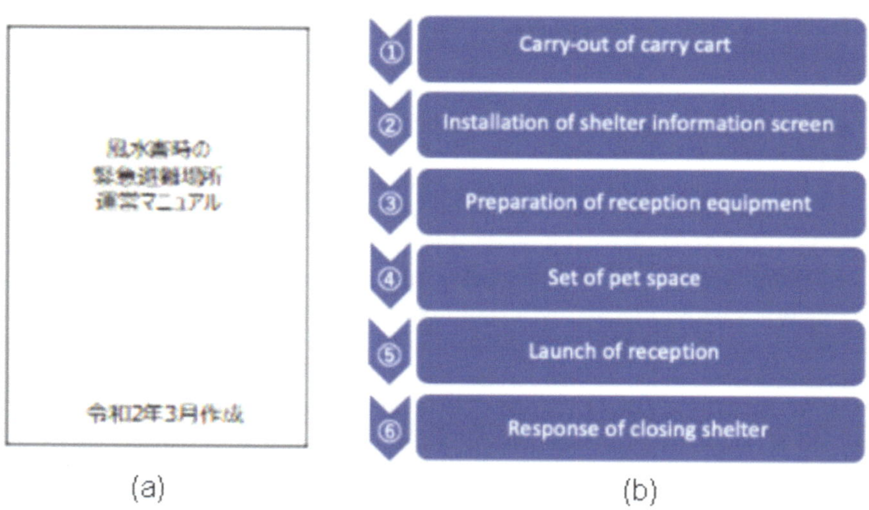

Figure 4.
(a) Emergency evacuation site operation manual in case of wind and flood damage (cover). (b) Flood damage kit (workflow).

was opened. The scenario, such as "for home-care patients with a positive PCR test, contact a medical institution. I started to carry it" was prepared in this experiment. Both teams need to report and share the situation with the ward headquarters that it is ready.

On the other hand, in the second experiment, the teams did not contact the ward headquarters, only the evacuation shelter was opened, and the differences between the manual and BOSS teams were analyzed.

Regarding the use of the facilities in the experiment, the first team was to use only the gymnasium, and the equipment such as classrooms, toilets were set up virtually.

The following five suggestions were used in the verification process. (1) For time, the minimum unit was 30 seconds = 0.5 minutes, and work-hours were recorded. (2) Regarding the operator's movement, a fixed video camera was installed, and the operator's movement was confirmed. (3) Regarding the leader's instructions, the activities were recorded with a voice recorder including the leader's statement. (4) Regarding the omissions of work, the manual and BOSS contents were compared with the contents of work performed during the verification test. (5) Regarding the participants' impressions and opinions, this research conducted a questionnaire survey and grasped the work's impressions.

5. Results of the first experiment

For the first experiment, the manual team had 23.5 minutes to complete the work, and the total number of work was 17. On the other hand, the BOSS team had 20.5 minutes to complete the work, and the total number of work was 24. The BOSS team carried out more works than the manual team.

The hypothesis is that manual teams do not take instructions from the leader and everyone tends to do the same work and duplicate the work. While on the other hand, the BOSS team leader gives instructions to the members, and each person works in parallel at the same time. It is conceivable that the work was achieved while being aware of the division of roles. Here, this research will verify the hypothesis by analyzing the leader's actions and each member while comparing the manual team and the BOSS team centering on the Gantt chart.

5.1 Gantt chart

Figure 5 shows the Gantt chart of three members in a manual team. The horizontal axis is the time (minutes), the vertical axis is the working name, yellow is the working time to complete a task by one person, orange is the work done by multiple people simultaneously. Gray represents the waiting time. Once all three members of the manual team finish the same work, they do the same work again in the second half, and there is much rework. However, time is running out.

On the other hand, **Figure 6** is a Gantt chart of three BOSS teams. Mr. C and Mr. D (like Mr. C and Ms. D for convenience in this research) have finished the work they started once and then moved on to the next work, and there is almost no rework. Incidentally, after 19 minutes, Mr. C seems to have accumulated from the start of the work because he was waiting for other members to finish their works. After that, contact sharing was started, and the work status was confirmed as a whole. Overall, the BOSS team did not see any rework in simple work and did not observe any waitings.

The big difference between the manual team leader and the BOSS team leader is that the manual team leader tends to do the same work as the member, but the

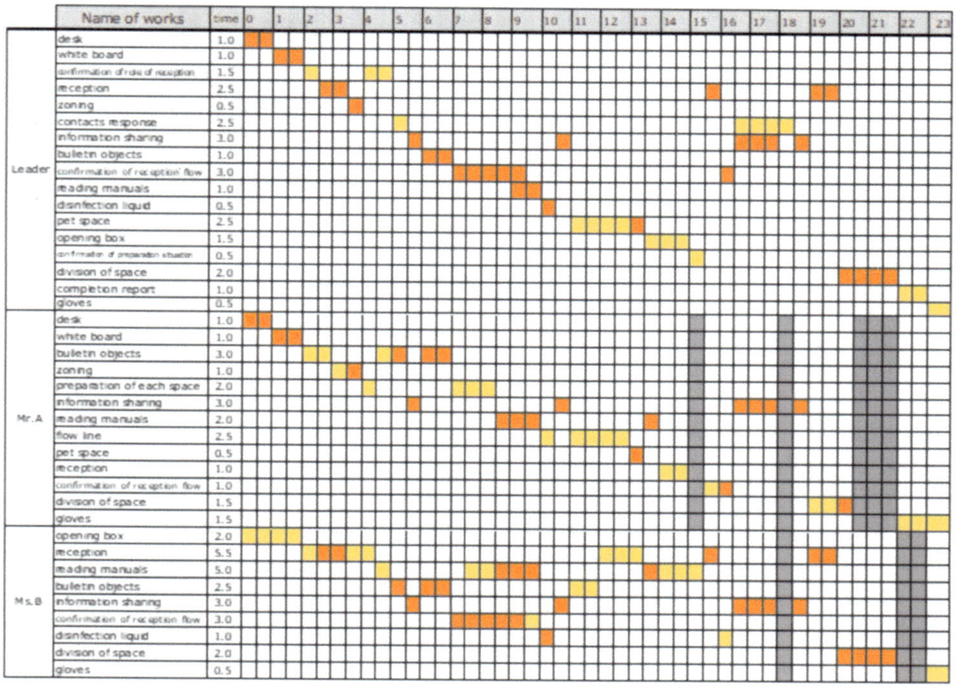

Figure 5.
Gantt chart of three people in a manual team (min.)

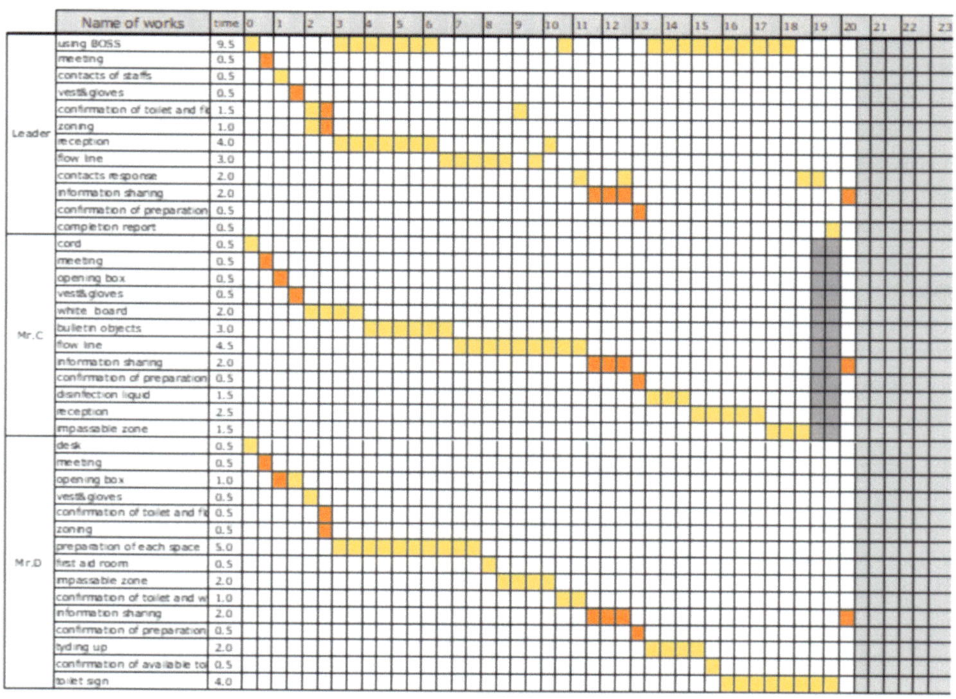

Figure 6.
Gantt chart of three people in a BOSS team (min.)

leader of the BOSS team instructs the members to work. The BOSS team leader tended to be around receptions to operate the BOSS, manage overall progress, and give instructions to members.

5.2 Waiting time

Next, analyze the waiting time. **Figure** 7 shows each team's waiting time. **Figure** 7(**a**) shows the manual team waiting time, and **Figure** 7(**b**) shows the BOSS team waiting time.

Two members observed the waiting time in the manual team due to waiting for the leader's instructions. On the other hand, since the leader responded while instructing each member of the BOSS team, there was almost no waiting time. Although Mr. C's waiting time was observed, he has completed all the instructed work and is waiting for other members' work to be completed.

Comparing the two teams shows that the manual team has a long waiting time than the BOSS team. The manual team did not receive any instructions from the leader and thought about what kind of action to take in each case centered on the leader. So, since there was no instruction, a waiting time occurred. Also, regarding the three members' actions, there was a tendency to simultaneously perform the same work, and no clear division of roles was observed. On the other hand, the BOSS team leader gave instructions to the members while confirming the workflow of the BOSS, and the members reported to the leader when the instructed work was over and received the next instruction. Since the leader instructed the work while checking the BOSS, the leader did not hesitate to wonder which work to carry out.

5.3 Number of works

When comparing the manual team and the BOSS team, the manual team members performed the same work with all three members without explicit instructions from the leader, so the manual team worked more than the BOSS team. Therefore, here, the number of works was focused on and compared.

Figure 8(a) shows the number of works performed by three manual team members. Here, the work was classified as "response work (simple work), contact/confirmation, and see manual." The works are directly related to the evacuation center establishment; contact/confirmation is contact with the ward headquarter. This result refers to sharing information and confirming the content of work.

Among the three members, the number of work done by the leader was the largest. The reason is that the leader carries out the response work by himself. It was also seen that three members were doing the same work simultaneously by multiple members.

Figure 8(b) shows the number of works performed by the three members of the BOSS teams. Like the manual team, the work was categorized into three types:

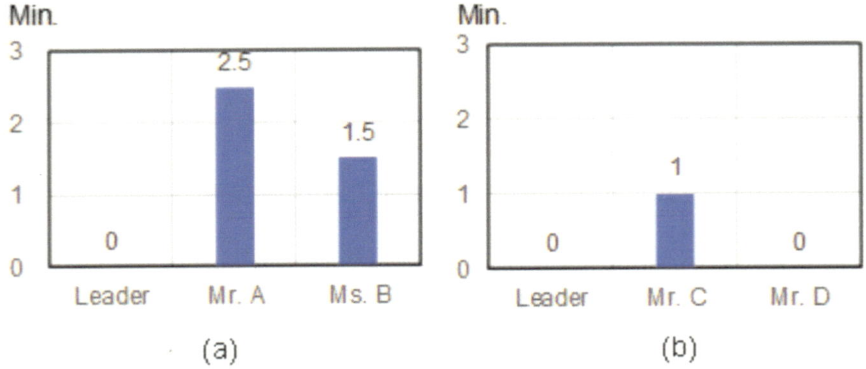

Figure 7.
(a) Manual team waiting time (unit:Minute). (b) Boss team waiting time (unit:Minute).

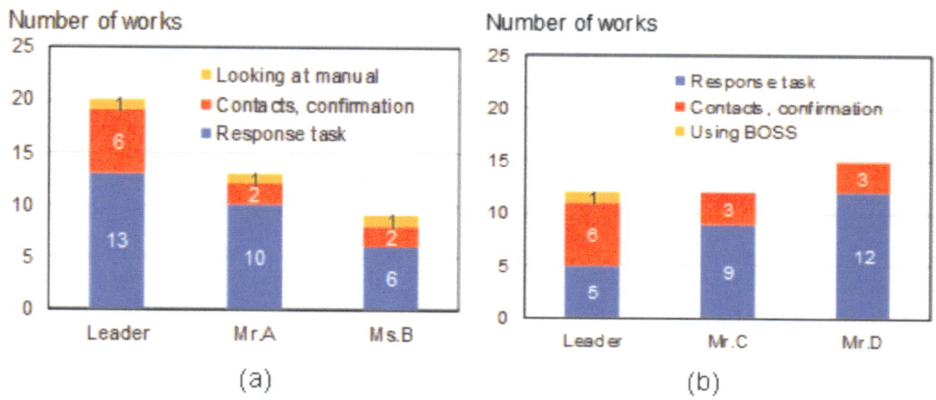

Figure 8.
(a) Number of works performed by 3 people in the manual team. (b) Number of works performed by 3 people in the BOSS team.

"response work (simple work), contact/confirmation (meeting), and BOSS operation." As a result, the number of tasks performed by the leader is the same as the number of tasks performed by Mr. C, but the number of tasks for simple tasks is the smallest. This result indicates that the leader mainly acted on instructing the members, and the members were performing the instructed work. The leader also operated the BOSS alone, shared the BOSS screen with the members, and managed the progress while telling the members the current position in the overall workflow.

5.4 Number of works by number of people

The manual team was observed to have three members doing the same work simultaneously, but here, this research analyzes to what extent multiple people were doing the same work.

Figure 9(a) shows the working time by the number of manual team members. The "1, 2, 3," on the horizontal axis means the number of people who simultaneously worked on the same task. According to this, the total number of working time by one person and the total number of working time by two persons are almost identical. The number of contacts/confirmations is more significant in the tasks performed by two people than in the tasks performed by one person. Besides, the time for simple work by two or three people is significant.

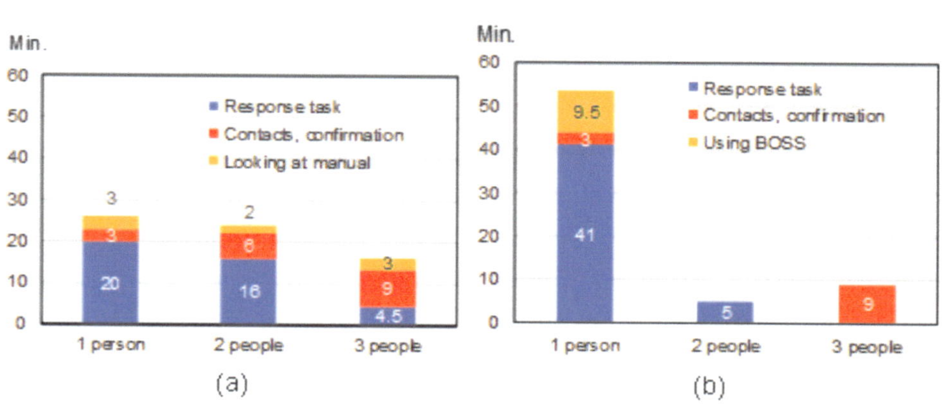

Figure 9.
(a) Working time by the number of the manual team. (b) Working time by the number of the BOSS team.

Figure 9(b) shows the work-hours by the number of members in the BOSS team. As shown in **Figure 9(a)**, while the manual team spends much time working with multiple people, members of the BOSS team almost always worked alone. It was also observed that multiple people did not perform simple work together because the three people's work was for sharing the situation, such as the meeting.

In other words, the manual team worked on a single task by multiple people simultaneously, but the BOSS teams simultaneously carried out different tasks in parallel.

5.5 Working time by business

Figure 10 shows the working time of each team. **Figure 10(a)** shows the working time by the manual team's work. The preparation for a reception is the time consuming, and three people work together. The next task that took a long time was information sharing. In terms of information sharing, it was often used as a time when the three members were considering what to do next. Also, "confirmation of reception flow" in the third was used to examine the reception flow. Regarding the top 5 works with the most work time, there was a strong tendency that three people were doing the same work simultaneously or with a certain amount of time.

On the other hand, **Figure 10(b)** shows the working time by the work of the BOSS team.

Like the manual team, it takes time to prepare for reception and share information. Since three people gather to share the situation, it means that three people work together, but for other things, there are many tasks that one person does, such as setting up "each space" and "flow line (reception)". The tasks performed by the three people were contact sharing, sanitary equipment, confirmation of preparation status, and meetings, all of which can be said to be tasks that should be performed by all.

In other words, it can be said that the BOSS team was working independently, except for those who were required to respond to multiple people.

Figure 11 shows the working time by both teams. Regarding the working time by business, it can be seen that the BOSS team generally performs various types of works in a short time. On the other hand, the manual team has a longer work time

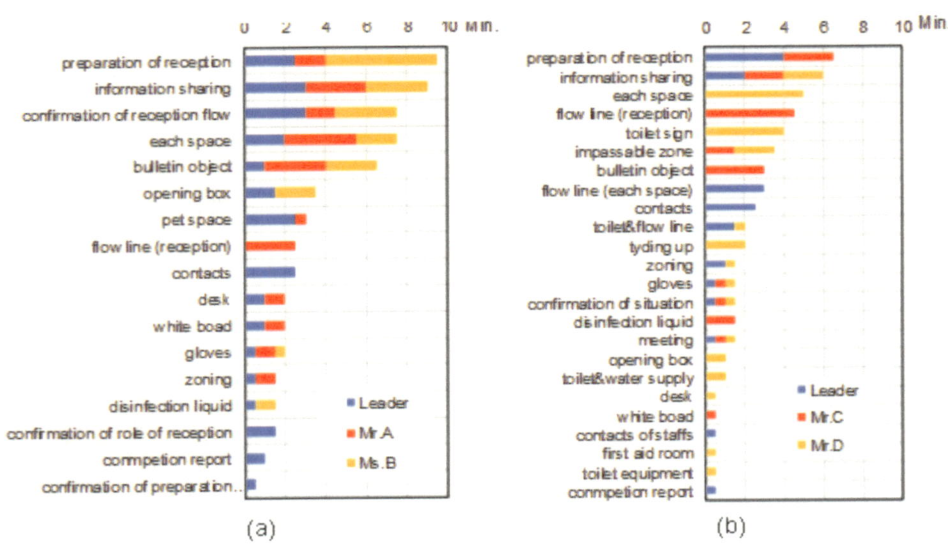

Figure 10.
(a) Work-hours by work of the manual team. (b) Work-hours by work of the BOSS team.

for each work because the manual team often works with multiple people, so the sum of the entire team's working time increases.

Regarding the kind of work, the manual team's work, but the BOSS team did not respond: "Confirmation of reception flow, the setting of pet space, confirmation of the role of reception." The manual team spends much time confirming the reception flow, and they were doing shelter opening work while discussing the next reception each time. While the pet space was set up according to the shelter opening kit, the BOSS team carried out the work according to the BOSS workflow because there was a note on the detailed sheet of the BOSS workflow.

The BOSS team did not set up a pet space because the set up of a pet space was not included in the workflow.

Also, the tasks that the BOSS team did were to display the toilet guidance and the impassable zone, clean up the opening kit, check the location of the toilet and water system, but the manual team did not carry out these tasks. In comparison, the number of operations performed by the BOSS team is larger than the manual team. All of these works are registered in the BOSS. In other words, if works are not registered in the BOSS, responders may omit the responses, so it is necessary to enhance the flowchart in the BOSS.

5.6 Omission of work

Table 4 shows the manual team's work during the verification test concerning the work content described in the manual. Many Xs indicate that the work was not performed and that many of the work had been omitted.

On the other hand, **Table 5** shows the BOSS contents and the BOSS team's work during the experiment. If both the start and end of the work are entered, it is considered that the work is completed. Besides, when nothing is inputted, it is regarded as non-performing work and is represented by X. The work indicated by the diagonal lines was excluded because it was set to be already performed in the experiment scenario. As a result, one work has been omitted.

5.7 Leader's statement

Table 6 shows the statements of the leaders of both teams. The manual team leader seemed to ask other members for their opinions and think while working

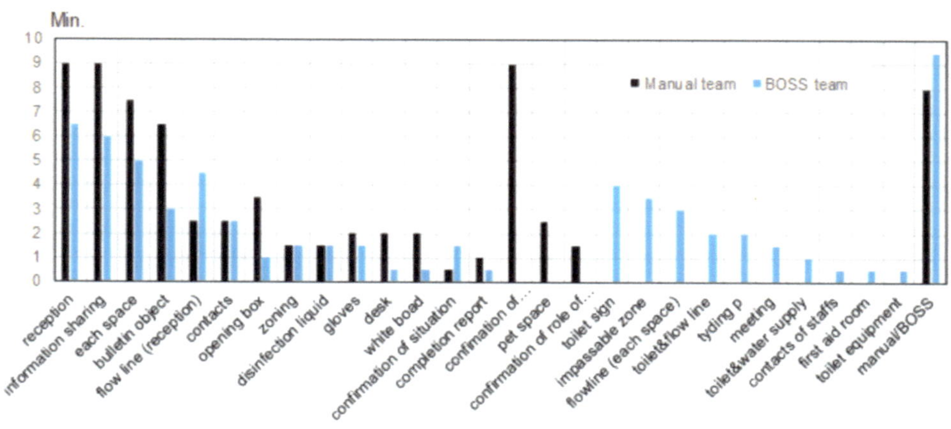

Figure 11.
Working time of each team by business (unit: Minutes).

Name of the works	Specific content	Whether or implement
Meeting	Confirmation of work to be performed on a time series basis	×
	Division of roles	×
Reception	Reception card	○
Improvement of acceptance environment	pet space	○
	Simple air mat in space for people who need consideration	×
	Toilet paper	×
	LED lantern	×
Confirmation of communication	Contact to ward headquarters	Δ(start×, end○)
Information guidance	Shelter information screen	○
	Arrow sign	○
	Reception	○
	Rules of evacuation life	○
Securing evacuation space exclusively for fevers	Separating evacuation space	○
	Separating toilets and flow lines	×
Confirmation of goods	Thermometer	○
	Disinfection liquid	○
	Mask	○
	Goggles	×
	Tape for social distancing	○

Table 4.
Manual contents and tasks performed by the manual team during the verification test.

together. Also, since he is speaking while considering the next action, he often asks the members questions or confirms the questions asked.

On the other hand, the BOSS team leader, while checking the workflow in the BOSS, thinks about who will do what and cuts it, so he gives instructions to the members. While confirming whether it has been done, the next work instruction is given.

Since the work contents are organized as a workflow in the BOSS, it is possible to decide the responses while looking at the overall image. So, there are a few omissions. Also, since all members can confirm the same workflow, it gives the impression that activities can be performed with the same purpose in one team.

6. Results of the second experiment

Based on the results of the first experiment, we conducted the second experiment.

6.1 Difference between the first and second experiments

6.1.1 BOSS contents and usage manual

Table 7 shows the contents of the manuals used by the manual team in the second experiment. In the first experiment, the conventional evacuation shelter

Name of the works	Whether or implement
Transport of necessary materials from the stockpiling warehouse	/
Confirmation of evacuation shelter safety and lifeline	/
Meeting	○
Opening box	○
Confirmation of the place of use in the facility	/
Bulletin object	×
Installation of a reception desk and shelter information board	○
Preparation of hygiene equipment for the operator	○
First aid room	○
Preparation of dense contactor space	○
Preparation of space for people with fever and poor physical condition	○
Preparation of space for people who need consideration	○
Preparation of space for healthy people	○
Disinfection liquid	○
Confirmation of the number of available toilets and equipment	○
Installation of guide signs for available toilets	○
Completion report	○

Table 5.
BOSS contents and tasks performed by the BOSS team during the verification test.

Manual team	BOSS team
-You have to secure a flow line for this. Where do you want to go?	-First of all, please Mr. C will make decisions about the classroom.
-I'd like to attach a little glue to the intervals when I was waiting at the shelter.	-First of all, Mr. D, please overhang the bulletin board.
-Mr. A, do you want to guide? At B's reception.	-Mr. C, is it OK to secure a space for those who need attention?
-Arc you going to separate it roughly? What do you do? Then, do you want to do it all, for the time being?	-Mr. D, may I ask you to check the toilet equipment?

Table 6.
Statements (instruction) of leaders to the staffs.

operation manual and COVID-19 countermeasure manual were used. However, in the second experiment, one manual that summarizes evacuation shelter operation in flood damage under the COVID-19 was used.

Table 8 shows the number of works registered with the BOSS in the first and second experiments.

Based on the first proof experiment, the content of work for the immediate response and the initial response was updated, and the number of works increased. In the initial response, four works were added between the first and second rounds, and in the immediate response, there were changes, including slight changes in work names, but 20 works were added overall. Specifically, the work was revised, as shown in **Table 9**. For one work, there are a variety of other works because the work was set to be performed sometimes in the same workflow. For example, there

Work Description	Specific Contents
Carry-out of carry cart	Transport from stocking warehouse
Installation of shelter information screen	Installation of shelter information screen
	Confirmation of reception place
Preparation of reception equipment	Non-contact thermometer
	Disinfection liquid
	Mask
	Gloves (if necessary)
	Face guard (if necessary)
	Pet card
	Cheek sheet
	Reception card
	User list for people who need consideration
	Health check-list
Pet space	Bulletin object
	Blue sheet

Table 7.
Contents of manual.

Phase	First experiment	Second experiment
Prevention	54	54
Preparedness	90	110
Initial response	28	32
Emergency response	66	66
Reconstruction/Recovery	6	6
Total	244	268

Table 8.
Number of operations in the BOSS for the first and second experiment.

is "(Periodical) Ventilation and disinfection" because the leader and members gradually became less aware of the infectious disease during the first experiment. They contacted each other and treated the evacuees politely. To maintain the basic principles against common infectious diseases (such as dense avoidance, hand washing, disinfection, ventilation.), tasks such as "Ventilation and disinfection" and "Thorough avoidance of three dense situations in evacuation shelters" were performed to raise awareness of infectious disease control regularly.

In the first experiment, the manual team carried out the "installation of pet space" in the immediately preceding response, but the BOSS team did not, and the reason was that the BOSS was not in good condition. Since it is considered, the task of "setting up pet space" was added to the BOSS.

6.2 Results of the second experiment

As a result of the second experiment, the manual team had 32.5 minutes to complete the work, and the total number of works was 10. On the other hand, the

Phase	Name of the works	First experiment	Second experiment
Preparedness	(Regularly) Ventilation and disinfection	—	4
	If a dense contact person who knows in advance arrives, it will be accepted into the accommodation space once	1	—
	If a dense contact person who knows beforehand arrives, it will be accepted into the accommodation space	—	2
	If the symptom worsens, request emergency transportation	2	—
	Physical condition management and follow-up	1	2
	Wearing masks properly and thoroughly disinfecting fingers		1
	Evacuee receptionist: Preparation for confirming the health status of evacuees	1	—
	Thorough avoidance of three Cs situation in evacuation shelter	—	2
	Pet space	—	1
	Transport request to medical institution	1	—
	Disinfection of frequently contacted bans (handrails, doorknobs, etc.)	—	2
	If the symptom worsens and the urgency is high, call 119 and contact the ward headquarters	—	4
	If the symptom worsens, request an emergency transportation	1	—
	Final confirmation of set location	—	1
	Situation management of dense contactor space	—	1
	Dense contact person transportation completion report	1	2
	Providing information to evacuees, updating from time to time	1	—
	Set of information equipment (TV, whiteboard) for evacuees	1	—
	Provision of information provision equipment (TV, whiteboard) to evacuees, provision of information *Updated thereafter	—	1
	Confirming the number of evacuees	—	4
	Installation of evacuation center information screens, etc.	—	1
	Thorough avoidance of three Cs situation in evacuation shelter	—	2
Initial response	(regularly) Ventilation and disinfection	—	1
	Confirming the number of evacuees	—	1
	Thorough avoidance of three Cs situation in evacuation shelters	—	2

Table 9.
Descriptions of works in the BOSS modified between the first and second experiment.

BOSS team had 36 minutes to complete the work, and the total number of works was 15. This result means that the BOSS team carried out more jobs because the manual team has 0.31 work/minute and the BOSS team has 0.42 work/minute in terms of unit time.

6.2.1 Gantt chart

Figure 12 is a Gantt chart showing the work performed by five people in the manual team and the required time. The horizontal axis represents the time taken (minutes). The vertical axis represents the work done. Yellow is the work time for one person, orange is the time performed by multiple people simultaneously, and the gray shows the waiting time.

In the first half, it can be seen that work once completed can be resumed later. The waiting time is seen in the middle stage and the latter half. Some works were completed in the intermediate stage, and the members were unsure what to respond to next. A member who finished early waited for other members to complete their work in the second half.

One characteristic of the manual team is that the leader took the initiative and delivered it to other members.

Figure 13 is a Gantt chart showing the work and time required by the five members of the BOSS team.

There is almost no rework, and each work is completed before moving on to the next work. Therefore, less waiting time is seen. In the latter half, some waiting time occurs. The time taken for all members to complete their works was defined as the work completion time by the team; the person who completed his work earlier waits for other members to finish theirs.

6.2.2 Waiting time

Figure 14 shows each team's waiting time. **Figure 14(a)** shows the waiting time for the manual team, and **Figure 14(b)** shows the waiting time for the BOSS team.

It could be seen that the manual team had a longer waiting time than the BOSS team. In particular, according to the Gantt chart (**Figure 12**), Mr. E, Ms. G, and Ms. H continue to wait without any instructions during the work.

Also, regarding the BOSS team's waiting time, although Mr. K had a waiting time in the middle stage, the waiting time was not long (**Figure 14(b)**).

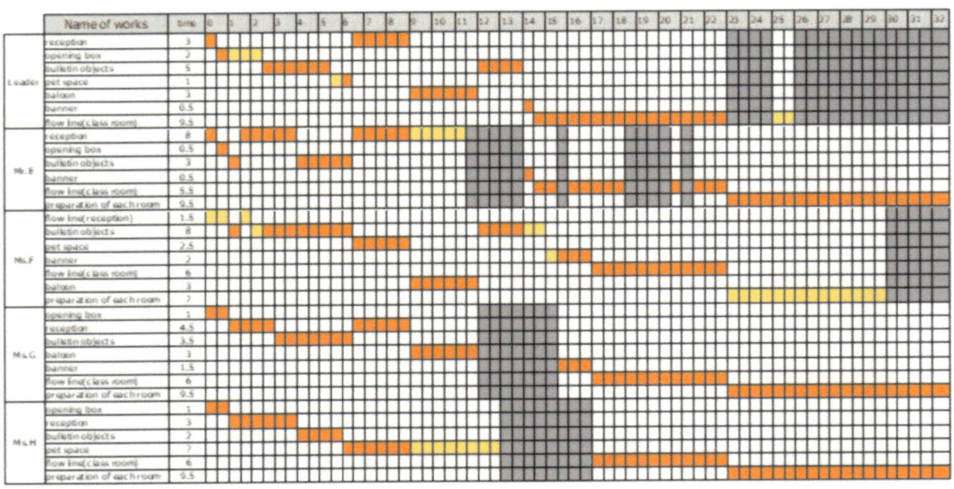

Figure 12.
Gantt chart of five people in the manual team (min.).

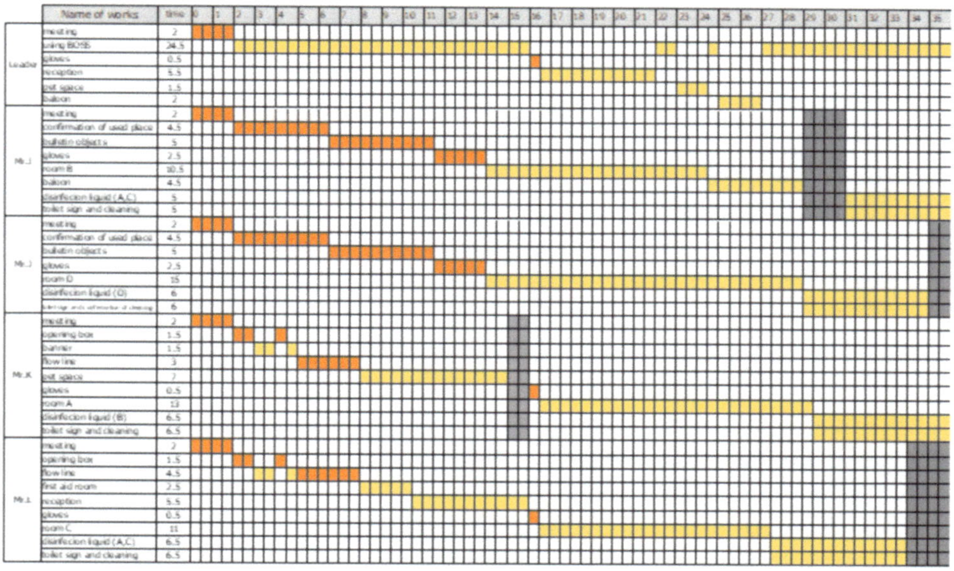

Figure 13.
Gantt chart of five people in the BOSS team (min.).

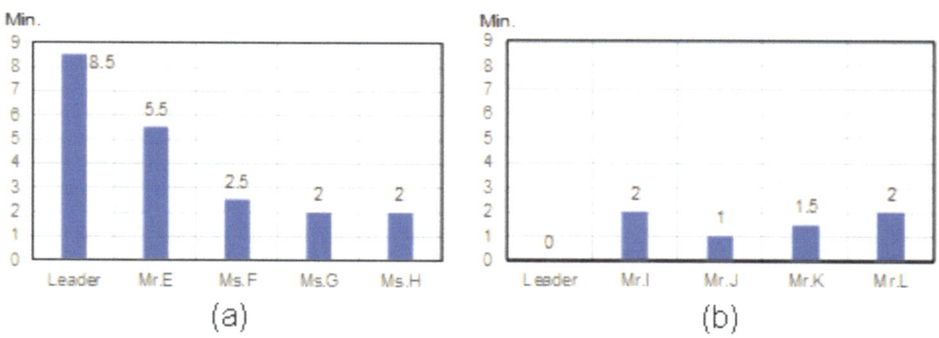

Figure 14.
(a) Waiting time of the manual team (unit:Minute). ***Figure 14(b)*** *Waiting time of the BOSS team (unit: Minute).*

6.2.3 Number of operations

Figure 15(a) shows the number of tasks performed by the five members of the manual team. The number of tasks performed by the leader and the number of response tasks (simple tasks) is the largest. The leaders and members are working while discussing for the next task. The leader was not looking at the manual, but Ms. G and Ms. H checked occasionally. This is because the manual team prepared the works a day before the experiment (in advance), and the leader took the initiative.

Looking at the Gantt chart, the manual team started the experiment without having a meeting to share the leader's response plan and instructions. In contrast, the BOSS team had time to hold a meeting with members to confirm the work content.

Figure 15(b) shows the number of operations performed by the BOSS teams. The number of tasks performed by the leader and the number of simple tasks is the

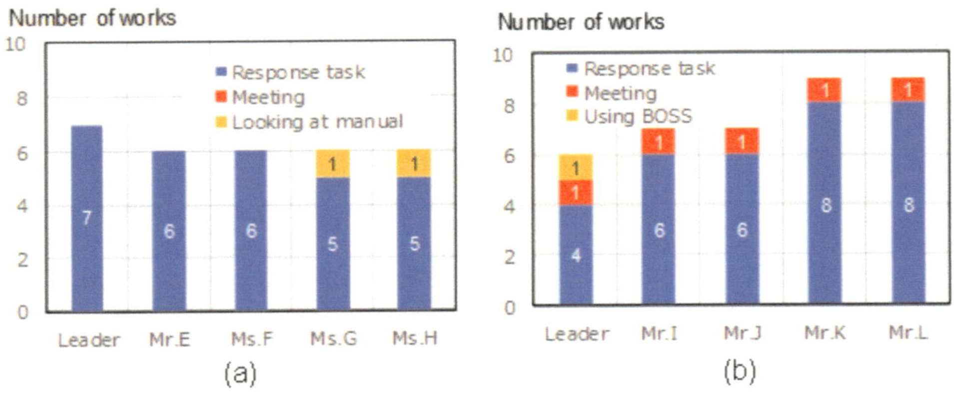

Figure 15.
(a) Number of tasks performed by 5 people in the manual team. (b) Number of tasks performed by 5 people in the BOSS team.

smallest. The leader instructed the members about the work contents by having a meeting with the members and only the leader performed the BOSS operations.

6.2.4 Working time by the number of people

The manual team received few instructions from the leader and worked while appropriately considering the members, and the leader took the initiative in responding. Hence, it is probable that multiple people worked. Therefore, this research analyzes how various people work at the same time.

Figure 16(a) shows the working time of each member of the manual team.

The same work was carried out by 4 members most often, and all members often did the same work. The response task (simple work) consisted of 1 to 5 people.

Figure 16(b) shows the working time of each member of the BOSS team. According to the table, one of the members did the most work. Although there are times when three members work at the same time, they did different works separately. The same type of work was not performed by four people simultaneously, and the work performed by all members at the same time was only the meeting.

6.2.5 Working time by business

Figure 17(a) shows the working time of each task performed by the manual team.

It took time to prepare the classroom and set the flow line to the classroom. This result includes the time required to move from the reception to each classroom.

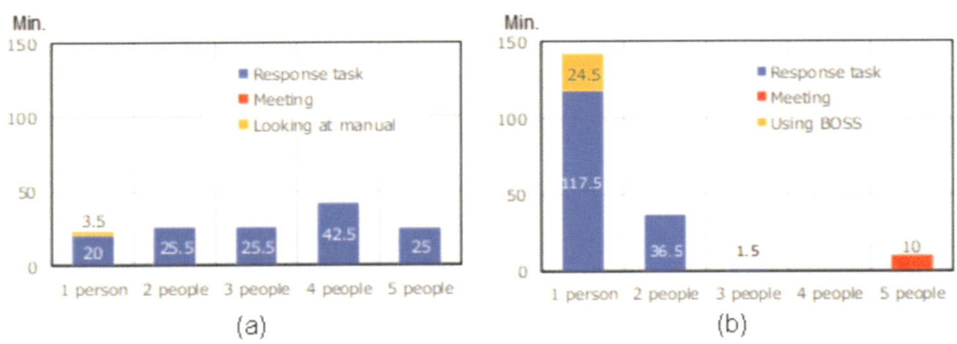

Figure 16.
(a) Working time by the number of the manual teams. (b) Working time by the number of the BOSS teams.

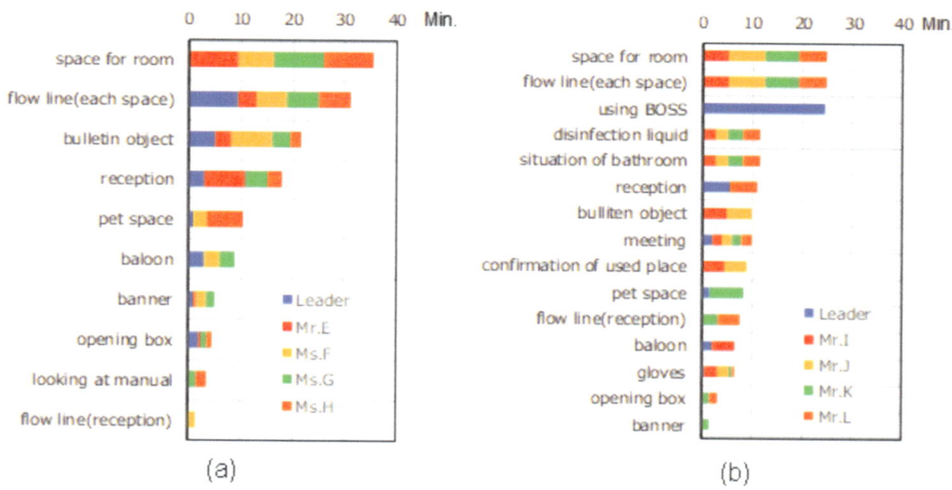

Figure 17.
(a) Work-hours by the manual team (unit: Minutes). (b) Work-hours by the BOSS team (unit: Minutes).

Besides, the leader is often involved in simple work and heads the classroom. All five people set the flow line and overhung the bulletin board. Furthermore, multiple people respond to the same work.

Figure 17(b) shows the working time of each task performed by the BOSS team. As with the manual team, it took time to prepare the classroom and set the classroom flow line. While operating the BOSS, this research focuses on grasping the work's progress by receiving instructions and work completion reports from other members. Furthermore, suppose multiple people do the same work. Work-hours are the same for the classroom preparation, setting of the flow line toward the classroom, installation of the disinfectant in each classroom, and confirmation of the toilets' status by the leader's instruction.

Figure 18 shows the working time of each task performed by both teams.

The manual team spends more time on each task.

On the other hand, the total number of tasks performed by the BOSS team is enormous: the BOSS team performs all functions carried out by the manual team. It then confirms the toilet's status and the facility use location not performed by the manual team. All of these tasks are included in the BOSS.

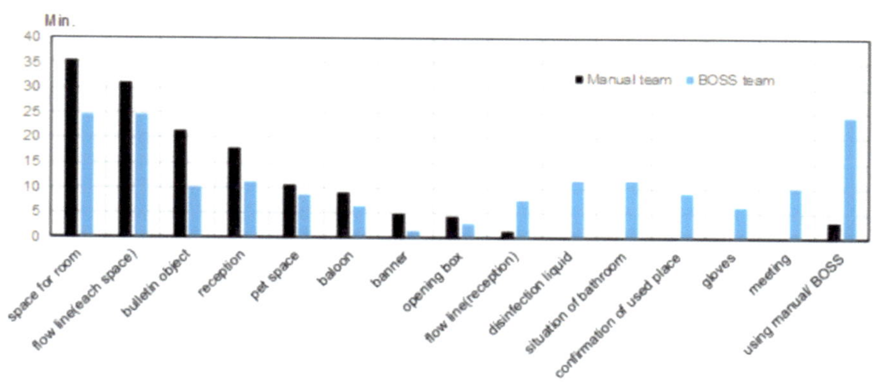

Figure 18.
Working time of each team by business (unit: Minutes).

6.2.6 Omission of work

Table 10 shows the manual's contents and the manual team's work during the experiment test. Gloves and face guards were not prepared. It was stated, "If necessary," and it may have been decided that it is not necessary.

Table 11 shows the BOSS contents and the work performed by the BOSS team during the verification experiment. If both the start and end of the work are entered, it means that the work is completed and represented as O. However, if the work is started but it is not completed, it is represented as △. All works in the BOSS contents were carried out, and no omissions occurred.

6.2.7 Leader's statement

Table 12 shows the statements of the leaders of both teams.

While the manual team leader performs the task, instructions were to the nearby member in the form of handing over the task. The leader is not aware of who should do the task. The team leader gives instructions while confirming what work should be done by other members.

On the other hand, the BOSS team leader specifies who will do what task and add a description of the specific work content. Besides, while confirming the completion of work, the next work instruction will be given.

7. Opinions of participants

Table 13 shows the opinions of the participants obtained from the questionnaire.

Participants' opinions as seen by their work can be regarded as subjective opinions, while visitors as objective opinions comparing both the manual and the BOSS team.

Name of the works	Specific contents	Remark
Carry-out of carry cart	Transport from stocking warehouse	O
Installation of shelter information screen	Installation of shelter information screen	O
	Confirmation of reception place	O
Preparation of reception equipment	Non-contact thermometer	O
	Disinfection liquid	O
	Mask	O
	Gloves_(if necessary)	×
	Face guard_(if necessary)	×
	Pet card	O
	Cheek sheet	O
	Reception card	O
	User list for people who need consideration	O
	Health check-list	O
Pet space	Bulletin object	O
	Blue sheet	O

Table 10.
Manual contents and tasks performed by the manual team during the verification test.

Name of the works	Remark
Transport of necessary materials from the stockpiling warehouse	○
Confirmation of evacuation shelter safety and lifeline	○
Meeting	○
Opening box	○
Confirmation of the place of use in the facility	○
Installation of evacuation center information screens, etc.	○
Bulletin object	○
Installation of a reception desk and shelter information board	○
Preparation of hygiene equipment for the operator	○
First aid room	○
Preparation of dense contactor space	○
Preparation of space for people with fever and poor physical condition	○
Preparation of space for people who need consideration	○
Preparation of space for healthy people	○
Disinfection liquid	○
Confirmation of the number of available toilets and equipment	○
Installation of guide signs for available toilets	○
Pet space	○
Final confirmation of set location	Δ
Completion report	Δ

Table 11.
The BOSS contents and tasks performed by the BOSS team during the verification test.

Manual team	BOSS team
-Let us create a reception first.	-Mr.K and Mr.L, please prepare the opening box.
-Let us put a mask on the entrance	-Next, Mr. K and Mr. L, please prepare the sanitation equipment of the operator. Prepare masks, gloves, face guards, and ponchos.
-The flow line is the one with the arrow. Is it okay if I put it on the floor?	

Table 12.
Statements of the leaders of both teams.

From these opinions, it is considered that the BOSS has a significant advantage since it can grasp the whole flow.

Since the work can be grasped in the form of a flow chart, it is easy to understand the work to be performed next, and the instructions of the leader are apparent. It can be said that the instructions can be given while prefetching the flow. Clear instructions from the leader lead to a sense of security for the staff who act upon receiving the instructions and can prevent confusion at the worksite.

8. The BOSS effect

As shown in **Table 14**, the effects of the BOSS are summarized in terms of quality, cost (burden), and period (working time).

Administrator
• As the person in charge of the command confirms, the next work to be done is precise, and if there is a shortage or mistake in the work I did, it will be pointed out immediately and as a position to engage in the shelter for the first time.
• It is easy to understand because it is organized into a workflow, and the activities arc organized. Also, since the start and end can be discriminated by color, there is a little omission of what to do.
• Because everyone sees one workflow, I felt that I could act with the same purpose. Also, I felt that the quickness and efficiency of contacting the headquarters increased, and that the collection of information at the headquarters was significant.
• While confirming the overall flow chart, I gave work instructions, and if there was room, I could look at two ahead and work based on the progress situation.
• The instructions from the leader of the BOSS team were clear, I have followed them strictly, there was no confusion about what to do at the site, and a smooth setup was possible.
Visitor
• The workflow of the evacuation shelter is chronological, so it is easy to grasp.
• The fact that the evacuation shelter's operation status is known at the headquarters, leads to a sense of security at both the headquarters and the evacuation shelter. It will lead to the simplification of regular reports and quick support from the headquarters.
• When preparing for the shelter's opening, the BOSS team seemed to be conducting instructions efficiently with the conductor looking ahead to a few hands.
• The staff is older than the leader, but the leader gave appropriate instructions.

Table 13.
Opinions of participants.

Category	Overview
Quality	• Q depends on (1) priority of work and (2) difficulty of work (necessity of qualification and experience). • Based on this, we will provide the necessary support concerning the effective allocation of staff. Concerning the assignment of tasks to support staff from other local governments, a system will be created to secure and allocate human resources in consideration of priority and difficulty.
Cost	• C depends on (1) operations that require a large number of human resources, and (2) equipment purchases and expenditures to residents. • It is a work that requires a large number of workers, such as the operation of shelters, management of supplies, emergency risk assessment of buildings, and building disaster investigation. The load analysis of the entire work is performed to determine the concentration and distribution of personnel. Secure financial resources for processing government expenditure (other than personal expenses). Or try to minimize spending. The evaluation of cost-effectiveness due to advanced measures is also related to this.
Delivery	• D depends on (1) process length, (2) stagnation between processes, and (3) gap between supply and demand. • To understand whether the work period ends in a short time frame or takes a long time frame to improve the bottleneck. For operations that require an extended period, such as restoration and reconstruction, building an organizational system based on this is possible. • For the waiting between processes, the difference between the end and the start of each process is discovered, and the shortage of personnel, materials/equipment, and information that cause the waiting are eliminated. • The demand difference between supply and demand is that the need disappears due to the difference between the demand generation timing and the actual supply time point. It is the situation that supplies cannot be kept due to the changing needs of supplies.

Table 14.
Opinions of the BOSS effect.

8.1 Improvement in work quality by increasing the number of works (Q: Quality)

In both the 1st and 2nd cases, the BOSS team performed more tasks than the manual team as a whole. In both cases, the BOSS team spent less time doing the same job with multiple people. Since many tasks were performed, it is considered that the number of functions that could be performed increased by sharing the jobs efficiently. Also, many jobs might have been performed because there were few omissions of works. For the BOSS team only, members' tasks include checking the status of the toilet and setting up an emergency room, which will improve the quality of life of evacuees.

8.2 Reduction of the burden on the operator (C: Cost)

The use of the BOSS clarified the instructions of the leader. Under clear instructions, it is considered that the staff can move quickly, and it is unlikely that there will be a difference in movement due to experience. Therefore, it will be easier to grasp the workflow of the whole work, and it will be possible for even inexperienced leaders to give appropriate instructions. Thus, the BOSS workflow, which is the axis of action, can be useful and can reduce the mental burden on the operator.

8.3 Reduction of working time (D: Period)

The BOSS team used fewer work-hours to complete their tasks. This is because the number of people engaged in one work is reduced by appropriately sharing the work under the leader's direction, and the work as a team is small. Also, in both cases, the BOSS team spent less time doing no work and had less rework. By eliminating waste and increasing efficiency, the overall working time could be shortened. The BOSS team's general work-hours for evacuation centers were shorter for the first experiment and longer in the second experiment than for the manual team. By doing so, the BOSS team achieved a reduction in overall working time, even though it does more work than the manual team. Although in the first appearance, the manual team had previously trained in the same workflow and remembered the workflow well, so it was impossible to see the effect of shortening the overall working time in the BOSS team.

9. Conclusions

This study compared and verified the disaster response process management system BOSS activities, or without BOSS, the manual for evacuation shelter management operations under COVID-19. Two experiments were conducted to clarify the effects and issues of using the BOSS.

As a result, by utilizing the BOSS, the leader gave instructions to the members to clarify their roles. The members shared the work efficiently without duplication, waiting for instructions, and what to do next. The BOSS team had little waiting time (retention) without hesitation about what to do, there were few omissions of work, and many kinds of works were completed during the experiment time compared to the manual team. The leader's remarks were instructed and commanded for the members. It was also found that even young leaders who have no actual disaster response experience can take the minimum response by checking the BOSS workflow.

It was also found that it is necessary to include the minimum required functions in the BOSS. It was also found that it is necessary to devise ways to mechanically check important functions matters on the system, such as creating a checklist.

As a future issue, it is possible that omissions will occur if the BOSS is not entered as a workflow.

Acknowledgements

We would like to thank members of Kawasaki City for the success of the experiments.

Author details

Arisa Yasui[1]*, Muneyoshi Numada[2] and Chaitanya Krishna[3]

1 Department of Civil Engineering, School of Engineering, The University of Tokyo, Japan

2 Interfaculty Initiative in Information Studies/Institute of Industrial Science, The University of Tokyo, Japan

3 Institute of Industrial Science, The University of Tokyo, Japan

*Address all correspondence to: antant36z@icloud.com

References

[1] Coibion, Olivier; Gorodnichenko, Yuriy; Weber, Michael; Weber, Michael (2020): The Cost of the COVID-19 Crisis: Lockdowns, Macroeconomic Expectations, and Consumer Spending, CESifo Working Paper, No. 8292, Center for Economic Studies and Ifo Institute (CESifo), Munich

[2] The Asahi Shinbun, Second wave of virus has now hit Japan, COVID-19 panel expert says, August 20th, 2020. (2020.8.27 access) http://www.asahi.com/ajw/articles/13652103

[3] Kazuko Ueyama et al. Current status of Niimi Public University as an evacuation center due to heavy rain disaster in July 2018 and implications for the future, Bulletin of Niimi Public University, 39, pp.185–187, 2018.

[4] Takashi Ito, Hiroshi Kawana, School as a "designated shelter" in case of disaster A case study in an elementary school located in the area affected by the earthquake and tsunami caused by the Great East Japan Earthquake, Bulletin of Faculty of Education, Ibaraki University, 65, pp.425. -435, 2016.

[5] Kouadio, I., K., Aljunid, S., Taro Kamigaki, T., Hammad, K., & Oshitani, H., Infectious diseases following natural disasters: prevention and control measures, Expert Review of Anti-infective Therapy, 10(1), pp.95–104, 2012.

[6] Izumikawa, K., Infection control after and during natural disaster, Acute Medicine & Surgery, 6, pp.5–11, 2019.

[7] Kawano, T., Tsugawa, Y., Nishiyama, K., Morita, H., Yamamura, O., & Hasegawa, K., Shelter crowding and increased incidence of acute respiratory infection in evacuees following the Great Eastern Japan Earthquake and tsunami, Epidemiology and Infection,144(4),pp.787–95, 2016.

[8] Ken Kanemi, Taro Kamigaki, Keiji Mimura, Hitoshi Oshiya, Surveillance of Infectious Diseases in Evacuation Areas in Miyagi Prefecture after the Great East Japan Earthquake, Jpn JP, 60(10), pp.659–664, 2013.

[9] Kenichi Goto and Fumio Oka, Pandemic of Infectious Gastroenteritis and Countermeasures against Infection in Kumamoto Earthquake Refuge, Journal of Infectious Diseases, 91(5), pp.790–795, 2017.

[10] Uckay, I., Sax, H., Harbarth, S., Bernard, L., & Pittet, D., Multi-resistant infections in repatriated patients after natural disasters: lessons learned from the 2004 tsunami for hospital infection control, Journal of Hospital Infection, 68, pp.1–8, 2008.

[11] Yee, E., L., Palacio, H., Atmar, R., L., et al, Widespread outbreak of norovirus gastroenteritis among evacuees of Hurricane Katrina residing in a large "megashelter" in Houston, Texas: lessons learned for prevention, Clinical Infectious Diseases, 44, pp.1032–1039, 2007.

[12] Mitsuhiro Higashida, Shigenori Tanabe, Haruo Hayashi, Preparation of crisis response manual focusing on effectiveness, Proceedings of the Institute of Social Safety Science, 10, pp.473–482, 2008.

[13] Yuta Yamada et al., To establish a method for creating a disaster response work manual based on normal work, Proceedings of the Institute of Social Safety Science, 10, pp.67–76, 2008.

[14] Takeyasu Suzuki and Masakazu Amami, Development of disaster response management system for local governments and its application to disaster response training, Proceedings of the Japan Society of Civil Engineers Earthquake Engineering, pp. 781–790, 2007.

[15] Munenari Inoguchi, Haruo Hayashi, Mitsuhiro Higashida, Development of disaster response work analysis method for defining requirements only by staff for construction of disaster response support system, Proceedings of the Institute of Social Safety Science, 8, pp.1–10, 2006.

[16] Shingo Nagamatsu and Kenji Koshiyama, How the disaster support staff of the local government were coordinated, Proceedings of the Institute of Social Safety Science, 29, pp.125–134, 2016.

[17] Shinji Akitomi et al., Consideration of medical treatment in the Great East Japan Earthquake based on the emergency support function, Proceedings of the Institute of Social Safety Science, 32, pp.1–7, 2018.

[18] Muneyoshi Numada and Kimio Meguro, Fundamental study for development of disaster prevention process system, Production research, 67 (2), pp.149–153, 2015.

[19] People and Disaster Prevention Future Center, Preliminary Checklist for Preventing Infection at Opening of Evacuation Center Ver.2-Guide Version-, DRI Special Issue, 2020. (2020.8.27 access) http://www.dri.ne.jp/wordpress/wp-content/uploads/sp_report_vol1_ver2.pdf

[20] JVOAD, New Coronavirus Evacuation Life Help Support Book, 2020.5.29. (2020.8.27 access) http://jvoad.jp/wp-content/uploads/2020/06/d9f7d81889f8c980e6d4958a150d7730.pdf

Chapter 8

Households' Adaptation to Climate Change Hazards in Semi-Arid Region of Mopani, South Africa

Musa Yusuf Jimoh, Peter Bikam, Hector Chikoore, James Chakwizira and Emaculate Ingwani

Abstract

New climate change realities are no longer a doubtful phenomenon, but realities to adapt and live with. Its cogent impacts and implications' dispositions pervade all sectors and geographic scales, making no sector or geographic area immune, nor any human endeavor spared from the associated adversities. The consequences of this emerging climate order are already manifesting, with narratives written beyond the alterations in temperature and precipitation, particularly in urban areas of semi-arid region of South Africa. The need to better understand and respond to the new climate change realities is particularly acute in this region. Thus, this chapter highlights the concept of adaptation as a fundamental component of managing climate change vulnerability, through identifying and providing insight in respect of some available climate change adaptation models and how these models fit within the premises and programmes of sustainable adaptation in semi-arid region with gaps identification. The efforts of governments within the global context are examined with households' individual adaptation strategies to climate change hazards in Mopani District. The factors hindering the success of sustainable urban climate change adaptation strategic framework and urban households' adaptive systems are also subjects of debate and constitute the concluding remarks to the chapter.

Keywords: households, adaptation, climate change, semi-arid, South Africa

1. Introduction

Adaptation to climate change hazard is attracting growing international attention as confidence in forecasts for climate change is rising [1]. Developing countries have unique adaptation needs because of high vulnerabilities and the tendencies to bear a significant share of global climate change costs [2]. The Intergovernmental Panel on Climate Change's (IPCC) Fourth Assessment Report noted that public recognition and concern about the global environmental issue of human-induced climate change has reached unprecedented heights. Research into the drivers, both natural and anthropogenic, the character and magnitude, their impact on human living conditions and ecosystems, and possible approaches to adaptation and

IntechOpen

mitigation, as well-as understanding of the complex relationships with ecosystems interacting with them, has also increased in recent years [3].

While anthropogenic greenhouse gas emissions, which aggravates climate change are mainly from rich industrialized countries, the consequences of which are projected to be relatively acute and more serious in developing countries particularly in semi-arid region of Africa, where, for instance, rise in temperature and reduction in precipitation are likely to result in high evaporation, with serious health related consequences [4, 5]. South Africa like many developing countries' national economies and employment heavily rely on climate-fed activities [6], coupled with high poverty levels, limited technological and weak institutional ability to adapt to climate change qualifies for classical case in which urban populations (children, elderly, persons with disabilities and women) are more susceptible to climate change adversities [7].

Nonetheless, climate change adaptation strategies and projects on one hand, still focus mainly on sustainable rural adaptation, without much attention on urban areas, especially small and medium towns, where there is increasing household vulnerability and climate change pressures [8]. Current literature on adaptation to climate change in urban areas are largely coastal and big city biased [9–11]. On the other hand the early years of international climate change studies' attention on adaptation as a strategy was compromised by mitigation and impacts [12]. In recent years, several models incorporate mitigation, as an anthropogenic intervention to the changing climate [3] and has rapidly escalated, while models that incorporate adaptation are still in their various stages of development, advancement and yet to reach maturity [13].

Inherently, it has become urgent to focus on approaches and instruments that assist with the reduction and reversal of the prevailing and unescapable climate change hazards, coupled with the need to maximize the immediate manifestation of the net benefits of adaptation [14]. As an essential policy response, local level and individual (including private) households' adaptation strategies to climate change needs to be apportioned the desired priority in climate change policy agendas at all levels and scales of governance.

This chapter aims through a holistic approach, to provide the highlights of the South African governments at several levels and scales of governance to advance adaption and mitigation urban household practices and interventions. This analysis and discussion is conducted within the global context of existing adaptation framework that incorporate the local level and individual households' (private) adaptive practices, efforts and initiatives. Furthermore, the chapter also identifies some of the key issues hindering the success of urban adaptation policies and interventions in the region.

In brief, the chapter places in perspective, the basic steps necessary for a more participatory urban management for sustainable households' adaptation to climate-related hazards in the semi-arid region of Mopani, South Africa.

2. Literature review

2.1 Adaptation, a fundamental component of climate change vulnerability

The new climate is no longer a doubtful global reality, but a phenomenon that we need to learn to live with for years to come [11]. Its disposition to leaving no facet of human endeavor immune from its negative externalities are unpredictable and presents very worrisome realities for the contemporary society and urban communities [15] largely manifesting beyond alterations in temperature and precipitation threatening the existence of humanity, particularly in Africa, and other developing countries [3, 16].

Adaptive ability to climate change hazard is considered a new field of endeavor, serving as a converging point for several experts, ranging from development experts, climate scientists, planners, disaster managers, and a host of other experts and disciplines/fields [17]. This has brought about divergent conceptual models to the study of vulnerability and adaptation, though addressing similar issues and emphasizing similar processes, but rather with different vocabularies [18]. The growth in the body of literature on the conceptual issue has brought about a confusing set of terminologies with unclear relationship [16, 19, 20]. However, notwithstanding the differences, the recognitions and understanding of the need to curtail the adversities of the phenomenon is the most crucial.

The frustrations from the present context of failure to successfully mitigate greenhouse gas emissions and curtail its associate developmental issues has resulted in adaptation becoming not only an inevitable strategy to frustrate vulnerability but also an integral social components for vulnerability assessment [16, 21]. However, this course of action is still in the trial periods of being considered relevant, particularly within science and policy contexts [22, 23].

Adaptation to climate change, is the "adjustment in natural or human systems in response to actual or expected climatic stimuli or their effects, which moderates harm or exploits beneficial opportunities" [24] as cited in [25]. It can take various dimensions from being reactive or anticipatory, private or public, and autonomous or planned [24], it can equally either be active or passive [26], spontaneous or prompted by alteration in conditions [27]. It is however a phenomenon that its success is hinges on the adoption of several co-and interdependent factors, including but not limited to human, technological y and policy matters.

However, many regions of the world, particularly Africa currently have limited access to these technologies, appropriate information and financial resources [28]. The cost-effective use of adaptation strategies will therefore depend upon the availability of financial resources, technology transfer, and cultural, educational, managerial, institutional, legal and regulatory practices, both applicable domestically and internationally [29, 30]. Hence "the need to consider indigenous knowledge system-based support and intervention", for effective climate change adaptation strategies as one of the under-studied and utilized adaptation and mitigation strategy especially in Africa and developing countries in general [31].

In this chapter we further argue that like all other cultures, in Mopani District people are essentially adaptive, while exposed to environmental variability and risky circumstances in the past. These events called to question the local people's adaptive capacity in respect to environmental variability and risk within the resources and technologies available options to them [32]. Therefore, to efficiently and adequately confront the prevailing and the potential climate change hazards, indigenous knowledge (IK) must be embraced, but be enhanced particularly beyond peoples' experienced coping option ranges [33, 34]. The development and adoption of IK notion has necessitated the paradigm shift from organic adaptation approach to scientific adaptation framework models that attempt to respond to gaps ranging from adaptation needs determination, to adaptation assessment and interventions. The ensuing section of this chapter attempts at the identification of some existing adaptation models from a historical lens perspective, to typologies of these models and gaps that characterized them as well-as offering suggestions for improvements.

2.2 Climate change adaptation models

Adaptation modeling field is wide, varied and is punctuated with largely unclear disciplinary/field boundaries [35]. The definition of what are its constituents is equally open to numerous interpretations, with tagging of several models as adaptation models added another confusing layer to the identity and boundaries controversies [36]. In several contexts of science, models are considered very essential and key in different fields, disciplines and specialties. For instance, Evolutionary models are very important in the biological sciences disciplines while the agent-based models are a dominant feature in the social sciences [37]. Models are painstakingly built, tested, compared and revised in light of practice and feed-back loop for future lessons [36].

On a general note a classification of models on climate change adaptation was further made in line with the existing ones by [36] who identified two distinct typologies or categories of adaptation models, these include:

a. Adaptation Centered Models (ACMs); and

b. Impact Centered Models (ICMs);

Over time, advancement in understanding the consequences of climate change and policy interpretations and the associated challenges has occasioned a shift in global priority in climate change policy [37–40]. At the onset, an undeviating cause–effect style prevailed, then climate situations forms the foundation upon which future climate impacts is estimated, which then outlines the needs for adaptation. With this linear concept, on one hand, adaptation to climate change is divorced from social activities and processes where needs are informed by scientific manipulations [41, 42] on one hand. On the other hand, a more comprehensive approach where the risk assessment is guided by management of past climatic hazards experience, and adaptation recommendation are determined by the option's probability to reduce the prevailing and future climate risks while synergizing with other policy objectives, and existing management activity [43]. The later concept is currently in vogue and has enjoyed patronage from researchers, academics and policy makers, informed by its openness and comprehensiveness. Upon the determination and assessment of the needs, the choice of the form of adaptation will be made from the following identified three adaptation options:

a. No-regrets adaptation options;

b. Proactive anticipatory adaptation; and

c. Win-win adaptation

These options are not new, but the policies in various forms of decision models about them in Mopani District like other municipalities is currently characterized by limited attention and priorities [44]. Thus, making the success of the municipal adaptation efforts to appear unsuccessful.

However, a probe into the available literature and survey analysis with respect to climate change adaptation and the various adopted models in the study area, revealed some essential issues. These are policy related issues; Climate change issue; and adaptation issues. These issues form the fundamentals upon which the following identified gaps are considered critical in the existing models. These include:

2.2.1 Gaps regarding climate change adaptation models

With respect to climate change adaptation and the various adopted models, survey revealed some essential issues related to policy; Climate change issue; and adaptation issues. These issues form the fundamentals upon which the following identified gaps are considered critical in the existing models. These include:

2.2.1.1 Gaps in relations to the current climate change adaptation models

Our findings revealed that many models on climate change adaptation, apart from being highly mathematical in nature, are based on methodological ideas that originate from the advanced economies [45], limiting their applicability in local African communities' context. This is because the assumptions upon which the models are largely based are alien to the prevailing realities in the region particularly in Semi-arid region of Mopani District in South Africa. In addition, several of these models are largely rural biased [46], or centered on metropolitan, big and coastal cities [47]. Similarly, some are rather infrastructure or sector-specific adaptation framework such as water, transport agriculture and energy sectors [48–52], while rather than local community based adaptation models, other models have focused on macro level postulations [53]. Hence the need for a flexible household-based conceptual framework model that is participatory and applicable at all levels of policy and decision making.

2.2.1.2 Gaps related to policies

Several studies have advocated for household-based climate change adaptation strategy to be anchored by municipal planning agency [54–56]. The study acknowledges that most local municipalities in the district are still relying on macro level climate change adaptation formulated policies from the national government. Despite that the impacts of climate change on both human and environment are well acknowledged in the various municipalities' planning instruments (Integrated Development Plans, Spatial Development Framework etc.), yet, little evidence exist to indicate the efforts to pragmatically and coherently address the challenges [44].

2.2.1.3 Gaps related to reporting climate change events

During data collection, our interactions with the community members, revealed that municipal governments were rather relying on reactive adaptation procedure rather than proactive. The delay in reporting of incidence of hazards have resulted in more costly, more devastating and sometimes unrepairable situations. Due to the devastating consequences often occasioned by late reporting of climate change emergencies, the climate change adaptation challenges are complex, dynamic and contextual, thereby requiring urgent attention by stakeholders.

For adaptation to be beneficial and cost effective, it should not be solely reactionary but rather proactive and anticipatory [57]. Changing climate is no longer in doubt so also is the likelihood of the trend to proceed to the coming century at an unprecedented rate in history, as projected [20, 58] with strong signals to a rising hazards for regions of such countries that are already water-stressed, like Mopani District, Limpopo province [5, 59] and other semi-arid regions of Africa are also projected.

Hence for effective communities adaptation, government at all levels have pivotal roles to play, particularly within the global context of adaptation framework as guided by the current regime. On this basis the chapter highlights the various steps taken by the South African governments (National, Provincial, District and Local Municipalities) with regard to adaptation needs of the citizens, through policies, program and projects interventions.

2.3 Efforts of South African governments within the global context of adaptation framework

In order to understand the roles of the various levels of government in South Africa in combatting the menace occasioned by climate change across the country, activities of government (National, Provincial, District as well-as local municipalities) regarding climate change adaptation were reviewed. This was assessed through the policies, strategies and legislations (Acts), and it was equally further done within the global context. The Republic of South Africa, being a signatory to Kyoto Protocol and a part of the United Nations Framework Convention on Climate Change (UNFCCC), has taken several initiatives (past and present) in striving to fulfil the expected obligations as regards the protection of citizens and the regional territory against the aggressiveness of climate change and its associated adversities. Some of these efforts as regards adaptation are highlighted in different eras in this section, and these include the following:

2.3.1 Apartheid era

The legal framework for managing disaster in South Africa preceding democratic rule, were largely administered by the Civil Protection Act No. 67 of 1977 [60]. The National Disaster Management Framework (NDMF) was initiated but was characterized by inadequacies following over a hundred lives lost to Lainsburg floods in the year 1981 alone [61]. In reaction to this incidence, out of various legislative and structural reforms that were put together to overhaul the system for proper integration of disaster management was the South Africa Constitution of 1996.

2.3.2 Post-apartheid era

The South Africa Constitution of 1996 marked the beginning of a prominent legislative and structural reforms of disaster management, by specifying the roles of the government at all levels in Part A Schedule 4 [62]. This necessitated the extension of the Civil Protection Act, the pioneer integrated policy on the management of disaster, called "the Green Paper on Disaster Management" [63].

The same era has witnessed active participation in various fora and conventions with respect to dealing with the challenges and opportunities that climate change presents since 1994. Commitment have been shown by the country to sustainable development with both active international participation and institutionalizing national frameworks. The frameworks include out of others: the Kyoto Protocol, the United Nations Framework Convention on Climate Change, the Cancun Agreement, the Convention on International Trade in Endangered Species of Wild Fauna and Flora, the Ramsah Convention on Wetlands of International Importance, the Basel Convention on the Control of Transboundary Movement of Hazardous Wastes and Their Disposal, and the Montreal Protocol for the Protection of the Ozone Layer [64] cit. in [65]. This suggests an involving participation of the country on climate change issues at the global realm.

Similarly, South Africa has at various times successfully instituted some climate change related legal frameworks that are either sector-prone (e.g. waste management, carbon tax, transport, energy efficiency, renewable energy and several others). Other related planning instrument is the Integrated Development Planning (IDP), through which short and medium-terms development objectives, strategies and programs are prepared as strategic plans for municipalities. It is a key instrument for guiding and informing fiscal allocation (budget), administration and decision making for service delivery and development within municipal's jurisdiction [66]. Subsequently, White Paper on Disaster Management was produced, emphasizing proactive and integrated approach in the management of disaster through public (stakeholders) participation and capacity building [67]. Targeting the creation of National Disaster Management Centre, enhance disaster prevention among the poor and disadvantaged zones, ensure adequate funding system and facilitate access to information (South African Government Gazette).

In 2002 Disaster Management Act 57 of 2002 was institutionalized, highlighting the guiding philosophies for disaster management and defined tasks [68]. The Act provides for the establishment of Intergovernmental Committee on Disaster Management, with powers to the Premier of the concerned Province and Local Government to select members. While at national level, the Minister of Cooperative Governance and Traditional Affairs is empowered to establish a National Disaster Management Advisory Forum with several governmental and non-governmental organizational representatives, traditional institutions and various professional, Sec 5. (1). Section 8. (1) Establishes a National Disaster Management Centre (NDMF) to form part of, at the same time functions within the control of the Minister under a state department of the public service. Provincial disaster management framework is instituted in Section 28. (1) of the Act not only to be established but also implement a disaster management framework aligned to the NDMF objectives and in consistent with the provisions of Act (No. 57,2002) and the NDMF, 33 (I). The local government is empowered under chapter 5 to appoint a disaster officer [62]. This gave birth to the establishment of Mopani District Disaster Management Centre at District Municipal level and the appointment of disaster manager in the five local municipalities in Mopani District as gathered during our field survey, they equally had plans for disaster management framework [69].

In 2011, the parliaments of South Africa adopted the National Climate Change Response Strategy (NCCRS). The policy document is generally anchored on some strategic priorities such as risk reduction and management; mitigation actions (with significant targeted outcomes); sectoral responses; policy and regulatory alignment.

The establishment of the National Climate Change Monitoring and Evaluation System came with the objectives of tracking South Africa's transition to a climate resilient society, by following-up on the country's transition to a lower carbon economy and by tracking climate finance. The benefits of the system include out of others, the provision of an evidence-based impacts and the vulnerabilities to climate change, and providing learning for the workability and otherwise of climate change response. This will inform the future responses to climate change as well-as facilitate the assessment of the impact and need for climate finance and institutionalizing national communication and biennial update reports. As promising as these objectives are, the M&E system has till now been struggling to find its rightful place, because of the disconnect between the municipalities and the grassroots where the data (for national communication and biennial reports) ought to be generated. Although the M&S system is substantially mitigation-focused not adaptation oriented, it still remains a viable too and mechanism for managing adaptation if well captured and harnessed.

The specific urban policy and planning that was institutionalized that "seek to influence the distributions and operations of investment and consumption processes in cities for the common good" [70] was the South African Integrated Urban Development Framework (IUDF) as approved in 2016 by Cabinet. Although attempt by various Government's Departments have in different ways attempted to address the challenges of urban areas since 1994 with significant achievements in areas such as service extension, municipal reform, urban renewal and economic infrastructure development, these efforts are largely viewed as inadequate [71]. Not so much achievements have been recorded in the mainstreaming of climate change to urban planning. The municipalities in Mopani District are still relying on the National Urban Policy without plans (currently) to have theirs that embrace the economic, political, social and environmental peculiarities of their respective areas. However, the adoption of the Paris Agreement as well as the New Urban Agenda, signaled a renewed motivation for action, particularly to mainstreaming climate change in Urban Policy.

The Mopani District Municipality in line with the National Disaster Management Act 2005 acknowledges the current and the potential climate change threats to both human and the environment. It equally recognizes the need for actions to mitigate, as well as prepare for the projected changes (adaptation) in the District. Consequent upon this, the district municipality in 2016 developed Vulnerability Assessment and Climate Change Response Plan to prioritize the development of Climate Change Response strategies. The Plan recognizes several numbers of ways that climate change will impact on human settlements across the district and thus identifies related indicators, sub-projects and actions for inclusion in the service delivery and the plans for budget implementation [72]. Our field survey revealed that the identified projects are held for paucity of funds.

The frustrations from the failure of municipalities to guarantee the protection of households through the implementation of a pragmatic actions have prompted private adaptation initiatives across the selected towns in the district to curtail the impacts of climate change. The section of this chapter succeeding the description of the study area and methods, addresses the various initiatives of households towards coping with climate change in the selected towns.

3. Study area and methods

3.1 Description of the study area

Located in the semi-arid region, the northern-most province (Limpopo) of South Africa, Mopani District Municipality is a category C municipality (**Figure 1**).

The district consist of five local Municipalities, including: Greater Giyani, (the district administrative seat), Maruleng, Greater Letaba, Ba-Phalaborwa and Greater Tzaneen. The municipality is situated on Longitudes: 29° 52′E to 31° 52′E and Latitudes: 23° 0′S to 24° 38′S, with 31° E as the central meridian. It covers 13,948.418 ha (10.2%) of the surface area of South Africa. It shares boundaries in the east with Mozambique, in the north, with Vhembe District Municipality through Thulamela & Makhado municipalities, while bordered in the south, by Mpumalanga province through Ehlanzeni District Municipality and, by Capricorn District Municipality to the west [72].

3.2 Climate of Mopani District

Being within the semi-arid region, the district is characterized by temperature that ranges from a high average of 21°C in the Mountainous areas with a very high average of 25°C in the dry low-veld areas of Kruger National Park. In the district Frost rarely occurs, while the monthly distribution of the average daily maximum temperatures indicates that the average midday temperatures for Mopani Rest Camp (Kruger) range from 23.7°C in June to 30°C in January. The region is the coldest during June when the mercury drops to 8°C on average during the night. The District falls within the Letaba Catchment area, which is 13 779 km^2 and has a mean annual

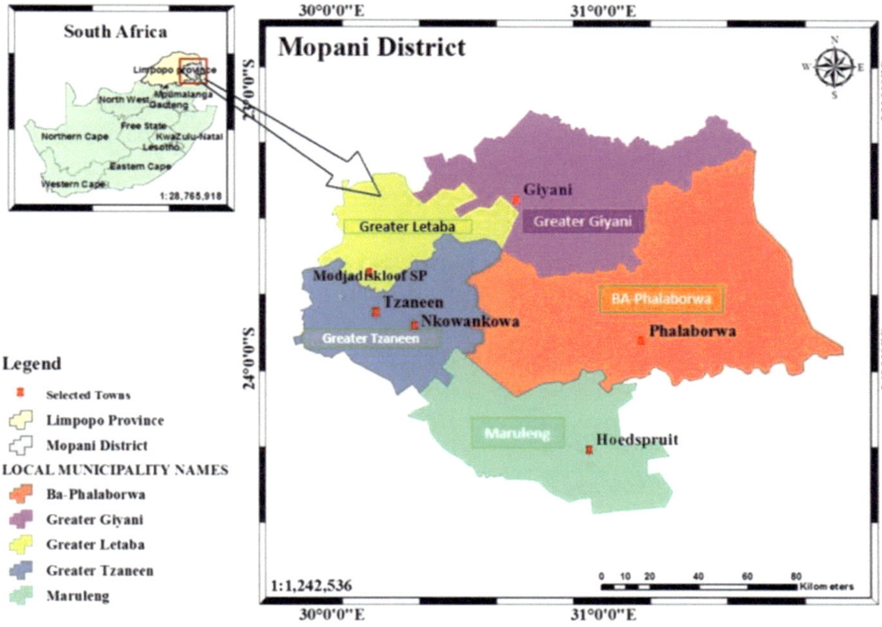

Figure 1.
Mopani District municipality showing the five local Municipalities within the context of Limpopo Province and South Africa Context.

precipitation of 612mm (Environmental Management Framework for the Olifants and Letaba river catchment areas, Report, 2009).

Mopani Rest Camp (Kruger) receives about 520 mm of rain per year, with most proportion (85%) of precipitation in Mopani District is received in mid-summer, while with the lowest (3 mm) is received in June and the highest (96 mm) in January [72]. The rainfall varies from the mountainous zones in the Drakensberg Mountains (2000 mm/a) and the dry low-veld in the Kruger National Park (400 mm/a). The district is situated within the Letaba Catchments area which has a 612 mm Mean annual precipitation.

Climate is recognized by the municipality to be changing, altering and resulting to rising temperature and reduced and erratic rainfall across the district, which is a reflection of the regional climate [58, 75]. The new climate pattern according to the district Integrated Development Plan has caused reduction of access to potable water, food security threats and increase health effects to poverty stricken communities [72]. Part of the strategies identified in the planning instrument for the mitigation of the emission of global warming causing-gases include: utilizing every space for plants, using alternative forms of energy and strict control against deforestation.

3.3 Methods

Consequent upon climate change challenges, households in Mopani have consistently adopted several private and individual strategies to adapt with the varying climate change extreme events. In line with this, we examine the individual household coping strategies to climate change related extreme events and hazards in six purposively selected small and medium-sized towns (Tzaneen, Nkowankowa, Hoedspruit, Modjadjiskloof, Phalaborwa and Giyani) in Mopani District. Sample size of 500 were estimated and drawn using multistage random sampling method, with proportional share to each towns. Guided by the focus of this chapter, data collection methods emphasizes direct/personal interviews, questionnaire and visual inspection/ transect walk in order to ensure a high rate of response. Though the study adopts mixed method, open-ended questions were minimized, and well-ordered, where necessary. Review of existing literature was used to complement the current research findings. The investigated variables were isolated because of their being the direct location-specific effects occasioned by climate change. These variables are categorized into three, these include: those strategies related to increased temperature; reduced water level (rainfalls); and incidence of flood. These variables were cross tabulated against the selected towns of respondents and are discussed as follow.

4. Result and discussion

4.1 Households' efforts towards adapting to climate change in the semi-arid region of Mopani District

Although efforts are on-going globally, regionally and nationally to reverse the trend in climate variability through research, treaties, collaborations, dialogues and other mechanisms, it is essential to appreciate that adaptation to the new climate change regime remains for now, the only realistic and sustainable option that is available [73, 74]. However, household's private adaptation strategy is becoming an increasingly important component to the urban setting, since the end to the failure of Municipal governments to effectively deal with adaptation to climate change in urban center is indeterminate.

4.1.1 Households' temperature coping strategies

Occurrence of heat waves as a result of rise in temperature is generally evident in the semi-arid region of South Africa [58] and particularly in Mopani District of Limpopo province [75]. According to [76], households' and municipal responses to cope with high temperature or heat waves can be undertaken via tree planting and several other strategies. In line with this understanding, household's individual rising temperature coping strategy in the selected towns in the district were identified. The strategies include tree planting, minimizing bush burning, preservation of water bodies, eco-friendly faming practice, Flower and Grass Planting, the use of Fan and Air conditioner and the creation of Parks and Gardens. **Table 1**, depicts the responses from households, on the preferred coping strategies for increasing temperature across the selected towns in Mopani, these are subsequently discussed.

Tree planting is one of the popular coping strategies and was recommended for mitigating the impact of high temperature [77]. This assertion was validated in the selected towns, with the results obtained from our investigation where 63.3% of households in Tzaneen, and 56.6% in Nkowankowa were in agreement with tree planting strategy to cope with heat waves. Hoedspruit accounted for 88.9%, Modjadjiskloof 50% while in Phalaborwa and Giyani 95.2% and 100% of households employed the strategy respectively. The findings suggests wide range of acceptability of tree planting as temperature coping strategy. The general acceptability of the strategy across the towns was adduced to its affordability and effectiveness as a coping strategy for increasing temperature.

An examination of the relevance of reducing bush burning as a strategy for coping with temperature in the selected towns in Mopani District was undertaken and the results of the respondents' answers to the strategy reveals that

Towns	Tree planting	Flower and grass planting	Create parks & garden	Minimize bush burning	preserve water bodies	Eco-friendly farming	Use of fans and air condition
Tzaneen	66.3	45.6	0.0	100.0	0.0	0.0	100.0
Nkowankowa	56.6	18.3	0.0	100.0	0.0	0.6	100.0
Hoedspruit	88.9	100.0	0.0	100.0	83.3	11.1	100.0
Modjadjiskloof	50.0	80.0	0.0	100.0	60.0	0.0	100.0
Phalaborwa	95.2	98.8	0.0	98.0	77.4	0.0	100.0
Giyani	100.0	51.5	0.0	100.0	2.3	0.0	100.0

Source: Authors' Field Data, 2019.

Table 1.
Temperature coping strategies across Mopani District.

in Nkowankowa, Tzaneen, Hoedspruit and Modjadjiskloof, minimizing bush burning was considered by every household, as an appropriate strategy, while in Phalaborwa, 2% of the entire households surveyed declined the choice of the strategy. The employment of bush burning minimization to curtail the impact of heatwaves at municipal level will be an acceptable and effective strategy that will make meaningful impacts across the district.

With respect to the respondents adopting the conservation of water bodies in their communities, **Table 1** shows that more than four in every five respondents in Phalaborwa, three in every five in Modjadjiskloof respectively endorsed the strategy to cope with temperature. However, the strategy only enjoyed the acceptability of only 2.3% respondents in Giyani. This strategy was equally unpopular in both Tzaneen and Nkowankowa. The reason for Hoedspruit, Phalaborwa and Modjadjiskloof in favor of this strategy was traced to the awareness of the benefits of the strategy among households, facilitated by NGOs and the respective municipalities. This result reflects that the municipalities of the two towns complied with the water conservation Act No. 36 [78].

In the narratives of current literature reviewed on the adoption of eco-friendly farming practices as temperature coping strategy, it shows that it is a promising strategy as reported by [79]. But the results of the acceptability test of the strategy in the selected towns show otherwise. For example, in Tzaneen, Modjadjiskloof, Phalaborwa and Giyani, no respondent indicated adopting it as a strategy. Only Hoedspruit accounted for 11.1%. The result reflects the economic activities of significant proportion of respondents from non-primary sources particularly agriculture. Thus, prescribing it as coping strategy for temperature in the district might not be very impactful to the majority of households.

However, According to [80] as cited in [81] Green infrastructure is very useful in contributing to mitigate the effects of hard surfacing by modifying ambient temperatures as well as creating recreational opportunities among other advantages. Our investigation revealed that every households in Hoedspruit town adopted the strategy, while 98.8%, 80% and 51.5% households in Phalaborwa, Modjadjisklooof and Giyani towns adopted the green infrastructure strategy respectively too. Flower and grass planting seems to be a widespread and suitable temperature coping strategy in the selected towns, except in Nkowankowa where only 18.3% of the household embraced the strategy.

The use of Fan and Air conditioner appeared to be a very satisfactory strategy that was favored by every households traversing the selected towns in Mopani. This was adduced to by the respondents that the former (fan) is affordable, accessible and environmentally friendly. However, while the latter (air conditioners) was enhanced by the stability of electricity, it does not only escalates the energy bills because of the increased loads resulting from cooling, but it equally exacerbates urban heat island in its own capacity. Thus [82] submit that for effective alleviation of urban warming and enhanced cooling, there is, as a necessity the need to reduce air-conditioning anthropogenic heat.

The responses obtained from the survey conducted on the creation of Neighbourhood Parks and Garden as a temperature coping strategy by the households across the selected towns is presented in **Table 1**. The result shows a consensus among the households that the siting and development of neighborhood parks and garden was the responsibility of the governments at different levels. This was reflected in households' responses where no household indicated creating Parks and Garden as a personal temperature coping approach. However, children who desire to recreate use available spaces like access road around them to play soccer, not minding risks involved.

4.1.2 Households' water scarcity coping strategies

With respect to water scarcity, the households were required to indicate the strategy they use during climate related drought or long heat waves that reduce the water quantity in their area. The variable used to capture the households' responses include: rain water harvesting, water embankment, use of storage tanks, water treatment to improve quality and use of water vendor service. These results are presented in **Figure 2**.

An examination of household coping strategies regarding change in water level in the selected towns, as summarized in **Figure 2**, shows that rainwater harvesting as a strategy was not popular among the households. The results shows that 5.1% respondents in Tzaneen and 5% in Nkowankowa adopted water harvesting as a strategy, while in Phalaborwa and Giyani both accounted for 5% and 3.1% respectively. However, both Modjadjiskloof and Hoedsrpruit towns did not use such a strategy because according to them, it is time consuming and that the quality of harvested water was most times compromised.

However, **Figure 2** shows the results of the examination of the use of storage tanks to cope with reducing water level. It was discovered that 100% and 81% of households in Hoedspruit and Phalaborwa respectively used the strategy to backup, to forestall the impacts of water shortages. In Tzaneen and Nkowankowa 54.4% and 51% of their respective household used same strategy. Similarly, in Modjadjiskloof

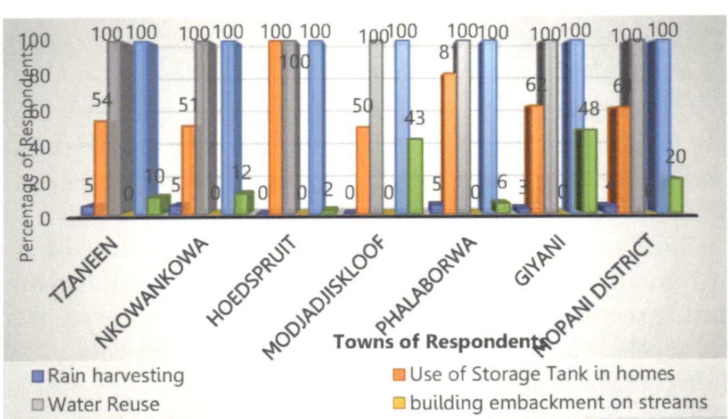

Figure 2.
Coping strategies for change in water level across towns and Mopani. Source: Author's Field Data, 2019.

and Giyani the households that used storage tanks were respectively 50% and 61.5%. On the average 61.1% of the respondents have used or still using storage tanks to adapt to reducing water level in their communities. The study implied that the storage of water in tanks is an acceptable strategy because water provision is not always at RDP level.

A significant indicator of health is water scarcity, which includes both its availability and quality [83]. Water use is beyond drinking, it is intimately linked to food security, sanitation and hygiene contributing to health burdens. Poor and vulnerable communities suffer the most from the adverse effects of climate change on water and health related issues and that the adaptation strategy which can effectively reduce the strain on water resources include wastewater recycling and reuse [84]. This was tested in the selected towns, and was found that when water became scarce, such as in 2016 and 2017 droughts periods in Limpopo province, most households turned to the re-use of water due to the scarcity of water for domestic and others uses. Our investigation further showed that 100% of the respondents re-use water as was advised by the Department of Water Affairs, when Limpopo Province, was declared a disaster province.

Water treatment was one of the variables we requested the households to give their response if they use such strategy. Although according to the science of water treatment which involved reverse osmosis etc, we were more interested in treatment such as water boiling, using aqua active bleaching agents such as hypochlorite to disinfect the water before use. The results in **Figure 2** shows that 100% of the respondents use non-complicated methods to treat their water when it becomes very scarce and necessary.

The general practice particularly in the peri-urban areas of the selected towns is that most of them buy water from water vendors who sell water in containers ranging from R5 to R25 depending on the quantity sold. The study showed that not all respondent were disposed to buying water from vendors maybe because some could not be guaranteed the quality of the water. Patronage of water vendors was common among those households who did not have stand pipes in their yards. However, in Modjadjiskloof and Giyani 43% and 48% respectively used water vendor services to cope with water scarcity (**Figure 2**).

4.1.3 Household strategy for flood control in Mopani

As rightly noted [85] that with increasing havoc of floods in the urban center, and its negative impacts particularly on the poorest and the most susceptible, effective coping strategies require the combination of protective infrastructure, nature-based approaches, and risk financing (insurance) schemes to curtail floods and cushion their adversities. Flash floods has resulted into several degrees of damages in South Africa [65] as well-as some parts of the selected towns in Mopani District Municipality [75]. This occurred at different times, frequencies and intensities. This phenomenon has in the past resulted in households loosing properties ranging from home assets to farm crops and farm produce. The results of the survey showed that with respect to flood control strategies, the most popular include the construction of embankment to prevent over flow of rivers, the use of Furrow around their house, building of walls to protect houses during flash floods, growing of lawns, removal of solid waste from the storm water drainages, re-enforcement of dwellings with stones and concretes.

Our findings suggest that building embankment around houses is a popular strategy particularly among those residing close or whose offices are in close proximity to rivers, along erosion line, or terrain threatening sites. Embankments are usually constructed by the community or the local municipality. One aspect of the embankment as a strategy to cope with floods is that it fends off water and shelters settlements from flooding. About 58% of respondents was recorded in Modjadjiskloof and 32% in Phalaborwa, Hoedspruit was 23%, while Giyani and Nkowankowa both depicted 27% and 36% accordingly. In a further probe to why majority did not adopt the strategy, respondents noted it to be an expensive option, which often failed when the construction was not done to structural specifications.

With respect to the use of Sandbags, as a strategy, in Giyani 24% of the households indicated its adoption as the option to protect their properties against flood. In Nkowankowa 17%, Tzaneen was 11%, while Modjadjiskloof households accounted for 22% that used sandbags. The households' justification for the use of sandbags as a coping strategy to protect against flooding was hinged on its affordability, ease of building and availability of the material components.

The use of furrow was equally investigated to ascertain whether or not is an acceptable strategy among the households in the district. The result indicates that 5% of Tzaneen residents are using Furrows around their properties, while about one in every four households in Nkowankowa adopted the same strategy. Households in Hoedspruit and Modjadjskloof that used the strategy accounted respectively for 16% and 36% and both Phalaborwa and Giyani accounted for 12% and 18%. The result suggest that the strategy was not embraced by the majority of the households across the selected towns. According to the respondents, the option was considered costly and not an effective strategy compared to others.

The proportion of households' that adopts the building of protective walls around their houses to cope with flood in the six selected towns indicates that this is a commonly used strategy in the study area. Both Modjadjiskloof and Nkowankowa used it as a strategy mostly. With 76% of its household, Modjadjiskloof recorded the highest proportion of household that used the protective walls as strategy, while 18% of the households in Nkowankowa used the strategy. These results was significantly influenced by the terrain of individual towns under consideration as towns with relatively low lying terrain recorded lower patronage of the strategy, while town with steep slope like Modjadjiskloof adopted it most.

According to [80], Green infrastructure is useful in curtailing surface runoff among other benefits [81]. From the results of analysis, households' response with respect to growing grasses to reduce the effects of floods in the selected towns revealed that 60% of households in Hoedspruit grew lawn to reduce the flow of surface run off that erodes the top soil. The study showed that 37% of the households in Modjadjiskloof and 36% in Tzaneen grew lawn to reduce erosion while 2.5% and 20% employed the same strategy in Nkowankowa and Giyani towns respectively. This strategy apart from protecting the surface top soil from erosion, it also keep a good ambient of the environment.

The respondents' answers to the cleaning and removal of waste from drainage channels and systems appeared an acceptable coping strategy across the selected towns in Mopani. 32% of households in Tzaneen do evacuate waste from drainages, while as low as 5% of Nkowankowa households used the strategy to avoid over flow of drainages. However, more than two out of every five Phalaborwa residents engaged in clearing of their drainages to prevent flooding. The study further shows that one tenth of Giyani household embraced the strategy as well. Further to this, drainage and stream channelization was popular, accounting for 25% of Tzaneen households, while one fifth of Hoedspruit households embraced drainage channelization in coping with the incidence of flood.

The use of concrete and stones by households to reinforce their housing foundation serves dual purposes as a way to stabilize the building as well-as safeguard it against any unexpected floods that can erode the building foundation.

About 88% of Hoedspruit household endorsed it, while 87% of the households in Modjadjiskloof as well-as 78% of them in Phalaborwa used it as a strategy to cope with floods. However, Tzaneen account for 40% of houses in this category, while Giyani town accounted for 34%.

Obviously without waiting endlessly for government, households across the district have taken creative initiatives to respond within the available resources at their disposal to climate change related hazards. However, households' capacities are limited by several factors, ranging from economic, social, and attitudinal. Unless pioneered, championed and facilitated by government, household adaptation may not achieve the desired goal. Although several factors collaborate to hamper the success of urban adaptation in the semi-arid region of Mopani, South Africa. These limiting factors are identified in the next section.

4.2 The factors hindering the success of urban adaptation strategies in the Mopani region

This section identifies the factors that inhibit the successes of urban adaptation to climate change hazards. Through our interactions with the households in the selected towns, the key informant (particularly the municipal staff and professionals) and other stakeholders, buttressed by the findings from the planning instruments (IDPs) of the five local municipalities in Mopani District, several inhibiting factors clogging the successes of urban climate change adaptation in these municipalities were uncovered. These out of others can be stratified into both internal and external factors. These are discussed as follow:

Internal factors are those factors that the local municipalities recognized as being within their mandates and powers, on one hand. These include but are not limited to paucity of fund, principally from budgetary allocation. Limited human capacity to embark on the required types of planning for integrated adaptation mainstreaming, compounded by the paucity of knowledge of adequate climate issues at the local municipal level. Higher competition that exist between the mandates of government, resulting in less priority being accorded to long-term planning issues (like climate change) in favor of short-term actions and gains. The Situation is further compounded by the South African need to tackle the backlogs of service amidst coping with both current and future needs of the people. Thus, rendering long-term interventions unattractive to politicians who run a short political tenure to execute. With long-term horizon nature of climate change projections, it contradicts with the short-term political and development programs of these municipalities.

In addition, system's failure manifest across the selected towns, for instance drainages and water ways blockages, absence of drainages in many instances, sewer leakages (like the case of Nkowankowa and Phalaborwa), and backlogs of service across the municipalities are clear indicators. Others factors include policy inadequacies resulting from municipal reliance on national policies (such as urban and other climate adaptation-related policies). The dichotomized land management and operational deficiencies where traditional institutions are in charge of unproclaim land with no responsibility to provide services. Absence of interface programs between the municipalities and the Universities and other research institutions for information and knowledge sharing as well-as research activities regarding climate change and urban development. There was equally no evidence to show collaborations with private sector (banks, insurance and individual philanthropists) on adaptation issues.

Furthermore, external factors include high poverty rate, low literacy level and unemployment. Lack of reliable and verifiable hazard incident reporting systems that can guarantee disaster hotspot identification and monitoring for early warning. Nevertheless, some of these identified factors (policy shortcomings, institutional weakness etc.), lack of political will plays a significant role.

5. Conclusion

There is no doubt that the new climate is here so also are the attendant hazard that we have to live with in decades to come. With the long-term nature of ongoing global mitigation efforts, adaptation remains the available strategy that must be collaboratively embraced to cope with climate change prone hazards in the urban centers of semi-arid region of South Africa.

Thus, we emphasize the need for a participatory urban management strategy for sustainable adaptation to climate-related hazards, while calling on Scholars to develop models of urban adaptation to climate change that may not necessarily be highly mathematical, but recognize the technological level, social and economic peculiarities of urban Africa, particularly in the semi-arid region of Mopani, South Africa.

The need to urgently review the procedure for reporting climate change hazards and emergencies to promote early warning system, should be revisited. Hazards reporting should be facilitated by the incorporation of instant reporting components in to the existing or a new reporting protocols. This chapter referred to this as "hotspot reporting and monitoring system", through the implementation and development of a mobile phone facilitated protocol that makes citizens the reporters of climate hazards.

It is therefore important to identify and simplify trends and carry out assessment of the effectiveness of prevailing and future policies that may be directed towards urban households' adaptation to climate change hazard in semi-arid region of Mopani South Africa for impactful delivery. In addition, such adaptation policies should be locally-driven and must address climate change as a multifaceted phenomenon and not limited as solely to being tackled as an environmental issue, while integrating local knowledge approaches.

Although, it may be uneasy to convince politicians to prioritize climate change (a long-term development agenda) over and above short tenure political agenda, conversations and strategies to encourage the implantation of long-term sustainable projects should be persuaded. But, because climate change phenomenon as well-as its related consequences are real and already manifesting [58], thus, research institutions, private sector (corporate organization) and NGOs are urged to assist in facilitating training of municipal staff and reorientation program for politicians, particularly by promoting the inclusion of climate change hazard management agenda in the political parties manifestoes while facilitating private adaptation strategies at community level.

Strategies like tree planting, urban greening, drainage channelization, and harmonization of the dichotomized land management in the district are some of the strategic window to curtail climate change hazards in the semi-arid region of Mopani South Africa.

Acknowledgements

University of Venda, Thohoyandou, South Africa is acknowledge for funding the research. The University of Ilorin is equally acknowledge for granting me the permission and sponsorship of the Ph.D program.

DOI: http://dx.doi.org/10.5772/intechopen.94759

Author details

Musa Yusuf Jimoh[1,2]*, Peter Bikam[1], Hector Chikoore[3], James Chakwizira[1] and Emaculate Ingwani[1]

1 Department of Urban and Regional Planning, University of Venda, Thohoyandou, Limpopo, South Africa

2 Department of Urban and Regional Planning, University of Ilorin, Ilorin, Nigeria

3 Units for Environmental Sciences and Management, North-west University, Vanderbijlpark, South Africa

*Address all correspondence to: jimoh.my@unilorin.edu.ng

References

[1] Mertz O, Halsnæs K, Olesen JE, Rasmussen K. Adaptation to climate change in developing countries. Environmental management. 2009 May 1;**43**(5):743-752

[2] Huq S, Reid H, Konate M, Rahman A, Sokona Y, Crick F. Mainstreaming adaptation to climate change in least developed countries (LDCs). Climate Policy. 2004 Jan 1;**4**(1):25-43

[3] Change IP. Climate change 2007: the physical science basis: summary for policymakers. Geneva: IPCC; 2007 Feb.

[4] Christensen JH, Christensen OB. A summary of the PRUDENCE model projections of changes in European climate by the end of this century. Climatic change. 2007 May 1;**81**(1):7-30

[5] Thompson AA, Matamale L, Kharidza SD. Impact of climate change on children's health in Limpopo Province, South Africa. International Journal of Environmental Research and Public Health. 2012 Mar;**9**(3):831-854

[6] Bryan E, Deressa TT, Gbetibouo GA, Ringler C. Adaptation to climate change in Ethiopia and South Africa: options and constraints. Environmental science & policy. 2009 Jun 1;**12**(4):413-426

[7] Yohe G, Tol RS. Indicators for social and economic coping capacity—moving toward a working definition of adaptive capacity. Global environmental change. 2002 Apr 1;**12**(1):25-40

[8] Thomas et al. (2009)

[9] Bigio A. Adapting to climate change and preparing for natural disasters in the coastal cities of North Africa. In: Urban Research Symposium. Cities and Climate Change: Responding to an Urgent Agenda; 2009 June. pp. 28-30

[10] Hallegatte S, Corfee-Morlot J. Understanding climate change impacts, vulnerability and adaptation at city scale: an introduction. Climatic Change. 2011 Jan 1;**104**(1):1-2

[11] Nhuan MT, Tue NT, Hue NT, Quy TD, Lieu TM. An indicator-based approach to quantifying the adaptive capacity of urban households: the case of Da Nang city, Central Vietnam. Urban Climate. 2016 Mar 1;**15**:60-69

[12] Kates RW. Cautionary tales: adaptation and the global poor. Climatic change. 2000 Apr 1;**45**(1):5-17

[13] Dickinson T. The compendium of adaptation models for climate change. Environment Canada; 2007.

[14] Berkhout F. Rationales for adaptation in EU climate change policies. Climate Policy. 2005 Jan 1;**5**(3):377-391

[15] Adger WN, Huq S, Brown K, Conway D, Hulme M. Adaptation to climate change in the developing world. Progress in development studies. 2003 Jul;**3**(3):179-195

[16] Burton I, Huq S, Lim B, Pilifosova O, Schipper EL. From impacts assessment to adaptation priorities: the shaping of adaptation policy. Climate policy. 2002 Jan 1;**2**(2-3):145-159

[17] Taylor M. The political ecology of climate change adaptation: Livelihoods, agrarian change and the conflicts of development. Routledge; 2014, Nov 17

[18] Dewulf A. Contrasting frames in policy debates on climate change adaptation. Wiley Interdisciplinary Reviews: Climate Change. 2013 Jul;**4**(4):321-330

[19] Smith JB, Schneider SH, Oppenheimer M, Yohe GW, Hare W, Mastrandrea MD, et al. Assessing dangerous climate change through an

update of the Intergovernmental Panel on Climate Change (IPCC) "reasons for concern". Proceedings of the national Academy of Sciences. 2009 Mar 17;**106**(11):4133-4137

[20] Adger WN, Arnell NW, Tompkins EL. Successful adaptation to climate change across scales. Global environmental change. 2005 Jul 1;**15**(2):77-86

[21] Schipper EL. Climate change adaptation and development: Exploring the linkages. Tyndall Centre for Climate Change Research Working Paper. 2007 Jul;**107**:13

[22] Klein RJ. Adaptation to climate variability and change: what is optimal and appropriate. Climate Change in the Mediterranean: Socio-Economic Perspectives of Impacts, Vulnerability and Adaptation. 2003;**32**

[23] Schipper L, Pelling M. Disaster risk, climate change and international development: scope for, and challenges to, integration. Disasters. 2006 Mar;**30**(1):19-38

[24] Parry ML, Canziani OF, Palutikof JP, Van Der Linden PJ, Hanson CE. IPCC, 2007: climate change 2007: impacts, adaptation and vulnerability. Contribution of working group II to the fourth assessment report of the intergovernmental panel on climate change. Cambridge Uni-versity Press, Cambridge, UK. 2007.

[25] Levina E, Tirpak D. Adaptation to climate change. Key Terms.

[26] Stakhiv, EZ, Policy Implications of Climate Change Impacts on Water Resources Management, Water Policy, 1(2), pp. 159-175, 1998

[27] Watson RT, Zinyowera MC, Moss RH. Climate change 1995. Impacts, adaptations and mitigation of climate change: Scientific-technical analyses. 1996.

[28] Collier P, Conway G, Venables T. Climate change and Africa. Oxford Review of Economic Policy. 2008;**24**(2):337-353

[29] Laukkonen J, Blanco PK, Lenhart J, Keiner M, Cavric B, Kinuthia-Njenga C. Combining climate change adaptation and mitigation measures at the local level. Habitat international. 2009 Jul 1;**33**(3):287-292

[30] Hannah L. A global conservation system for climate-change adaptation. Conservation Biology. 2010;**24**(1):70-77

[31] Brook T. Collaboration: Japanese agents and local elites in wartime China. Harvard University Press; 2005

[32] Nyong A, Adesina F, Elasha BO. The value of indigenous knowledge in climate change mitigation and adaptation strategies in the African Sahel. Mitigation and Adaptation strategies for global Change. 2007 Jun 1;**12**(5):787-797

[33] Newton J, Paci CJ, Ogden A. Climate change and natural hazards in northern Canada: integrating indigenous perspectives with government policy. InMitigation of Natural Hazards and Disasters: International Perspectives 2005 (pp. 209-239). Springer. Dordrecht.

[34] Adger WN, Brown K. Vulnerability and resilience to environmental change: ecological and social perspectives. London: Wiley-Blackwell; 2009 Feb 11

[35] Miller F, Osbahr H, Boyd E, Thomalla F, Bharwani S, Ziervogel G, et al. Resilience and vulnerability: complementary or conflicting concepts? Ecology and Society. 2010 Sep 1;**15**(3)

[36] Dickinson T. The compendium of adaptation models for climate change. In: Environment Canada (EC). 2008

[37] Frigg R. Hartmann S. In: Models in science. 2006

[38] Krankina ON, Dixon RK, Kirilenko AP, Kobak KI. Global climate change adaptation: examples from Russian boreal forests. Climatic Change. 1997 May 1;**36**(1-2):197-215

[39] Lewis JI. China's strategic priorities in international climate change negotiations. Washington Quarterly. 2008 Jan 1;**31**(1):155-174

[40] Thaker J, Leiserowitz A. Shifting discourses of climate change in India. Climatic Change. 2014 Mar 1;**123**(2):107-119

[41] Carter JG, Cavan G, Connelly A, Guy S, Handley J, Kazmierczak A. Climate change and the city: Building capacity for urban adaptation. Progress in planning. 2015 Jan 1;**95**:1-66

[42] Adger WN, Arnell NW, Tompkins EL. Successful adaptation to climate change across scales. Global environmental change. 2005 Jul 1;**15**(2):77-86

[43] Huq S, Reid H, Konate M, Rahman A, Sokona Y, Crick F. Mainstreaming adaptation to climate change in least developed countries (LDCs). Climate Policy. 2004 Jan 1;**4**(1):25-43

[44] Pieterse A, du Toit J, van Niekerk W. Climate change adaptation mainstreaming in the planning instruments of two South African local municipalities. Development Southern Africa. 2020 May 13:1-6.

[45] Füssel HM. Adaptation planning for climate change: concepts, assessment approaches, and key lessons. Sustainability science. 2007 Oct 1;**2**(2):265-275

[46] Bharwani S. Understanding complex behavior and decision making using ethnographic knowledge elicitation tools (KnETs). Social Science Computer Review. 2006 Feb;**24**(1):78-105

[47] Kithiia J, Dowling R. An integrated city-level planning process to address the impacts of climate change in Kenya: The case of Mombasa. Cities. 2010 Dec 1;**27**(6):466-475

[48] Draper AJ, Jenkins MW, Kirby KW, Lund JR, Howitt RE. Economic-engineering optimization for California water management. Journal of water resources planning and management. 2003 May;**129**(3):155-164

[49] Jenkins MW, Lund JR, Howitt RE, Draper AJ, Msangi SM, Tanaka SK, et al. Optimization of California's water supply system: results and insights. Journal of Water Resources Planning and Management. 2004 Jul;**130**(4):271-280

[50] Lin QG, Huang GH, Bass B. An energy systems modelling approach for the planning of power generation: a North American case study. International journal of computer applications in technology. 2005 Jan 1;**22**(2-3):151-159

[51] Callaway, J., and Hellmuth, M. Climate risk management for development: economic considerations', presented at Stern Review Workshop on "Economics of Adaptation", 2006 May, available from http://www.sternreview.org.uk.

[52] deBruin KC, Dellink RB, Tol RS. AD-DICE: An implementation of adaptation in the DICE model. Working Papers FNU-126

[53] Mendelsohn R, Nordhaus W. The impact of global warming on agriculture: Reply. The American Economic Review. 1996 Dec 1;**86**(5):1312-1315

[54] Adger WN, Agrawal S, Mirza MM, Conde C, O'brien

KL, Pulhin J, Pulwarty R, Smit B, Takahashi K. Assessment of adaptation practices, options, constraints and capacity.

[55] Huq S, Kovats S, Reid H. Satterthwaite D. Reducing risks to cities from disasters and climate change.

[56] Milly PC, Betancourt J, Falkenmark M, Hirsch RM, Kundzewicz ZW, Lettenmaier DP, et al. Stationarity is dead: Whither water management? Earth. 2008;**4**:20

[57] Bouwer LM, Aerts JC. Financing climate change adaptation. Disasters. 2006 Mar;**30**(1):49-63

[58] Chikoore H. Drought in southern Africa (Doctoral dissertation, UNIVERSITY OF ZULULAND South Africa).

[59] Niang I, Osman-Elasha B, Githeko A, Yanda PZ, Medany M, Vogel A, Boko M, Tabo R, Nyong A. Africa Climate Change 2007: Impacts, Adaptation and Vulnerability: Contribution of Working Group II to the Fourth Assessment Report of the Intergovernmental Panel on Climate Change..

[60] Zuma BM, Luyt CD, Tandlich R, Chirenda T. Flood disaster management in South Africa: legislative framework and current challenges. INTECH Open Access Publisher; 2012 Sep 10

[61] Zuma BM, Luyt CD, Tandlich R, Chirenda T. Flood disaster management in South Africa: legislative framework and current challenges. IntechOpen; 2012 Sep 10.

[62] South African Local Government Association (SALGA). Disaster Risk Management Status Assessment at Municipalities in South Africa. South Africa: Pretoria; 2011

[63] South Africa Republic. Green Paper Disaster Management. Vol. 1998. Government Printer: Pretoria; 1998

[64] Montmasson-Clair G. Green economy policy framework and employment opportunity: A South African case study. Trade and Industrial Policy Strategies (TIPS), Working Paper Series. 2012 Aug;2.

[65] Lethoko MX. Inclusion of climate change strategies in municipal Integrated Development Plans: A case from seven municipalities in Limpopo Province, South Africa. Jàmbá: Journal of Disaster Risk Studies. 2016;**8**(3)

[66] Department of Provincial and Local government, IDP guide pack, Government Printers, Pretoria. 200b

[67] South African Government Gazette. Pretoria: Government Printer. 1999.

[68] South Africa Republic. Disaster Management Act No. 57 of 2002. Pretoria: Government Printer.2003

[69] Jimoh MY, Bikam P, Chikoore H. The Influence of Socioeconomic Factors on Households' Vulnerability to Climate Change in Semiarid Towns of Mopani, South Africa. Climate. 2021 Jan;9(1):13.

[70] Pillay U. Urban policy in post-apartheid South Africa: context, evolution and future directions. InUrban Forum 2008 Jun 1 (Vol. 19, No. 2, p. 109). Springer Netherlands.

[71] Nel E, Rogerson CM. Evolving local economic development policy and practice in South Africa with special reference to smaller urban centres. InUrban Forum 2007 Jun 1 (Vol. 18, No. 2, pp. 1-11). Springer Netherlands.

[72] Mopani District Municipality Demographic (2016). https://municipalities.co.za/demographic/128/mopani-district-municipality

[73] Barreca A, Clay K, Deschenes O, Greenstone M, Shapiro JS. Adapting to climate change: The remarkable decline in the US temperature-mortality

relationship over the twentieth century. Journal of Political Economy. 2016 Feb 1;124(1):105-159.

[74] Zacharias M, Naresh Kumar S, Singh SD, DN SR, Aggarwal PK. Assessment of impacts of climate change on rice and wheat in the Indo-Gangetic plains.

[75] Munyai RB, Musyoki A, Nethengwe NS. An assessment of flood vulnerability and adaptation: A case study of Hamutsha-Muungamunwe village, Makhado municipality. Jàmbá: Journal of Disaster Risk Studies. 2019;**11**(2):1-8

[76] Hayley L, David S. Local Authority Responses to Climate Change in South Africa: The Challenges of Transboundary Governance. Sustainability. 2018;**2018**(10):2542. DOI: 10.3390/su10072542 www.mdpi.com/journal/sustainability

[77] Saxe H, Cannell MG, Johnsen Ø, Ryan MG, Vourlitis G. Tree and forest functioning in response to global warming. New phytologist. 2001 Mar;**149**(3):369-399

[78] Republic of South Africa. National Water Act 36 of 1998 [online] Available at www.gov.za (accessed 20.08.2020)

[79] Vincent K, Cull T, Chanika D, Hamazakaza P, Joubert A, Macome E, et al. Farmers' responses to climate variability and change in southern Africa–is it coping or adaptation? Climate and Development. 2013 Jul 1;**5**(3):194-205

[80] Pataki DE, Boone CG, Hogue TS, Jenerette GD, McFadden JP, Pincetl S. Socio-ecohydrology and the urban water challenge. Ecohydrology. 2011 Mar;**4**(2):341-347

[81] Matthews T, Lo AY, Byrne JA. Reconceptualizing green infrastructure for climate change adaptation: Barriers to adoption and drivers for uptake by spatial planners. Landscape and Urban Planning. 2015 Jun 1;**138**:155-163

[82] Kikegawa Y, Genchi Y, Kondo H, Hanaki K (2006) Impacts of city-block-scale countermeasures against urban heat-island phenomena upon a building's energy-consumption for air-conditioning. Appl Energy **83**:649-668

[83] DeNicola E, Aburizaiza OS, Siddique A, Khwaja H, Carpenter DO. Climate change and water scarcity: The case of Saudi Arabia. Annals of global health. 2015 May 1;**81**(3):342-353

[84] Jongman B. Effective adaptation to rising flood risk. Nature communications. 2018 May 29;**9**(1):1-3

[85] Congedo L, Macchi S. The demographic dimension of climate change vulnerability: exploring the relation between population growth and urban sprawl in Dar es Salaam. Current Opinion in Environmental Sustainability. 2015 Apr 1;**13**:1-10

Chapter 9

Flooding and Its Impact on Education

Edson Munsaka and Sebia Mutasa

Abstract

Within academia there is ongoing discussion over what constitutes natural disaster or what does not. Floods, which in recent years have taken the world by surprise, come into the discussion too. While that is the case, its impact on education systems is least discussed, if ever, yet literature acknowledges floods as one of the most devastating disasters ever recorded in human history. This Chapter, while attempting to examine the impact flooding has on education systems in Africa, it also explores whether flood sits well in the category of natural disaster. Furthermore, the writers also critically examine and interrogates adequacy of states responses to prevent flooding affecting education systems. The Chapter further explores whether flooding and its related impact on the education system is a disaster risk governance failure.

Keywords: flooding, education, governance, hazards

1. Introduction

Within academia there is ongoing discussion over what constitutes natural disaster or what does not. Floods, which in recent years have taken the world by surprise, come into the discussion too. While that is the case, its impact on education systems is least discussed, if ever, yet literature acknowledges floods as one of the most devastating disasters ever recorded in human history globally. This Chapter, while attempting to examine the impact flooding has on education systems in Africa, it also explores whether flood sits well in the category of natural disaster. Furthermore, the writers also critically examine and interrogates adequacy of states responses to prevent flooding affecting education systems. The Chapter further explores whether flooding and its related impact on the education system is a disaster risk governance failure prior the conclusion section.

2. The hazard: floods and its invention

Floods come in various forms and are living testimonies of the conflict between urban development and weather-related vulnerabilities. According to Van Niekerk and Nemakonde [1], fluvial floods, flash floods and glacial lake outburst floods make the list. For Sub-Saharan Africa the main floods that inflict the region come from periods of high intensity rainfall emanating from tropical cyclones and storm surges [2]. For instance, the 2018 to 2019 South Indian cyclone was a fluvial flood that resulted in flood damage of exponential levels in Africa [3]. This was caused mainly by cyclone Idai which affected Zimbabwe, Malawi and Mozambique leaving a trail of destructions and deaths [4]. As for flash floods, these are of a short duration caused by high intensity rainfall that quickly floods the smaller basins. Africa is overwhelmed with periods of extreme rainfall and recurrent floods which may be associated with El Nino events which leave a lot of destructions in all sectors of the economies [5]. However, it is a misnomer to suggest that it is only the African continent that is prone to floods. In his article titled, 'The Floods: a man-made disaster?', [6] warns the Army Corps of Engineers in America that its concrete navigation structures in the Mississippi River were intensifying floods, and that its plans to build more wingdikes and weirs would exacerbate a severe and growing problem. Grunwald [6] clearly shows the human hand in the creation of flooding condition. It therefore follows that, while there are naturally caused floods that would have occurred from time to time that natural systems could mostly handle, our development of a lot of the world's landscape and our consumptive lifestyles have led to not only in the increase in flood occurrences, but also cause floods. Flood as natural event, relates to God's venting of anger and flood as man-made implies man's failure to mitigate and prepare adequately for hazard to reduce its impact on infrastructure on which livelihoods depend on.

3. Associated flood benefits

Tockner and Stanford [7] opine that floods have always been an important part of nature's regeneration process providing several benefits to countries in the developing world. To that end, some people choose to live in flood hazard prone areas and accept the high levels of risks because of the benefits they access from such areas [8]. The benefits of living in flood prone areas outweigh the risks involved for the vulnerable communities as the floods bring silt to nourish the deltas and to fertilise crops and seasonal fisheries [9]. Studies on floods in Tanzania's Rufiji River Delta by Sandberg found out that it allowed for post flood agriculture which is of benefits to the community in that area as they grow cotton, peas and maize [10]. In case of Bangladesh, flood has noted to be necessary for fertilising, irrigating the fields and enabling fish to spawn [9]. In such scenarios, and as the populations grow, governments are obliged to provide services such as education infrastructure to the inhabitants despite the threat posed by flooding.

The value of flooding to mankind is not only limited to soil enrichment for agricultural purposes. Flood water absorbed underground recharges the underground aquifers to supply fresh water to rivers, wells, dams, and lakes from flood water to the extent that many countries are dependent on aquifers for fresh water. Floods are an alternating source of that fresh water supply [11]. Ecosystems also depend on flood water which carries and deposits nutrients rich in sediments that support both the plant and animal life of wetlands [12]. In this way, floods become a pull factor incentivising communities to choose to live in flood hazard prone areas, and accept the high levels of risks because of the economic benefits they access from such areas [8].

4. Flood as a destroyer

In view of the context of the here and now benefits of floods, the negative impact of floods on education, despite being real, tend to be masked and yet have a terrible impact on the same. A survey by the Asian Disaster Preparedness Center in 2002, in Cambodia speaks to this observation. The survey in question sought to identify the impacts of disasters on the education sector and the findings revealed that floods were one of the factors disrupting study program accomplishment and thus affecting the quality of current education, particularly in provinces which are prone to floods and where schools were constructed without proper flood resilience [13]. In the context of Zimbabwe, soon after gaining political independence in 1980, the introduced reforms in the education system that focused on the principle of 'Education for all' were adopted. Education for all principle embraced the practical principle of increasing the number of schools by building schools in marginalised areas and disadvantaged urban centres [14]. Incidentally some of the marginalised areas were in flood prone areas. As [14] further argues, the government involved local communities to help support schools through providing labour for moulding bricks and other resources. The emphasis was not so much on quality and cost effectiveness of the education system, but on accessibility to education. The Fast land reform programme in Zimbabwe did not help the situation either. People settled themselves under the disguise of black empowerment through access to land and they settled where ever they felt like. Some communities settled in flood prone areas. Owing to the population expansion in those places illegally occupied, the government had no option but to follow up with the construction of schools, commonly known as satellite schools. By allowing people to settle in such fragile spaces, in one way or the other, it helped to easy political tension that had started building in the country due to socio-economic hardships. However, like in the post-independence era quality of infrastructure in the context of resilience to shocks such as floods is not a priority. When floods eventually strike, because they impact communities in different degrees depending on the vulnerability of that community, the impact on education is least to be acknowledged.

The extent floods affect livelihoods and human life is well documented in literature. In fact, quantifying such loss has been the responders' priority. While that is the case, literature has it that one's political, social and economic status plays a significant role when it comes to exposure to risk of flooding. Poorest communities are the most vulnerable as they live in the most threatened locations [15] they lack the means to live in less vulnerable communities [16]. This construction of flooding implies that one's exposure to flooding is a social construction, thus in a way implying that if it is socially constructed, then flood fits well in the discussion of it as being man-made. Cann *et al.,* [17] are quick to remind us that floods affect the quality of drinking water thereby bringing with them water related infectious diseases such as cholera and hepatitis which will be compounded by damaged or overwhelmed sanitation system. Flood as destroyer has also been noted by other authors such as [18] who argue that it destroy agriculture, transport infrastructure, communication infrastructure, service buildings, and social facilities. Such loss compounds the deterioration of the social and economic lives of people as well as harming the national economy. While literature envisage an increase in flood threats due to various reasons and creations both natural and man-made [19], the extent to which it affects education systems globally has not received favourable attention. More often, after a floods we hear and read about the level of the impact on the socio-economic well-being of children if not mainly on economic related benefits of floods; As [13] have observed, the impacts on children's access and right to quality education have received little attention This has resulted in a lack of studies on the effects of floods on school children and infrastructure in most countries including in Zimbabwe.

5. Floods and its impact on education systems

In the education sector floods leave a trail of destructions which may results in children's education getting to a level where it cannot be salvaged. Schooling maybe cancelled, children may drop out of school and school absenteeism may occur if school buildings are used as evacuation centres. A case in point is on Cambodia floods that happens at beginning of academic year from July –December, and children and teachers fail to go to school because of damaged roads and having to travel across rivers becomes dangerous. Using boats increase the cost of getting to schools which parents fail to meet [20]. This is supported by [21] who argued that the most depressing effects of floods is to be found in the affected areas, as the students have to wade through the flooded fields or board canoes that are dangerously rowed through the floating water. Living conditions in evacuation centres, limited space in schools having taken in more students and limited teaching resources for teachers also have a psychological effect on children [22]. The destruction of school infrastructure by cyclone Idai in Zimbabwe in Chimanimani district bares testimony.

There has also been recognition in practice that schools are normally designated as evacuation centres by government authorities. While classroom offer relatively large space for the multitude in need, evacuates bring their animals into the evacuation centres and use the buildings to house their animals. In a situation where the evacuation centre is a school, as was the case in Cambodia [20], animals destroy the school infrastructure. This leaves the schools in pathetic situations and disheartens educationist. Such an outcome tends to lead to brain drain as teachers may find it difficult to take up teaching jobs in the affected areas thus causing shortages of qualified teaching personnel [21]. In the absence of qualified teaching personnel, coupled with unattractive and dilapidated learning infrastructure impede on the quality of education offered to students which in turn affect the performance of students. When schools are closed and stay closed longer to flood disturbances, female learners are further exposed to other risks such as early marriages.

6. Flood: governance inaptitude

It is important for the different stakeholders in a country to mobilise each other in order to develop different tools to manage floods. UNDP [23] in [24] defines governance as 'the different ways in which governments, private sector and in general all individuals and institutions in a society organise themselves to manage their common affairs'. In relation to floods risk governance, governance then refers to the structural context in which various actors with a role in the development and implementation of flood risk management policies act and interact [24]. The level

Case 1: Flood risk governance Cyclone Idai

The 2018–2019 Indian Ocean cyclone resulted in a level of flood damage previously unseen in Africa [3]. The main cause was cyclone Idai which affected Mozambique and Zimbabwe. It commenced in March 2019 as a tropical depression over Malawi which caused widespread flooding affecting almost a million people. This moved back out to sea forming cyclone Idai which hit the east coast of Mozambique before dissipating in the eastern Zimbabwe which is 200 km from away on the 14th of March and slowly moved to hit Chimanamani at about 7 pm the following day on a Friday.

Zimbabwe had more lead time to prepare for the cyclone and reduce the potential damage compared to Mozambique. Despite this relative advantage, it was hit the hardest when compared to Mozambique that had far less casualties, environmental and infrastructure destruction. Indeed, security favours those who are prepared. The Metrological Service in Zimbabwe had warned of the impending threat two (2) days before the cyclone landed. Chimanamani District was severely punished, with loses amounting to millions of US dollars, unimaginable environmental damage and loss of lives.

Case 2: Mozambique Foods in 2000 and 2007

Mozambique has a total of ninety three rivers of various sizes, and seasonal regimes cross Mozambique coastal plain [25]. The flooding of the Zambezi in 2000 affected 4.5 million people and approximately 800 died [26]. However, in the major flood of 2007 no more than 300,000 people were affected though water levels were as high as they were in 2000. This showed that lessons learned by the government, national and international NGOs from the 2000 flood disaster had paid off. It had led to improved warning system, establishment of protocols for disaster response, awareness-raising campaigns among the population, training of local government institutions and improved coordination among all stakeholders. Many of those that had been affected in the 2000 floods had been relocated to higher and safer areas.

Vyas-Doorgapersad and Lukamba [27] gave the same sentiments on how Mozambique has improved in its flood risk management from 2000 floods citing the 2010 floods. According to [28] in [27], 'the Mozambique government had gone to great lengths to implement disaster risk reduction measures in the aftermath of the floods in 2000–2001. It had updated the contingency plans, prepared emergency site plans, conducted simulation exercises and pre-positioned supplies. The efforts paid as shown in the level of destruction during cyclone Idai which was not as bad as that inflicted in Zimbabwe, yet it still had higher destructive power from the Indian ocean when it made landfall in Mozambique. This shows the power of the flood risk reduction measures that have been put in place so far by the Mozambique government and its partners.

and trail of destruction caused by a hazard is largely defined type of governance in existence.

While the two case studies of two neighbourly countries do not detail the trail of destruction related to the school infrastructure and holistic education system, it has relevance to the discussion in place. Literature has identified a lot of weakness and some strength in governance issues pertaining to flood risks in countries in Africa, particularly Sub-Sahara Africa. Van Niekerk and Nemakonde [1, 27] identified that a number of countries have governance challenges to effectively respond to disasters and manage risk reduction measures because they lack pro-active measures from the government side. This is attested by Zimbabwe's response to cyclone Idai as penned by [29, 30] who identified capacity and policy gaps around coordinating response, civil and social protection, humanitarian assistance, development planning and management and land policies. This could have affected the proper dissemination of educational information on disaster as well as the threat that was posed by the cyclone Idai. A lot could have been done by the countries involved judging by the time-lines of the events, which gave ample time to alert the communities of the impending threat.

7. Flood risk management and education continuity

A number of frameworks have been devised to understand how flood impact education systems as giving directions on initiatives for the protection of children during disasters and for education continuity. The Hyogo Framework for Action [31] which recognised the necessity for including disaster risk assessment, disaster preparedness programs and activities that minimise disaster impacts in schools clearly comes to mind. The Hyogo Framework for Action (HFA) 2005–2015 was adopted in 2005 by 168 member states including Zimbabwe, to build resilient nations and communities through substantial reduction in disaster losses by 2015 [32, 33]. It was the primary global framework for DRR to give critical guidance to all nations in their efforts to reduce risk [34]. As such, five key indicators were formulated to guide nations towards a more disaster resilient society. Closely related to disaster risk reduction (DRR) education integration is Priority 3 which reads "Use knowledge, innovation and education to build a culture of safety and resilience at all levels" [35], p. 8. This was to be implemented by integrating DRR knowledge in relevant sections of school curriculum, including local risk assessment and disaster preparedness programs in schools and institutions of higher learning, and implementing programs and activities in schools, that teach learners how to minimise the effects of hazards [5, 36]. In response to that call countries like Zimbabwe have made some strides in integrating disaster education in school curriculum [37]. Hence although, HFA lifespan has ended, it remains the rock on which her successor the Sendai Framework for Disaster Risk Reduction (SFDRR) (2015–2030) [38] is built on. Unlike its predecessor which had a ten year life span, the SFDRR has duration of 15 years. Priority 2 of the framework guides the strengthening of disaster risk governance to manage disasters risk through education. This priority encourages member states to formulate policies and legal frameworks relating to DRR education that are within their capacity, of all facets of their government to address key elements of DRR education. Importantly, it acknowledges schools as critically important facilities and calls for the implementation of structural, non-structural and functional disaster risk prevention and reduction measures. All the itemised intentions of the framework in question speak to the value of the education system. The Comprehensive School Safety Framework (CSS) provides a structure that can enhance school safety, strengthen disaster risk education, identify priorities to enhance students safety at school and ensure continued access to primary education for students following a catastrophic disaster [39]. It also allows for collaboration among the different stakeholders with a focus on aligning the education sector and disaster management policy [40, 41]. Not to be outdone and noted by [34] the Sustainable Development Goals also speaks to the value of education. These initiatives advanced and prioritised children's continued access to education, the safety of school sites and using education to assist countries in improving disaster risk reduction efforts. However, where governance issues slake flood is one of the hazards proving to be a barrier of these efforts.

8. Education: a right in disaster situation?

A number of global forums have enacted policy framework to acknowledge the importance of education to children. Sustainable Development Goal 4 stresses the importance of education by promoting an inclusive and quality education for all and lifelong learning as sustainability goal [21]. Wisner et al. [26] sees education as Children's right and this is supported by Article 28 of the United Nations Children Act that recognises that a child has a right to education. While these global commitments need to be applauded, there is also need to appreciate that these have not kept pace with the huge numbers of children affected [21]. Anecdotal evidence indicates that floods have potential to slow down and hinder the progress towards the achievements of the MDGs.

9. Approaching flood

Educational continuity is being threatened by floods and other disasters due to the adverse effects of climate change on countries in Africa. It is expected that children will bear a disproportionate share of the impacts of floods both in the immediate and long-term as documented by many researches [42]. Floods impact on education sector in different ways which include destruction of buildings and infrastructure, function of institutional and organisational structures as well as the wellbeing of individuals and communities [43]. Chang et al. [21] penned that, damaged schools disrupt hard won education right, and when instructional time is lost, ultimately quality of education drops, when there are no plans for alternative locations and students are denied continuous schooling, many will never be able to catch up and will drop out permanently [21]. The disruption of education due to flooding is common issues worldwide.

This is where governments and their development partners could make it priority to harness the capabilities of each community to respond to flood threats, since during a flood the first responders are active community members. Wisner et al. [26] penned that schools are more than just the site for educating students, there is more to a school that being classrooms that house the students for their lesson and these include recognising the symbolic, cultural, economic and political significance of the schools within communities [43].

10. Conclusion

Understanding floods as either natural or man-made is critical to not only intended interventions but also in locating education infrastructure. A number of studies around the world argue for the integration of DRR into education sector policy at multiple levels of government and stress importance of specific and strong local implementation based on national guidelines [44–46]. Post-flood educational continuity need to deeply engage with the physical, institutional and organisational context of the schools, as nuanced understanding of the vulnerabilities and capacities of school stakeholders must be central to strategic practice. Therefore, enabling environment and policy, strengthening communication and co-ordination between and among school stakeholders and governments as well as integrating DRR into education sector policies are key for averting flood induced school learning disruptions. Such an approach will ensure that building substandard and weak structures which are not resilient to flooding are eradicated. Such measures may include, raising the ground floor and adding floor levels, improving drainage systems and irrigation channels as well as promoting safe storage of teaching and learning equipment and supplies. A School disaster management framework that promotes standard the adoption of a flexible education calendar, taking cognisant of the need of adjusted exam schedules is paramount for promoting risk reduction and resilience education system. However, all may be seen to waste where flood is a disaster risk governance failure.

Acknowledgements

We are deeply grateful to National University of Science and Technology (NUST) and Cambridge School for allowing us to compile this paper using their time and library resources.

Conflict of interests

The authors declare that they have no conflict of interests.

Author details

Edson Munsaka[1,2]* and Sebia Mutasa[3]

1 National University of Science and Technology, Bulawayo, Zimbabwe

2 Faculty of Humanities, University of Johannesburg, Johannesburg, South Africa

3 Cambridge International School, Gaborone, Botswana

*Address all correspondence to: emunsaka6911@gmail.com

References

[1] Van Niekerk, D., and Nemakonde, L. D. Natural Hazards and Their Governance in Sub-Saharan Africa. Printed from theOxford Research Encyclopaedia, Natural Hazard Science [Internt]. 2017. Available from: naturalhazardscience.oxfordre,com [Accessed: 2020-09-12]

[2] Kundzewicz, Z. W., Kanae, S., Seneviratne, S.I., Handmer, J., Nicholls, N., Peduzzi, P., Mach, K. Hydrological Science Journal: Flood risk and climate change: Global and regional perspectives. 2014:59 (1), 1-28

[3] Masters, J. Africa's Hurricane Katrina: Tropical Cyclone Idai causes an extreme catastrophe. Weather Underground [Internet]. 2019. Available from: http://www.wunderground.com/cat6/Africas-Hurrican-Katrina-Tropical-Cyclone-Idai-Cause-Extreme-Catastrophe [Accessed: 2020-09-08]

[4] Lumbroso, D.Flood Risk Management in Africa. Journal of Flood Risk Management[Internet]. 2020. Available from: https://doi.org/10.1111/jfrs.12612 [Accessed: 2020-08-23]

[5] Field, B.C. Managing the risk of extreme events and disaster to advice climate change. American Journal of Public Health: 2012: 96 (3)

[6] Grunwald, M.l. The Floods: A Man-Made Disaster? [Internet]. 2008. Available from: http://content.time.com/time/nation/article/0,8599,1818040,00.html [Accessed: 2020-08-27]

[7] Tockner, K. and Stanford, J. A. Riverine flood plains: present state and future trends. Environmental conservation, 2002: 29 (3) 308-330

[8] Loayza, N.V, Olaberria, E., Rigolini, J., and Christiaensen, L. Natural Disaster and growth, going beyond the Average. World Development 2012:40(7)

[9] Schmuck, H. Rural Flooding, the Zambezi River in Mozambique [Internet]. 2012. Available from: https://reliefweb.int/report/mozambique/mozambique-floods-final-report-31[Accessed: 2020-09-22]

[10] Sandberg, A. The Impact of the Stiegler's Gorge Dam on Rufiji Flood Plain Agriculture. BRALUP Service Paper 74/2, March 1974.

[11] Bubeck P., Otto A., and Weichselgartner J. Societal impacts of flood hazards. In : Oxford Research Encyclopedia on Natural Hazards Science; 2017.

[12] Clarkson, B. R , Ausseil, A-G.,E., and Gerbeaux, P. Wetland ecosystem services. In Dymond, J.R editor. Ecosystem services in New Zealand, Lincoln: Manaaki Whenua Press; 2013. p.191-202

[13] Masese A, Neyole E and Ombachi, N. Worrying deluge: influence on household's wellbeing and access to education in lower Nyando basin, Kisumu county, Kenya. International Journal of Multidisciplinary Academic Research. 2016:4(5) ISSN 2309-3218

[14] Kanyongo Y. G. Zimbabwe's public education system reforms: Successes and challenges School of Education. International Education Journal. 2005: 6(1): 65-74. ISSN 1443-1475

[15] : Hudson, P., Bubeck, P., Pham, M., Hagedoorn, L., Lasage, R., and Haer, T. (2017) The Impacts of Flooding on Wellbeing and the Role of Ecosystem-Based Adaptation. Global Resilience Partnership

[16] Kim, J. Y., Rich and Poor: opportunities and challenges in the age of destruction. The World Bank. 2018

[17] Cann, K. F., Thomas, D., Salman, R. L., Wyn-Jones, A. P., and Kay, D. Extreme water- related weather events and water borne disease[Internet]. 2013. Available from: pubmed-ncbi. nlm.nih.gov/228777498 [Accessed: 2020-09-05]

[18] Parvin, G. A., Shimi, A. C., and Biswas. Flood in a changing climate: the Impact on livelihood and how the rural poor cope in Bangladesh. Climate 2016: 4(4) 1-15

[19] , MONRE. 2011. Climate change and sea level rise scenarios for Vietnam. Natural Resources- Environment and Mapping Publishing House. Ha Noi, Vietnam.

[20] Ardales, G. Y., Espaldone, M. V. O., Lasco, R. D., Quimbo, M. A. T., and Zamora, O. B. Impacts of Floods on Public Schools in Municipalities of Los Banos and Bay, Laguna. Philippines. Journal of Nature Studies .2016:15 (1) 19-40

[21] Chang, M. S., Shahneela, Khatoon, Z., and Ali Shah, S. G., Flood Disasters and Its Impacts on Child Education in Sindh: A case study of 2010 Flood. International Journal of Advanced Research. 2013: 1(3) 329-344

[22] Bangay, C. and Blum. N. Education responses to climate change and quality: Two parts of the same agenda? International Journal of Educational Development. 2010: 20, 359-368.

[23] UNDP. Capacity Development for Disaster Risk Reduction[Internet]. 2010. Available from: https://int.search.myway.com/search/GGmain.jhtml?p2=%5ECQO%5Exdm256%5ETTAB03%5EZW&ptb=E8EFD835-457F-4E86-85C2- 8F3BCA72F555&n=7867175a&ln=en&si=sl4&tpr=hpsb&trs=wtt&brwsid=&searchfor=+UNDP+%282010%29.+Capacity+Development+for+Disaster+Risk+Reduction.+&st=tab[Accessed:2020-08-19]

[24] UNISDR. Global Assessment Report 2015. Making Development Sustainable: The Future of Disaster Risk Management[Internet]. 2015. Available from: https://www.preventionweb.net/english/hyogo/gar/2015/en/gar-pdf/GAR2015_EN.pdf[Accessed:2020-09-23]

[25] Winser, B. Flood Prevention and Mitigation in the People's Republic of Mozambique, Disaster's. Disaster. 1979. 3, 293-306. http://dx.doi.org/10.1111/j.1467-7717.1979.tb00155.x

[26] Wisner, B., Blaikie, P., Cannon, T., and Davis, I. At Risk: Natural Hazards, People Vulnerability and Disaster 1st edition. London: Routledge. 2004. DOI: 10.4324/9780203428764

[27] Vysa-Doorgapersad, S and Lukumba, T. M. Disaster risk reduction policy for Sustainable Development in the Southern Africa Development Community: A Policy Perspective. Journal of Public Administration. 2012: 47 (4)

[28] Chilundo, D. Report on the SADC disaster risk reduction and preparedness plan planning workshop. Gaborone: SADC; 2010.

[29] Chatiza, K. Cyclone IDAI in Zimbabwe: An analysis of policy Implications for post-disaster institutional development to strengthen disaster risk management. www.oxfam.org November 2019

[30] Olowu, D. The Hyogo Framework for Action and Implications for Disaster Management and Reduction in Africa' Jàmbá: Journal of Disaster Risk Studies. 2010: 3(1): 303-320

[31] UNISDR Japan: National Progress Report on the Implementation of the Hyogo Framework of Action [Internet]. 2011. Available from: http://www.preventionweb.net/files19650_jpn_NationalHFA progress [Accessed: 2020-09-23]

[32] UNISDR. Global Assessment Report on Disaster Risk Reduction. London: United Nations. 2011a

[33] SADC. Sub-Regional Assessment on Mainstreaming and Implementing Disaster Risk Reduction Measures in Southern Africa[Internet]. 2013. Available from: http://www.gisportal.sadc:int/sadc/media_dl=571[Accessed: 2020-09-23]

[34] Mutasaa,S., and Munsaka, E. Botswana and international policies on the inclusion of disaster risk reduction in the school curriculum: Exploring the missing link. International Journal of Disaster Risk Reduction. 2019:40:1-7

[35] UNISDR. Towards a Culture of Prevention; Disaster Risk Reduction Begins In School. Good Practice and Lessons Learned. Geneva: UNISDR; 2007

[36] Tran, H.P. Regional Analysis on DRR Education in Asia Pacific Region: In the Context of Priority 3 of the Hyogo Framework for Action, Bangkok: UNISDR; 2009.

[37] Department of Civil Protection; Cyclone IDAI situational report as at 19th March 2019, Harare: Ministry of Local Government[Internet] 2019. Available from:http://www.drmzim.org/wp-content/uploads/2019/03/CYCLONE-IDAI SITUATIONAL-REPORT-19-March-2019.pdf[Accessed:2020-09-23]

[38] UNISDR. Indicators to monitor global targets of the Sendai Framework for Disaster Risk Reduction 2015-2030: A technical review. Geneva: United Nations Office for Disaster Risk Reduction;2015

[39] Shah, A. A., Ye, J., Lu, P., Ullah, R., Alishah, S. I., Fahad, S., and Naz, S. School Flood Emergency Preparedness in Khyber Pakhtumkhwa Province, Pakistan. International Journal for Disaster Risk Science. 2018: https://doi.org/10.1007/s13753-018-0175-8

[40] GADRRRES. Comprehensive School Safety Framework[Internet]. 2017. Available from:http://gadrrres.net/uploads/files/resources/CSS-Framework-2017.pdf [Accessed: 2020-09-25]

[41] Sakurai, A., and Shaw, R. Implications of 3:11 for disaster education and education for sustainable development in Japan: Fostering resilient communities after the triple disaster. London: Taylor and Francis; 2015:36-51

[42] Save the Children Sri Lanka. Promoting Children Friendly and Child Centered Disaster Risk Reduction in Sri Lanka, Colombo: Save the Children in Sri Lanka; 2005

[43] Forino, G and Von Meding, J. Post – Disaster Educational Continuity in Urban Flood. University of Newcastle, Newcastle; 2018

[44] Cadag, J. R, D., Petal, M., Luna, E., Gaillard, J. C., Pambid, L., and Santos, V.G. Hidden disasters: Recurrent flooding impacts on educational continuity in the Philippines. International Journal of Disaster Risk Reduction. 2017: DOI: 10.1016/j.ijdrr.2017.07.016

[45] Alam, K, Saadi, S, Hossain, K, Rahman, K A, Hossain, M, Al Rashed, A, Azad, A., Jasmeen, T., Neelim, A, Kabir, S, and Islam, T. Disaster resilient primary education in Bangladesh. Problems, priorities and actions for disaster risk management in primary education, Final Study Report, Save the Children UK; 2010.

[46] Amri, A., Bird, D., Ronan, K. R., and Towel, B. Disaster risk reduction education in Indonesia: Challenges and recommendations for scaling up. Natural Hazards and Earth System Science.2017:17(4): 595-612.

Chapter 10

Adoption of Conservation Agriculture as a Disaster Risk Reduction Tool in Chivi District, Zimbabwe

Jestina Chineka, Nthaduleni Samuel Nethengwe and Hector Chikoore

Abstract

Drought tops the list of disasters affecting southern Africa. In Zimbabwe droughts recur, leaving approximately three million people food insecure. Hence the adoption of sustainable adaptation strategies to drought becomes imperative. Conservation Agriculture (CA), has been successfully adopted in southern Africa to avert drought shocks among other agricultural challenges. Despite the success of CA in some regions, its effectiveness in semi-arid parts of Zimbabwe has been widely contested. However the effectiveness of a new technology, in the face of disasters depends on its adoption, reflecting its strength and usefulness. This chapter seeks to evaluate the adoption of CA in the semi-arid Chivi District of Zimbabwe and unpack factors affecting CA adoption to provide baseline data to policy makers in Zimbabwe and other similar environments. The chapter is based on data elicited from a survey held across Chivi District in Zimbabwe.

Keywords: adaptation, adoption, conservation agriculture, disaster risk, drought

1. Introduction

Climatic disasters such as drought have become a concern in Africa. Agriculture productivity in southern Africa is declining due to these disasters [1]. Zimbabwe is not an exceptional, [2] note that agricultural yields in Zimbabwe are averaging less a tonne per hectare, resulting in protracted food insecurities despite farmers having large pieces of land. Drought effects are felt in most parts of Zimbabwe where rainfall patterns have become erratic [3]. Hence communities are in dire need of effective, long-term strategies to cope. With recurrent droughts and current climate change projections, the future of food security is not only hinged on productivity and availability of food reserves but on addressing the challenges posed by climatic risks such as drought. Resilience of agricultural technologies is critical in communities where agriculture is the backbone such as in rural Zimbabwe.

CA is one humanitarian initiative introduced in Chivi District, to curb the effects of drought and ensure food security. CA is an agricultural system which seeks to conserve water and soil through its main principles of zero to minimal tillage, crop rotation and mulching. CA has been hailed globally for its ability to increase agricultural productivity under diverse climatic conditions. The same CA project in Chivi, was implemented in Zambia and increased crop yields by 240 to 400% [4]. In Kenya, Ghana and Malawi agricultural profitability increased [5–7]. Despite all this success CA project in Chivi has been characterized by conflict and contestations and its adoption has been very slow [8, 9]. It is within this breadth that this chapter seeks to assess CA adoption in Chivi and establish the weaker lines within the CA project.

IPCC's climate change projections predicting an increase in temperatures and acute rainfall shortages in southern Africa of between 1.5°C to 2.5°C under the 2.0°C GWL and 10 to 20% reduction in precipitation, it is crucial to draw sustainable adaptation strategies and improve resilience in rural communities, which are more vulnerable [10]. This research also unveils factors affecting the adoption of CA and enhance its effectiveness as an adaptation strategy to drought.

The effectiveness of a new technology depends on its adoption and also the project's adoption levels reflect on its strength thus convenience and usefulness in the user's interpersonal networks [11]. Adoption is defined "as the extent to which farmers put into practice a new innovation, given adequate information about the technology and the potential benefits" [12]. The Tradeoffs model inform that farmers are rational beings and only adopt a new system of agriculture if it's more viable [13]. This chapter sought to evaluate the adoption of CA in Chivi.

2. Methodology

The data used in this chapter was elicited from 140 household questionnaires administered across 16 wards of Chivi District and focus group discussions held in six wards of Chivi district. This data was also supported by data from key informant interviews held with three Non-Governmental Organisations (NGOs) and 16 Agricultural Research and Extension (AREX) officials.

Data capturing was organized in Microsoft Excel 2013 and later transferred to Stastical Package for Social Science (SPSS). Prior to the analysis, captured data was coded according to the levels of measurement. This allowed for uni- and bivariate data analyses. Data analysis was done using SPSS version 22. Chi square and Cramer's V value were calculated and analysis was set at 0.05 confidence level. In order to describe and identify relationships that must be taken into account and characterise CA project in Chivi District, frequency tables and bar graphs were generated (univariate analysis). Frequency distributions described the number of times the different attributes of a variable were observed in a sample. This allowed for the comparison of different variables. Statistical tests of significance were conducted on the levels of awareness and general perceptions in order to explore independent variables e.g. gender; age; level of education differences. Chi-square tests was used to calculate significant differences in different demographic groups on their adoption and practices in the Conservation Agriculture project [14]. A 95% level of significance was used, which is most commonly used in social research [15].

Cramer's V test was used to measure the strength of relationships. It measures the strength of relationship for any size of contingency table, and it offers good norming values from zero to one (0–1) for relative comparison of the strength of correlation regardless of the table size. For Cramer's V, 0.0 to 0.30, the strength is considered no relationship to weak; for Cramer's V, 0.31 to 0.70, the strength is considered moderate relationship; while for Cramer's V from 0.71 to 1.0, the strength of the relationship is considered strong [16].

For qualitative data analysis, Archive of Technology, Life world and Language. Text interpretation (Atlas.ti 8) was used to analyse data from the household questionnaires, focus group discussions and key informant interviews. Tools such as Co oc was used for comparisons using the occurrence frequency, Co-code Doc Table for numeric analysis as well as Networks and Report tool for visual and text analysis. Results from Atlas.ti 8 were used to compliment data from SPSS. Results were presented as graphs, charts, visuals and narratives.

3. Results and discussion

3.1 CA adoption in Chivi

The physical adoption of the CA project was measured based on project adoption records of NGOs operating in Chivi district.

Only 30% of households in Chivi are practicing CA, refer to **Figure 1**. AREX officials and Focus Group Discussants showed that CA started as early as 1995 in some wards such as Ward 10 but became more popular from 2008 when the government of Zimbabwe formalized it. This implies that the project has been long operating in the District, despite low adoption percentage. However, CA benefits are normally realized at least after 10 years of practice [17]. Hence a 30% adoption is not that low, considering that the project is formally slightly over a decade in most wards. After seeing the benefits more farmers are likely to adopt CA. However data on CA adoption trends did not support this. Key informants confirmed a decline in adoption trend over the years in all wards. In ward 21 of the 300 farmers who initially adopted CA in 2008 only 80 are currently practicing it. Of interest is that Ward 21 was listed as the third highest adopter of CA in the District by NGOs. This gives a gloomy picture to the sustainability of CA as a drought risk reduction tool in the District.

3.2 Extension of CA plots

To get an insight into the spatial adoption of CA and the long term plans of farmers on CA, plot sizes were also assessed. Key informants showed that farmers under the main NGO, CARE increased their demo plots from the 18 mother demo plots of 1 hectare to 180 baby demo plots across its 12 wards. However the questionnaire survey showed that 100% of CA farmers are still working on demonstration plots in groups and have not adopted the full CA package onto their individual

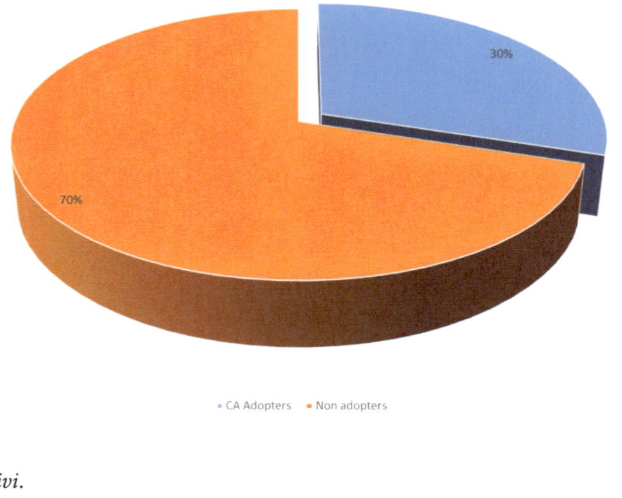

Figure 1.
CA adoption in Chivi.

plots. However 100% admitted to have adopted at least one of the CA principles and are using them in their conventional agriculture system. 52% of these farmers adopted planting on time, 80% crop rotation and 38% use of small grains. No CA farmers have adopted planting basins and mulching onto their traditional systems. NGOs supported these findings and added that planting basins and mulching principles are the most unpopular. These two principles could be the hindrance to effective adoption of CA as a disaster risk reduction tool in Chivi.

3.3 Social buy-in into CA

The social acceptance of the CA project was assessed to get the level of social acceptance of the project. Social discourse and verbatim around the CA project was used as indicators. 72% of participating groups under Focus group discussions described their role in CA as beneficiaries, refer to **Figure 2** below and very few had an active verbatim concerning their role under CA.

Verbatim assessment by Wards showed that only 28% of participant Wards had a positive view about their role in CA. Ward 21 and 24 showed an active role in CA (**Figure 3**). Ward 10 besides it being the first ward to be introduced to CA in 1995, over two decades ago it showed a passive role in CA project.

Focus group participants of about 72% admitted to being passive beneficiaries of CA and had no active or decisive role in the project. The community described NGOs as the "owners" of the project while AREX officials were described as "trainers". Throughout the whole cycle from its formulation to implementation community members are passive participants. On the discussion surrounding difference between CA and the conventional farming, 77% of participants showed that there is no difference in terms of benefits, this contradicted the views of NGOs and AREX officials, whom most of them pointed out the difference in yields per hectare in which CA has better yields. Social discourse on CA project was characterised by undertones of disassociation and negativity. The community coined the main principle of CA, zero tillage "Dhiga ufe" meaning dig and die, alluding to its labour intensive nature. Commenting on CA benefits the focus group participants who did not adopt the initiative said,

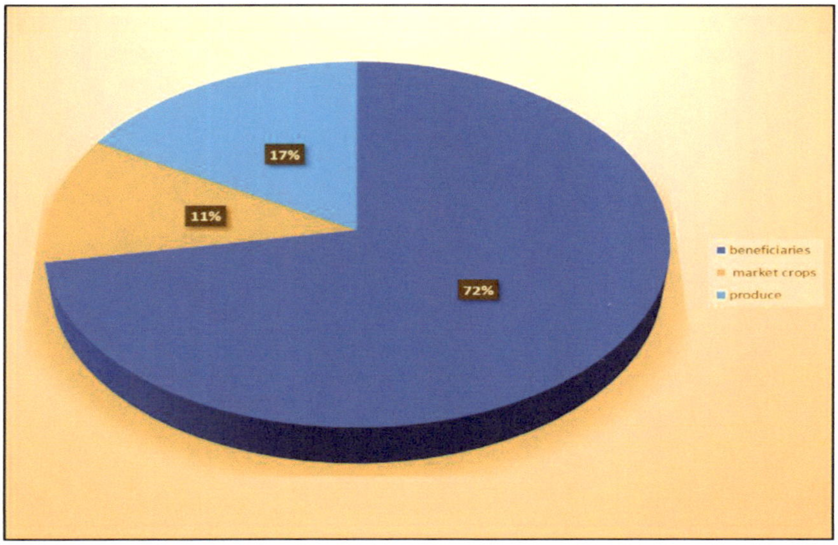

Figure 2.
Role of Chivi community in CA.

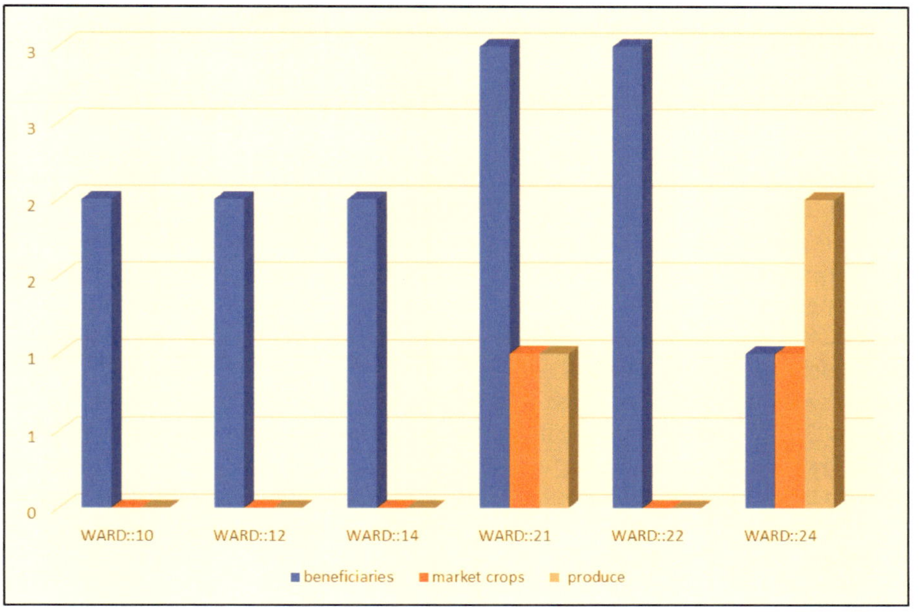

Figure 3.
Role of Chivi community in CA by Wards.

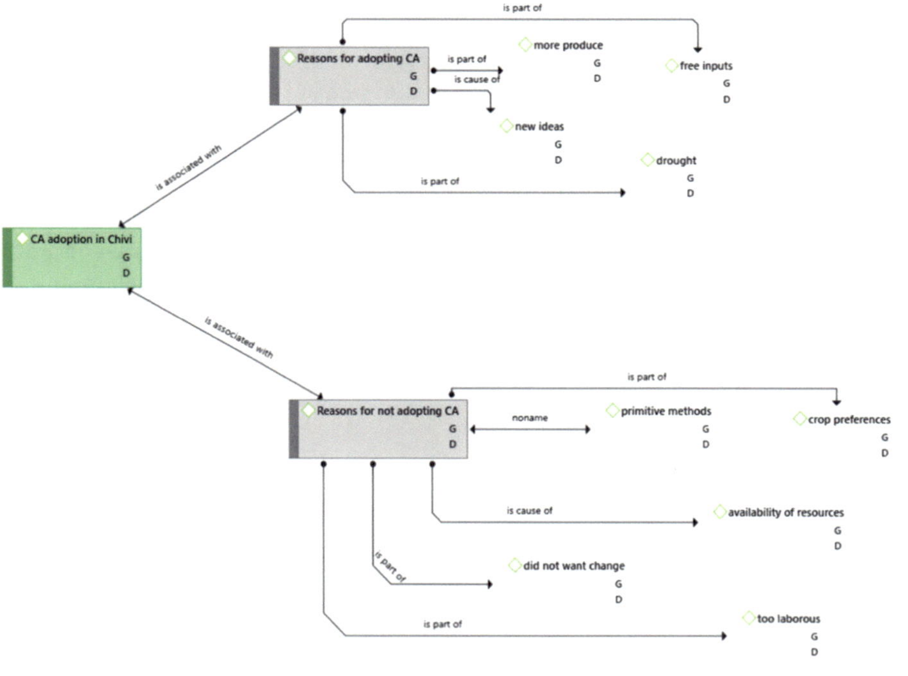

Figure 4.
Reasons behind CA adoption in Chivi District.

"Hapana akamborarama nedhiga udye, gore rezhara tose toforera mukomondera".

A Shona translation to no one has ever survived on CA, during drought, we all queue for food relief. This shows the community's negative view of CA as a disaster risk reduction tool.

Results on community buy-in showed that 80% of AREX officials described Chivi CA project buy-in as low. "Reluctant" and "not eager" were the most commonly used adverbs to describe community buy-in. Drought was also mentioned as a barrier to community buy in. The benefits of CA are said to be less visible due to recurrent droughts. Chivi community through focus group discussions also confirmed low buy into the CA project but had different reasons, refer to **Figure 4**.

The focus group participants who adopted CA in Chivi, did it for diverse reasons, 34% adopted for new farming ideas, 32% adapted to curb drought effects, 17% for free inputs and 17% to increase production. The participants who did not adopt CA had also their own reasons, 34% did not see the need as they have enough resources to continue with conventional farming, 34% mentioned use of primitive farming and labour intensive methods, and 32% did not prefer small grains and changing their traditional farming system. All, 100% of participants who did not adopt CA were aware of the challenges faced in agricultural production however they did not see CA as the solution to their challenges. The interesting argument was that CA is affected by drought the same way as conventional agriculture system. This was also indirectly brought up by AREX and NGOs interviews. They attributed the negative attitude of farmers towards CA to lack of tangible benefits which are being washed away by recurrent drought in Chivi.

3.4 Factors affecting CA adoption

Variables of human capital were assessed through a questionnaire administered to household heads to unpack factors affecting the effective adoption of CA project as a drought risk reduction tool in Chivi. The impact of human capital on CA adoption were tracked using the demographic characteristics of household questionnaire participants such gender, age, level of education, marital status, employment status and household incomes. Adoption of a new agriculture technology does not only depend on the nature of the technology but also its intended users. Hence the heterogeneity of farmers and their demographic structure influence the adoption of a new innovation [18]. Gender is an important characteristic in the adoption of CA considering the associated gender roles and dynamics especially in rural communities. Age is influential in new technology adoption. Adoption of new technology declines with age [19].

Chi-square ($\chi 2$) test was used to associate demographic characteristics of participants with CA adoption. The findings showed a relationship between gender and CA adoption, refer to **Table 1**.

Variable	Chi-square	df*	p-value	Cramer's V
Gender	6.056a	1	.014*	.209^0
Age	1.601a	3	.659	.107^0
Level of education	3.493a	3	.322	.159^0
Marital status	.280	2	.869	.0610
Employment status	2.249a	3	.522	.127^0

$^*P < 0.05$.
$^{**}P < 0.01$.
$^{***}P < 0.001$.
^{0}No relationship to weak.
^{1}Moderate relationship.
^{2}Strong relationship.

Table 1.
Human capital and Conservation Agriculture.

The Chi-square ($\chi 2$) analysis revealed a significant association of gender and being a conservation farmer ($p < 0.05$). However a Crammer test classified the relationship weak. These findings are supported by other surveys done in Chivi, which showed that women constitute the majority of communal small holder farmers [20, 21]. This was also supported by the key informant interviews held with NGOs operating in Chivi. NGOs target women in their CA projects, hence more women have adopted the project. However with more women involved in CA, the project ought to have a special design tailor-made to suit women's gender roles and their often tight work schedules for sustainability. The CA activity plan used in Chivi contradicts this, CA project activities run throughout the year [22]. CA project in Chivi, runs concurrently with the conventional agriculture, the main and traditional agriculture system practiced by every farmer as well as livestock farming. Therefore time could also be the barrier to effective CA adoption and would certainly hinder its effectiveness as a drought adaptation tool.

The relationships between age, level of education, marital and employment status and CA adoption were found statistically insignificant. On age the findings showed that more participants were in the active age group, the 30 to 50 year age group, followed by 51 to 60 year group, then the 60 and above. Despite these findings of the Chi square tests, it is also important to note that the age structure of Chivi highlighted a community operating in a poor economic environment, considering that the active population is fully engaged in small holder farming as opposed to the norm that active population often work off the family compounds in towns and cities. It also showed that small holder farming is a major source of livelihood in this community. Therefore there is a need for sound agricultural innovations to boost livelihoods and curb drought effects. Chivi age structure consisting of a higher percentage of the active population, shows that the community is not negatively affected by new technologies which is ideal for effective information dissemination critical in CA adoption [23]. Younger farmers make long-term plans in their operations and acquire necessary skills and knowledge better than old farmers.

The level of education is also of paramount importance to information dissemination, comprehension of information and querying of information sources. Majority of participants had secondary education with a 58.6%. Participants with primary education were 37.1%. The least participants had a tertiary qualification about 1.4% followed by those who never attained any formal education with 2.9%. The findings shows that Chivi District comprises of a literate population. This means that Chivi community is very much aware of their environment and if given adequate information on CA, it can comprehend it and make informed decisions on adopting or not adopting. In this case low adoption of CA might be more to do with the applicability or feasibility of the project design and assets other than human capital.

Social dimensions such as marital status are also of importance in the adoption of a new agricultural system [24]. Marital status and gender are critical in decision making, especially in crucial issues such as adoption of a new farming system. Dimensions such as gender roles in decision making roles and land ownership come into play. Majority of the household heads who participated in this study were married, with a 58%. Single participants constituted 18% whilst 17% were divorcees and 7% widows. This married to non-married ratio of 58:43% is a true reflection of the marital status in Chivi. According to census report population of widows and divorcees is rising due to factors such as prevalent HIV/AIDS and economic hardships [20]. In a social structure such as this there is a need for developmental projects such as CA to strengthen weak social networks and support the existing ones and to avoid project domains that create or exacerbate social tensions.

Amount	Frequency	Percent
≤$227	97	69.3
≥$228	43	30.7
Total	140	100

Table 2.
Chivi monthly household incomes.

Despite an insignificant statistical link between CA adoption and marital status, inherent gender dynamics in marital status of a rural society such as Chivi needs a closer scrutiny. The strength of gender roles in decision making and land ownership might not be visible on the ground but has a huge indirect influence on adoption of an agricultural innovation such as CA [25]. These gender roles are well-defined in Chivi, a predominantly rural district with only 30 out 32 rural wards [20, 24].

An interview with key informants on gender gaps in Chivi also acknowledged existent gender gaps. NGOs had problems with the registering Chivi women to a CA sister project of Nutritional gardens. Most women would register into this project under their husbands' names some of which divorced them and some not even in the community, working either in the cities or outside the country. This shows that even though women are the producers they are not the decision makers nor practical land owners. This becomes a bit complicated when they have to make crucial and life changing decisions such as changing the farming system from conventional plough system to CA. There is need to mainstream gender into a CA project, lest it might affect the sustainability of CA or further widen the gender gaps that already exists in agriculture.

Financial capacity as well as off field commitments also influence the adoption of CA technology. Employment status of participants revealed that most household heads who participated in this survey were unemployed. At least 59% of the participants were unemployed. Very few participants were employed with a 3.6% and about 28% were self-employed while 10% were on pension. The findings showed that the rate of employment in Chivi is very low and people who are employed work outside the District [8, 21].

Household heads' monthly incomes were also analysed. The incomes were categorised using the United Nations (UN) poverty datum line of US$1.90 per individual per day [26]. This was calculated for a 30 day month and further multiplied by 4 which is the average household size for Chivi District [20], refer to **Table 2**.

Most participants, thus 69.3% had a monthly income below the poverty datum line while only 30.7% of the households were out of the poverty threshold. This supports the UN (2019)'s assertion that sub-Saharan Africa has most of the people living below the poverty line together with South Asia. In line with these findings, focus group discussions also raised an interesting argument on CA impact in the community. Participants who adopted CA from its inception mentioned free inputs as one major reason which made them buy into the project while the non-adopters argued that CA project had blocked the issuing of free drought relief food by NGOs. These arguments speak to the high levels of poverty in the community. Hence for CA technology to be acceptable in the community it has to prove itself as a viable income generating project.

4. Conclusion and recommendations

This study notes that adoption of the CA project in Chivi is low, with some farmers withdrawing from the project in ward such as 21 and 22. For a strategy to be effective in the light of a disaster risk, it should cover all the affected people. Farmers are also still working on demo plots. Failure to transfer CA to their own plots and resorting to increase demo plots reflects that farmers do appreciate CA but they are barriers impeding CA practice on their own farming plots. Despite low adoption of the project, some CA practices have been adopted into the traditional farming systems. All farmers have at least adopted one CA principle and have incorporated it into their conventional agriculture system. This supports the view that CA has the potential as a drought adaptation tool. Crop rotation and use of small grains are the most adopted strategies, while mulching and planting basins are the least. These least adopted principles could be the barriers to effective adoption of CA project and effective adaptation to drought. Other barriers to CA adoption noted is the vulnerability of the CA system under a recurring drought environment. This support the assertion that farmers opt for a more viable agricultural option [13]. Hence CA option might be effective at smaller scale and farmers do not perceive the benefits at a larger scale. There is need for CA to be practiced on larger plots for tangible benefits and also for its high production during a better season to increase resilience of farmers during lean periods. The social discourse on CA is too negative, CA project officials need to involve the community in decision making as well as incorporate their local knowledge so as to build some sense of proprietorship and avoid knowledge contestations. On all factors affecting CA, gender and finance had the most significant impact. Gender dimensions such as overburden on women, land ownership and critical decision making powers need to be closely assessed and mainstreamed in the CA project. Finances heavily affect Chivi community where the majority of household heads are not employed and most households live on monthly incomes below the UN poverty datum line. CA concepts outside the project seem to be welcomed by Chivi community, hence for it to be an effective tool in drought risk reduction, more support has to be given to the community and NGOs need to take time in capacity building before rolling out the support. Future research can also look into finding common grounds in building resilient communities such as, blending indigenous and scientific knowledge into adaptation strategies as well as modifying agricultural extension models in rural areas.

Acknowledgements

I would like to acknowledge the support from NGOs working in Chivi District such as CARE international and Christian Care. The Department of Lands and Agriculture, ARREX officials at Chivi office for assisting with all the data which was used in this survey. The entire Chivi community for their willingness to share information. Last but not least, special thanks to the University of Venda for funding this research.

Conflict of interest

The authors declare no conflict of interest.

Author details

Jestina Chineka[1*], Nthaduleni Samuel Nethengwe[1] and Hector Chikoore[2]

1 Department of Geography and Geo-Information Sciences, University of Venda, Thohoyandou, South Africa

2 Unit for Environmental Sciences and Management, North-West University, Vanderbijlpark, South Africa

*Address all correspondence to: jessiench@gmail.com

References

[1] FAO. FAO–ADAPT. Framework programme on climate change adaptation. Food and Agricultural Organization of the United Nations, Rome [Internet]. 2011. Available from: http://www.fao.org/3/i2316e/i2316e00.pdf [Accessed 2020 June 12]

[2] Marongwe LS, Nyagumbo I, Kwazira K, Kassam A, Friedrich T. Conservation agriculture and sustainable crop intensification: a Zimbabwe Case Study: Food and Agriculture Organization of the United Nations (FAO) [Internet]. 2012. Available from: https://www.cabdirect.org/cabdirect/abstract/20133092187 [Accessed 2020 May 12]

[3] Unganai L, Mason SJ. Long-range predictability of Zimbabwe summer rainfall. International Journal of Climatology: A Journal of the Royal Meteorological Society. 2002;**22**(9):1091-1103. DOI: 10.1002/joc.786

[4] FAO. Policy on Gender Mainstreaming [Internet]. 2011. Available from: http://www.fao.org/climatechange/en [Accessed 2016 June 10]

[5] Kaumbutho P, Kienzle J. Conservation agriculture as practised in Kenya: two case studies. Nairobi, Kenya: African Conservation Tillage Network (ACT); Centre de Coopération Internationale de Recherche Agronomique pour le Développement (CIRAD); Food and Agriculture Organization of the United Nations (FAO); 2007.

[6] Boahen P, Dartey BA, Dogbe GD, Boadi EA, Triomphe B, Daamgard-Larsen S, et al. Conservation agriculture as practiced in Ghana. Nairobi: African Conservation Tillage Network/Paris. Agricultural Research for Development/Rome: Food and Agriculture Organization of the United Nations; 2007

[7] Mloza-Banda H, Nanthambwe S. Conservation agriculture programmes and projects in Malawi: Impacts and lessons. 2010.

[8] Nhodo L, Gukurume S, Mafongoya O. Contestations and conflicting lifeworlds in conservation farming practices in Zimbabwe: the experiences of peasant smallholder farmers in Chivi south district in Masvingo. Russian Journal of Agricultural and Socio-Economic Sciences. 2013;**16**(4)

[9] Gukurume S. Climate change, variability and sustainable agriculture in Zimbabwe's rural communities. Russian Journal of Agricultural and Socio-Economic Sciences. 2013;**14**(2)

[10] Maúre G, Pinto I, Ndebele-Murisa M, Muthige M, Lennard C, Nikulin G, et al. The southern African climate under 1.5 C and 2 C of global warming as simulated by CORDEX regional climate models. Environmental Research Letters. 2018;**13**(6):065002. DOI: 10.1088/1748-9326/ab9927

[11] Giesing, I. 2003. Information Technology adoption. Masters dissertation. South Africa. University of Pretoria.

[12] Ntshangase NL, Muroyiwa B, Sibanda M. Farmers' perceptions and factors influencing the adoption of no-till conservation agriculture by small-scale farmers in Zashuke, KwaZulu-Natal Province. Sustainability. 2018;**10**(2):555. DOI: 10.3390/su10020555

[13] Antle, J. and Valdivia. R, 2011. What is the TOA-MD Model? Basic Concepts and an Example. Agricultural and Resource Economics. Oregon State University.

[14] Babbie E, Mouton J. The practice of social research. Cape Town: Oxford University Press; 2010

[15] Fielding J, Gilbert N, Gilbert GN. Understanding social statistics. London: Sage Publications; 2006

[16] Essien HE. A Correlational Study of Active Learning, Academic Proficiency and Completion Rates of African American Students Enrolled in Developmental Mathematics Courses. Chicago, U.S.A.: Doctor of Education in Higher Education and Organizational Change faculty of Benedictine University; 2015

[17] Kassam A, Friedrich T, Shaxson F, Bartz H, Mello I, Kienzle J, et al. The spread of conservation agriculture: Policy and institutional support for adoption and uptake. Field Actions Science Reports The journal of field actions. 2014;7: 1-12. Available from: https://journals.openedition.org/factsreports/3720 [Accessed 2020 May 12]

[18] National Research Council. Publicly funded agricultural research and the changing structure of US agriculture: National Academies Press; 2002. DOI: 10.17226/10211

[19] Sunding D, Zilberman D. The agricultural innovation process: research and technology adoption in a changing agricultural sector. Handbooks in Economics. 2001;**18**(1A):207-262. DOI: 10.1.1.316.188&rep=rep1&type=pdf

[20] Zimbabwe National Statistics Agency. Census 2012. Preliminary Report [Internet]. 2012. Available from: http://archive.kubatana.net/html/archive/econ/121220zimstats.asp [Accessed 2020 May 12]

[21] Mudzonga E, editor. Farmers' adaptation to climate change in Chivi district of Zimbabwe. TRAPCA trade policy research forum. Arusha: Tansania; 2012

[22] Zimbabwe Conservation Agriculture Task Force. Farming for the Future A Guide to Conservation Agriculture in Zimbabwe[Internet]. 2009. Available from: http://www.foodgrainsbank.ca/uploads/Farming%20for%20the%20Future%20-20A%20Guide%20to%20Conservation%20Agriculture%20in%20Zimbabwe.pdf [Accessed 2017 August 15]

[23] FAO. Conservation Agriculture: Training guide for extension agents and farmers in Eastern Europe and Central Asia, Rome [Internet]. 2019. Available from: http://www.fao.org/3/i7154en/i7154en.pdf [Accessed 2020 June 14]

[24] Chineka J, Musyoki A, Kori E, Chikoore H. Gender mainstreaming: A lasting solution to disaster risk reduction. Jàmbá: Journal of Disaster Risk Studies. 2019;**11**(3):1-6. DOI: doi.org/10.4102/jamba.v11i3.723

[25] Nyanga PH. Factors influencing adoption and area under conservation agriculture: A mixed methods approach. Sustainable Agriculture Research. 2012;**1**(2):27-40. DOI: 10.22004/ag.econ.231353

[26] United Nations. Ending poverty [Internet]. 2019. Available from: https://www.un.org/en/sections/issues-depth/poverty/ [Accessed 2019 November 10]

Chapter 11

Disaster Relief Supply Management

Yusen Ye and Hong Yan

Abstract

Disaster relief supplies (DRS) play a vital role in natural disaster rescue and relief operations. Often DRS management is initiated and supported by the government, yet the related cost issues have not been fully emphasized. In the face of highly uncertain disaster locations and timing, these supplies are usually prepositioned without proper consumption, which causes enormous waste in practice both economically and environmentally. This chapter highlights the potential to bring the reverse logistics strategies in conventional business practice into DRS management. Incorporating the reverse flow of removed relief items with DRS supply chain management not only benefits in cost reduction and environmental protection, but also enhance the daily management and quality control of DRS. Relying on social trust and efficient marketing network provided by government coordination and international cooperation, the stable quality level and relatively integrated inventories of the removed DRS can achieve economies of scale in the reverse supply chain operations. This chapter aims to develop an understanding of DRS reverse logistics, which energizes the responsible management of DRS for economic, social, and environmental sustainability.

Keywords: disaster relief supply, relief material, reverse logistics, marketing strategy, resale, coordination

1. Introduction - What is disaster relief supply?

1.1 Disaster and relief

Natural disasters pose a great threat to human lives and property in all aspects. The frequency, intensity, and severity of disasters, are in a fast-growing trend year by year. According to the World Disasters Report (2018) of the International Federation of Red Cross and Red Crescent Societies (IFRC), a decade's number of 3751 disasters triggered by natural hazards had affected 2 billion people and estimated $1658 billion cost of damages. Floods, earthquakes, storms, tsunamis, wildfires, volcanic eruptions, and landslides are on the list of most common natural disasters over the last ten years (2008–2017). Some of them occur frequently in a year, such as floods (average 152) and storms (average 101), while some are not that frequent but extremely deadly, such as earthquakes (average 30) that caused 351,968 deaths and some 49% of the total [1]. There has been a big increase in the number of responses to disasters this decade. Alone with this situation, a large amount of disaster rescue and relief supplies has experienced tremendous growth, particularly in those major disasters such as forest fire, earthquake, typhoon, and floods.

Disaster relief supplies (DRS), including life necessities (e.g., food and drinks), living security items (e.g., clothes and shelters), medical supplies (e.g., medicines and healthcare products), and life-saving tools and equipment (digging tools, emergency lights, and large equipment), in response to disasters and crisis, play critical roles in saving lives and reconstructing communities in all recovering stages. Different supplies have different characteristics in terms of material lifetime and demand urgency and therefore need different management mechanisms. First, the demand for DRS is unpredictable due to the unexpected event at unexpected locations in most of the situations. Second, the amount, type, and emergency of demand are highly uncertain due to the different damage levels and geographic conditions. Third, the timely efficiency of the availability and delivery of DRS is extremely important in the relief process. Lastly, the demand and supply of these materials and equipment are mandatory and have strong social value. Furthermore, due to the low economic or market value of DRS, government agencies are responsible for the collection, purchasing, storing, delivery, and coordination of materials most of the time and situations.

1.2 Disaster relief supply management

DRS have different properties and thus are normally stored and managed in different ways. Under the central government regulations, there are typically two different management DRS management systems, centralized or decentralized, covering a country [2]. In a centralized system, such as that in China, DRS are properly arranged in different governmental levels and integrated by the central government level. A decentralized system, such as that in the US, DRS are physically handled at the state level.

In terms of material items, different supplies are held in different places. The immediate life necessities and living security items, including food, drink, basic digging tools, and simple clothes and shelters are normally stored in grassroots level facilities. These materials are of low value but have crucial importance in the initial rescuing period. Medical supplies, including medicines, dressing, first aids, surgical supports, are kept in the local hospitals and clinics. These supplies are vital in the first few days of life-saving and pain relief and need to be available in time. Other expensive medicines, surgical equipment, and special tools are available, and most of them are in actual daily uses, in large hospitals in large towns and cities. Large life-saving mechanical tools, equipment, and transportation modes are stored in the main facilities. These storages are all integrated into a largely complete network covering a whole region or a country.

Because of the uncertainty and emergency of disaster events, the demand for DRS usually cannot be satisfied by the prepositioned supplies in almost all cases. In these situations, private sectors, including non-government organizations (NGO), industrial organizations, and international organizations, collect and coordinate a large number of supplies. These supplies are planned and prepared by government agencies or obtained from all aspects of donations with various types, different amounts, and diversified conditions. They could be stored in anywhere of the country, nearby or far away. To properly manage these supplies is never an easy and simple task.

1.3 Management issues

Since the DRS are of much different market values, high uncertainty of demand time and amount, and involve multiple operations players, following management issues are frequently observed in rescue and relief practices.

1. The basic DRS of living necessities and first hand rescuing tools stored in the first line local warehouses are of low market value. They are prepared by government agencies and managed by grassroots staff. These people only obtain some basic training and spend a very limited time on material management. This implies that the daily management operations are not in a professional and even a bureaucratic manner. For instance, tons of bags of rice and flour were found expired and deteriorated in the warehouses of some affected towns several years after the 2008 Sichuan earthquake due to the lack of regular monitory and management. It is hard to have these grassroots organizations efficiently coordinated with other organizations to provide support for the first time.

2. The time efficiency of disaster relief operations is extremely important, but the operations capability is usually very limited, particularly in those sudden-onset disasters such as earthquakes, forest fires, or tsunami. The 2008 Sichuan earthquake in China lasted for less than a few minutes but took away around 70,000 lives, hundreds of thousands injured and a million houses crashed. Even it is possible to forecast some disasters such as typhoons and hurricanes, the precise information about the development and impact levels are hard to be obtained. Therefore, it is difficult to prepare the time availability of DRS.

3. The high uncertainties of disaster determine that the information about the timing, location, damage assessment, relief demand, and resource availability is limited, inaccurate, hysteretic, and even confusing. Besides, the demand is ever-changing due to the changing disaster environment, such as secondary disasters and volunteer coordination. Within 48 hours after the Sichuan earthquake in 2008 China, the number of rescue workers and volunteers surging into the earthquake centre town was four times more than the affected residents. Heavy rains and continuous aftershocks largely accelerated the running out of food and water. Also, broken infrastructure, limited professional manpower, poor communication, weak coordination, and unanticipated natural conditions together make the resource allocation extremely difficult.

4. Multiple relief sectors are always involved with complex coordination issues, particularly in the large natural disaster. Disaster relief operations involve thousands of converging supply chains to be coordinated in the rescuing and relief processes. However, it is often observed that these participants are too smart to be coordinated but want to be the coordinator. Many volunteers are keen to work in the sites exposed by major media, while the urgent needs in corners are often overlooked. Inefficiencies, duplications, and overlap in the management of DRS in the sites are frequently caused by poor information communication, unprofessional operations, or even language understanding. Furthermore, the scarcity of resources also poses inevitable allocation guideline issues about racial, class, ethical, and moral implications [3]. Furthermore, it is often confusing in rules, regulations, and other legal issues about social welfare in a life and death situation [4]. Poor coordination reduces the delivery time efficiency, pushes up the inventory and transportation cost, and results in huge economic and environmental waste. The phenomena are witnessed in almost all disasters like the 2008 Sichuan earthquake, the 2010 Haiti earthquake, and the 2011 Tohoku earthquake.

5. A disaster, such as an earthquake, flood, or tsunami, occurs once for years, without any regular pattern or warning signal. But once it happens, particularly in the early response periods, a large number of people and money are

needed. Relief groups, such as local community organizations, volunteer groups, or individuals are often short of budget to support their rescuing and relief operations. The regular budget provided by the government is minimal and only support the basic regular management. Disaster preparedness, including material, professional training, and community education, is often of the budget shortage. Large funds and donations are largely obtained after the disaster rather than in the earlier pre-disaster stages. The Brookings Institution investigates the hysteretic issues of donor funding, finding that the loss percentage could reach 7–28% per dollar [5]. The Federal Emergency Management Agency (FEMA) declares that every $1 the Federal Government invests in mitigation saves taxpayers an average of $6 in future spending [6], which shows that pre-disaster investment indeed works.

6. Both natural and social environments in the disaster area are always complicated and changing. Locally prepositioned disaster relief inventories are most likely damaged and the nearby procurements are extremely difficult. A large amount of donated relief supplies moves in the disaster relief area need to be carefully inspected, classified, and coordinated. Some of these supplies are often found not much helpful or even useless. In the US 1988 earthquake rescuing process, more than 1000 tons of medical supplies, donated from different organizations, were destroyed or discarded due to the conformity of related regulations. Therefore, complex and ever-changing DRS policies need to be designed based on a combined effect of the political, economic, demographic, and environmental realities of disaster-prone areas. It is reported that many developing countries have strong bureaucracy, weak economic capability, poor transportation network, and slow information systems. New issues appear without any signal, but require disaster rescuing and relieving organizations to undertake non-regular tasks. During the 2002 African flood crisis, most of the disaster-affected countries refused to accept the food provided by the World Food Programme (WFP) as they found that these food products were genetically modified (GM) and forced the WFP to replace them with the non-GM products [7].

2. DRS resale and refill

2.1 High availability and huge waste

Since the ultimate goal of managing disaster relief inventories is to guarantee the availability of emergency supplies, studies on inventory pre-positioning mostly focus on the problems of material type and quantity and the accessibility for rescue and relief operations. Yet, another critical issue of DRS management is to ensure the quality level of these supplies. These disaster supplies and the related inventory management issues are significantly different from those of commercial business products. First of all, the prepositioning of DRS is mostly initiated, organized, and managed by the government. In China, 25 national-level warehouses for disaster relief purposes have been established in different regions, more than 300 provincial/city-level and about 2500 county-level disaster relief warehouses cover the most population of major cities and towns. The state budget on the essential purchase of disaster relief supplies reaches 1.35 billion RMB in the year 2016. Enormous money has been spent to ensure an available supply source when a disaster occurs.

However, disasters are highly unpredictable so that estimating the potential future needs for rescue and relief activities is very difficult. We can never tell when an earthquake comes, what the affected population is, how much the infrastructure is destroyed, and how long the influence can last. To assume social responsibility, government agencies must keep high inventory level as rescue insurance against destructive catastrophes. On the other hand, a region is unlikely to be frequently struck by catastrophic event within a year. Even on the earthquake belt area, earthquakes occur at different places. It is then often observed that DRS are stored in idle for a long time. In practice, many warehouses are suffering great waste because of the expiry dates of some essential items, such as food, water, and medicine. Due to a lack of regular monitory and proper management, a considerable amount is stored out of date and discarded afterward. After the Sichuan Earthquake in 2008, many warehouses for the anti-disaster purpose have been set up in neighboring areas. In the following years, there have been successively reported that emergency supplies like expired food, water, and sanitation articles, have run to enormous waste. Furthermore, most local warehouses are not well operated, even without professional regular monitory. It results in a surprising number of financial losses every year.

In addition, when emergency supplies are needed, what is worse is that the quality of some short-life items can be at risk if without any turnover of the inventories. When an earthquake strikes, people are usually stuck in extreme situations of necessity shortages. Food, water, medicine, tents, and communication equipment are in desperate need. Emergency supplies need to be replenished with fresh items regularly to guarantee their quality availability in case of providing stale goods to affected people. Nevertheless, unqualified delivery of unfresh or even expired relief goods has been witnessed in the response operations to almost, if not all, disasters and crises. Because these consumable materials have a relatively short lifetime but are stored in a large volume, once out of date, they must be destroyed or discarded and then cause huge waste.

2.2 Resale and refill strategy

It is then clear that the DRS management should be considered from a business perspective, to work with upstream producers, suppliers and logistics service providers on product manufacturing, procurement planning, inventory policy implementing and logistics service operations. We need to define and examine the responsiveness of product value to changes in time to find out the turning point of the quality change. From the operations management point of view, a value deterioration function of an emergency supply item can be constructed to determine the optimal timing to remove old inventories for resale and replenish with new items, provided the trade-off among the logistics cost, the holding cost and the value depreciation.

Removed packaged food and bottled drinks, especially those dried items stored for disaster relief, are in good condition and might be welcomed to a secondary market. Although the freshness degrades over time, the food still has acceptable quality, clothes and essential rescue tools are almost at the same quality levels as new. From a business perspective, reselling the old inventories with markdown prices to a legitimate secondary market not only guarantees accurate access to available emergency supplies but also benefits the economy and environment by reducing tremendous waste. Therefore, managing the DRS is far more than to maintain the right type, quantity, and time availability of them. To design a detailed plan for monitoring the supplies in the inventory, understanding the secondary market, selling them at the right time points, and refilling the inventory has great potential.

The key point of the resale and refill strategy is to properly estimate the time point of turnovers. Long-time storage causes the quality level and commercial value of these items decreasing while too frequent replenishment can directly push up the logistics cost. Therefore, we need to identify the best timing to refresh old items. It depends on the deterioration of product quality and market value over time. Different than the fresh product, users mainly estimate the freshness of packaged food from expiry date information because the value deterioration is difficult to see in appearance. Thus, how product value perishes with age is another question to answer.

To start the resale and refill strategy, we first need to investigate the perishability in the quality of emergency supplies over time. Generally, the value of package food is depreciated very little at the initial period but suffers a drastic reduction when approaching the end of the lifetime. To recognize the turning point of the deterioration in product value, we consider the time responsiveness of value to show the responsive sensibility of the value deterioration to changes in time. Then, we can define the benefit function of DRS and conduct numerical analysis based on the historical data to manage the optimal replenishment timing to remove old items and replenish them with fresh ones. Our research reveals that there always exists a maximum benefit point throughout the product's lifetime for each item type.

Furthermore, to manage DRS in the business process, we need to work closely with product suppliers and logistics service providers. It is essentially a disaster relief supply chain management problem that aims at reducing the operational cost and enhancing the social and humanity effectiveness and efficiency. A sufficient cooperation and coordination scheme promises an overall successful implementation for the resale and refill strategy even it is not profit-oriented. It gives room for product manufacturers in material purchasing, production planning, and quality control, for disaster relief material merchandiser in product searching, purchasing and storage planning, for transportation service providers in earlier programming, routing design and labour arrangement in each refresh cycle.

Lastly, perhaps the most critically, the business perspective based operations can significantly improve the DRS management level. The government initiated and leaded DRS inventory, particularly in those local and first-line warehouses, does not have much incentive of economic revenue, and even neglects the management cost. A large number of local warehouses are assigned to local government agencies as a small part of their daily duty and are thus managed not at a professional standard. It inevitably causes undermined quality control problem and accelerates the product deterioration. With a proper revenue-sharing scheme, the local government is encouraged to assign officers specifically to be in charge of the DRS management. The job duties are clearly defined so that economic key performance indexes would be designed. They are responsible for secondary market research, various marketing operations, communication with product suppliers, and logistics service providers. This resale and refill initiative has been tested and implemented in a town in China for more than three years on a small scale. The practices are currently under research from both regulation and sociological aspects, to study the related issues and evaluate the effects. But the initial implementation experience shows that the quality of the operations is largely enhanced, the management cost is significantly reduced, and grassroots government agencies generally perform positive to tasks.

3. DRS reverse logistics

This section first introduces the scope and purpose of DRS reverse logistics. Then, we briefly describe the concept of conventional reverse logistics and identify the operational difficulties. Afterwards, we discuss the operational aspects of DRS reverse logistics for practical implementation.

3.1 Scope of DRS reverse logistics

The outbreak of coronavirus disease (COVID-19) from the beginning of the year 2020 has once again emphasized the significant importance of relief supply chain management as the critical items used for personal protection, diagnosis, and clinical care are facing a severe global shortage. On March 3, 2020, the World Health Organization (WHO) has issued a call for industry and governments to increase manufacturing by 40% to meet rising global demand in response to the shortage of personal protective equipment endangering health workers worldwide. Such predicament that produces additional human sufferings forces the relief supply chains to expand their capabilities in preparedness and response. While satisfying the increasing beneficiary needs and wants that generated by highly influential but low frequent disasters and crises, humanitarian decision-makers should think the reverse flow of diverse relief materials in "green" terms to quality-efficiently and cost-effectively manage the relief inventories at least cost to the environment.

There are four classes of DRS items and wastes including the following. (1) Most of the fundamental living necessities, including food, drink, simple tool, and basic medicine stored for disaster rescue and relief purposes are consumable. Take the massively reserved food commodities as an example. The "food waste hierarchy" has been widely discussed in relevant food-management research [8, 9], which indicates that the reuse option for surplus food. (2) The post-disaster relief materials, including medical support, disinfection purpose equipment, and medicine, tent, and clothes, have a longer lifetime but the quality and market quality also deteriorate. (3) Capital items and equipment, including large pieces of machinery used for life-savings, high-value medical equipment, advanced communication equipment, and large electrical generators, are usually expensive in investment but risky in technological obsolescence with relatively low-frequent demand. (4) All types of waste generated in the rescue and relief operations bring harm to the environment as well as increases the overall operational costs, which should be controlled at a minimum level. **Table 1** summarizes the types of DRS reverse logistics.

3.2 Purpose of DRS reverse logistics

DRS reverse logistics is about the recovery management of all sorts of relief supplies that are sent to the reverse flows, together with the green marketing strategies to ensure the efficient and cost-effective reuse and recycling of materials. Successful recovery management of relief materials can largely reduce the waste in need of proper disposal.

Hiding behind the humanitarian goals with highest priorities, the reverse logistics of relief supplies that fall short somewhere along the relief supply chain, or require removals from the points of consumption is not fully highlighted by the academic research or practical applications. However, the public has seen successive

Type	Example
Living necessities	Food, drink
Post-disaster relief materials	Shelters, clothes
Capital items and equipment	Life-saving machinery, medical equipment
All types of waste	Debris, packaging

Table 1.
Types of DRS reverse logistics.

reports about expired living necessities and rusty tools stored in emergency warehouses, spoiled food and out of date sanitation articles sent to the affected population, useless donations rushing into the disaster areas, etc. In the face of quality concerns, cost implications, and social responsibilities, managing reverse logistics of relief supplies should not be far from the decision-making core of humanitarian operations. While many operational issues, such as the pre-positioning, the procurement and the location-allocation of relief supplies, have attracted extensive attention in both academia and practice, the changing scope and significance of reverse logistics have not received much academic attention.

This is especially important in the humanitarian context. As "social" inventories, relief supplies should be handled in a responsible manner for economic, social, and environmental sustainability, not only due to limited budget and intensive public attention but rather on the macro-level protection of the degraded environment which is often one of the causes for humanitarian assistance [10].

3.3 Conventional reverse logistics

Reverse logistics or reverse supply chain has gained more and more attention in society through the whole product life due to the seriously declining natural environment. The essential role of the conventional reverse logistics is to recover the value of the products dropped out at different stages of the forward supply chain, including the production wastes in the manufacturing process, logistics damage in the transportation and inventory process, sample wear or unsold in the retailing stages, and most likely the end-of-life products after the consuming, so that physically discarded materials are reduced to the minimum level.

The operations in reverse logistics cover reuse, remarketing, repair, refurbishing, remanufacturing, cannibalization, recycling, and disposal. There are several difficulties in reverse logistics operations. The first one is the high operational cost in the inspection and classification of the returned products with different conditions, collection, and transportation of items with scattered locations, small amounts and uncertain timing, and remanufacturing and resale operations. In particular, the research reveals that moving a product back in reverse transportation may cost 5 to 9 times more than placing it in the original direction. To identify reusable product needs to inspect, classify, collect, repack, and allocate to the reuse locations where the related operational cost is much higher than to make a new product. Furthermore, special packaging material and process may need products collected but without preliminary package for reuse, remarketing, or even discarding, which causes more than double the cost of the forward product supply chain.

The second point is the economies of scale that are the most critical in all stages of any product supply chain, particularly logistics operations. A product or its component may get out of the forward supply chain due to different reasons in a random pattern. Therefore, all operations activities may be more or less unique. When a product can be at its end of use stage at any time, any place and with any quality conditions, which raises the issues of small collection and further inspection. To achieve the scale of operations, Walmart retailing network designed centralized returning centers and saved a considerable cost on product returns. In fact, the shortage of economies of scale can be seen in all aspects of the product reverse logistics, from recollection to inventory, from transportation to remanufacturing, and from recycling to remarketing.

The third point lies in the insufficient attention given by business management due to many business strategic reasons. These remanufactured products are of reduced quality and thus have a lower market value, but maybe even competitive to the original products. The market demand for these remanufactured products is hard to be identified and confirmed even with the low price. In general, these remanufactured products, from the remanufacturing process to marketing promotion, are always given the lowest priority in the company strategical plan, with limited financial resource allocation and other related supports. The management information system for a product supply chain designed for its regular business seldom serves the part of the reverse logistics. Even a professional third-party logistics service provider would allocate its capacity to its forward logistics operations since the service revenue for reverse logistics is lower than normal.

3.4 DRS reverse logistics

To save the natural environment, the concept and operations of the reverse logistics should be considered and implemented in the disaster relief materials management. The DRS reverse logistics operations deals with a diversity of materials and waste from food surplus to medical surplus, consumable goods to durable goods, end-of-use product to end-of-life product, and packaging waste to debris waste, that exposes itself to the management challenges in the process of reusing (including the maintaining services), repairing, refurbishing, recycling and disposal. Some of the activities must be carried out under difficult situations simultaneously with disaster relief operations (e.g., removing unsolicited donations or debris waste), some of the decisions should take serious consideration of the social and environmental impacts in advance (e.g., resale options for the products with market values). Most importantly, given budget constraints and cost burdens, it lacks incentives, or even impossible for the DRS management and suppliers to operate a wide range of reverse systems simply with moral requirements or enforced regulations. Innovative and cost-effective business models for recapturing value or benefits from closing material loops are much more attractive for both academic research and practical applications [11, 12].

To bring the Right relief supplies with the Right quality and Right quantity at the Right time to the Right sites to serve the Right beneficiaries, called 6R rule in disaster rescue and relief operations [13], is well discussed and implemented in practice. However, a rich body of knowledge documented in commercial inventory management research provides limited insights in disastrous conditions. Comparing with the conventional practice, the DRS reverse logistics has at least several advantages in product value, operational cost, economies of scale in operations management, and management strategical consideration.

The products or parts handled in the conventional reverse logistics are of low value and not directly useful. But, most of the goods stored for disaster rescue and relief purposes are consumable [10]. Take the massively reserved food commodities as an example. The "food waste hierarchy" has been widely discussed in relevant food-management research, which indicates that the reuse option for surplus food, which refers to the removal of food from its intended supply chain, is highly recommended if the removed food is still fit for human consumption. Since the prevention of product depreciation is of top priority, food inventories should be updated and replenished with fresh items periodically before going expired. When removed, DRS managers and their suppliers could have a set of marketing choices. Transferring to legitimate secondary markets with markdown prices or selling with promotions and discounts in the supplier's primary market are both cost-effective options, dependent on the residual value of the food. In addition, donating them to other ongoing humanitarian programs or local food aid services for the poor population could be taken into consideration.

Capital items and rescue and relief equipment are normally kept in a higher level of organization centers, which also provide regular maintenance and normal repairing after operations. Quality guaranteed recovery management of these relief items should be incorporated into the overall planning because it may have a profound impact on the next disaster response operations. There is a chance to develop "sharing economy" markets between the suppliers and the equipment management organizations to seek the pooling of costs and risks, which lies within effective communication of information, proper design of cooperation schemes, and corresponding regulations or insurance to avoid moral hazard, bilaterally. Currently, there is not sufficient attention in either academic or business aspects addressing the recovery management issues of capital items and equipment in the humanitarian context.

The DRS items are in relatively concentrated storage and have about the same conditions. This feature gives the economies of scale in all operations of reverse logistics. The inspection process conducted before the removal can be implemented in the regular standards based on the statistical sampling instead of individual checking done in the conventional reverse logistics. These items can be assigned for reuse or remarketing simply at a lower price. The transportation and logistics in moving these items can be properly pre-arranged since the moving timing, purpose, and destination are all planed previously. The logistics cost is low due to the economies of scale. The operations can achieve high time efficiency for loading, transportation, unloading, refurbishing, remanufacturing, disposal, and others in the batch. This is the largest distinction comparing to the operations in reverse logistics in business supply chain management.

DRS are prepared, organized, and managed mainly by the government or semi-government agencies. This gives much flexibility and large room for DRS reverse logistics operations management. The fundamental purpose of DRS reverse logistics is to recover material value such that the operational cost can be saved and to minimize the waste so that the natural environment can be protected. Unlike the reverse logistics activities led by commercial organizations, the operations strategy of DRS reverse logistics is not restricted by other strategical considerations, such as profit maximization or market cannibalization.

4. Government, NGO, and private sector cooperation

Any product supply chain has a core player who coordinates or lead the whole process such that the business benefit and operations efficiency is maximized. Management responsibility of the DRS reverse logistics is inevitably carried by the government agencies who are currently in charge of the DRS management, from strategy development to operations implementation, from information system establishment to coordination with other organizations. It would significantly extend the lifetime of these items and reveal the utilization level of all supplies to achieve a much better social value of DRS.

4.1 Coordination and leadership

A business profit-oriented supply chain may be leaded or coordinated by a core player with a certain strong power, exemplified by a product leading company who own the key technology such as Boeing, or a retailing company who operates the huge retailing network with a comprehensive product range such as Walmart, or a trading company who has a complete network connecting retailers and manufacturers such as Li & Fung. A non-profit-oriented DRS management can only be initiated and operated by a government agency that has the legal authority and social reputation as its core power.

The DRS reverse logistics starts from the beginning of these material design and manufacturing. The most distinctive feature of DRS is "available just in case". In other words, most of the consumable goods and even equipment must be well prepared but may not be used and would eventually move to the reverse logistics procedures. Therefore, it is critical to coordinate with the item suppliers in the product design stages. A consumable product for disaster relief purposes must be designed easily for inventory holding management and possibly for the secondary market. For example, a tent manufacturer slightly downgrades its product for disaster relief purpose, and buy them back with a predetermined lower price at the planned time for the secondary market. These tents are stored in the warehouses with perfect conditions, thus have almost the same market value as new since most of these tents are not for frequent use or even in single-use on the tourist market. Given the specified usage and characteristics of DRS, it is necessary to coordinate the multiple business partners in the whole product supply chain, from product manufacturing to transportation, not only to ensure the quality and functions needed but also to leave the possibility for reverse logistics.

During the rescue and relief process after the disaster event occurs, a large amount of relief items rushes into the sites from everywhere. In addition to the leading role of the DRS management in life-saving and reconstruction, government agency also needs to pay attention to material collection, screening, inspection, allocation, proper use, waste control, and surplus items reverse. Different organizations, individual volunteers, international rescue teams and military groups are all converging for life rescuing, but the precise material delivery, proper use, and waste generation issues are often ignored, even by the rescue authority. In the 2008 Sichuan earthquake, a huge amount of different materials was discarded, about one million tents were found useless and eventually reclaimed by a central government level authority.

Rescue and relief information systems play a vital role in DRS reverse logistics operations coordination. DRS are delivered to the disaster site in different rescue and relief stages. Alone with the time moving forward, the demand and material information should be more and more clear, but yet properly and timely shared by different organizations. Besides, a large amount of false information and rumors are spreading in different ways. A legal and authorized government information platform is needed to provide the real information of demand, supply, management, and reverse logistics.

4.2 Purchasing for donation

To encourage the reuse of DRS items, we propose a "Purchasing-as-Donation" marketing strategy for selling relief material surplus in a cause-related marketplace. These organizations or individual consumers, retailers, traders, original suppliers, and donors. For every relief surplus item sold out, the income is calculated as a cash donation to the DRS management. In addition to the attractive price, the market is also promoted by the environment-friendly concept and social responsibility. The original suppliers who buy back their products and send them to other markets can also be rewarded by further purchasing for further DRS inventory. Traders buying and reselling these DRS products are offered further business opportunities in the future.

This strategy takes advantage of other reusing choices in many ways. Putting aside the economic value that purchasing-as-donation offers, it enables the DRS management to get access to the people in need with alternative sources, not limited to the population supported by aid programs. By using monetary transfers, the proportion of inappropriate deliveries would be controlled at a minimum level.

The purchasing-as-donation model also has positive impacts on inventory operations management. As the poor inventory management is always to be blamed for the enormous waste of relief supplies, DRS managers are forced to take good care of their relief inventories before transferring them to the cause-related consumers. On the other hand, selling relief surplus products by using online platforms and information technology almost eliminates the searching cost for potential recipients, and more importantly, discloses the inventory and donation information to the public in case of corruption.

To support the purchasing-as-donation strategy, it is important to set up related policies and regulations. These policies and regulations involve the various rights of personal or organization donations, rules for international rescue teams, reallocation of donated items or cash, and other aspects. Another important support task is to coordinate other markets, including the international market and related recycling companies to fully recapture the value of the surplus materials.

4.3 E-commerce platform

The diffusion of ownership has complicated the disaster relief supply chains. There could involve a variety of relief material providers, including governments at all levels, local and international social organizations, and private sectors. The Internet revolution has provided new platforms and innovative technologies that facilitate their communication needs in acquiring, delivering, and recycling relief supplies. Equipped with well-established information and delivery systems, the fast-growing e-commerce platforms are enjoying evolving engagement in disaster relief supply chains in the following two aspects.

1. **Using their information systems to acquire relief supplies (forward-flow) and sell relief surplus products (reverse-flow).**

Developing sources for the acquisition of relief items to satisfy the surging demand of beneficiaries during disaster relief operations is a key task of disaster planning [14]. Meanwhile, once the relief supplies are removed and replenished with fresh items, it is important to find a cost-effective and environmentally friendly way to ensure the appropriate reuse of relief surplus products. The e-commerce platforms, such as Amazon.com, JD.com, and Alibaba.com, cannot only act as the suppliers of a wide range of relief materials but also provide marketplace services to assist the reverse flow of relief surplus products.

During the recent outbreak of coronavirus disease (COVID-19), one of the leading online retailing E-platform in China, JD.com, has launched an information platform for emergency resource sharing, which serves over 15,000 customers around the nation, including thousands of government agencies and health care institutions. Based on the information platform, the company has provided, on aggregate, over 660 million pieces of relief items within the first month of the pandemic. On the other hand, selling relief surplus products by using online platforms and information technology enables the humanitarian organizations to get access to the people in need with alternative sources, whereby the proportion of inappropriate deliveries would be controlled at a minimum level. Because poor inventory management is always to blame for the great waste of relief supplies, selling relief surplus products via transparent online platforms also forces the humanitarian organizations to take good care of their relief inventories before transferring them to the re-users.

2. **Using their delivery system to distribute the products.**

Disaster relief is about 80% of logistics it would follow [7]. As relief supplies are very likely to be distributed in challenging situations while in presence of compromised infrastructure, destroyed or blocked roads and limited manpower and resources, a widely-covered supply chain network (including the supplier partnerships, warehousing, and transportation systems and personnel resources) with well-established logistics mandates and information system will save a lot of time in delivery. Many e-commerce platforms have their own self-supporting logistics service system like Amazon.com and JD.com [15], laying the groundwork for a highly efficient distribution network. To combat the pandemic COVID-19, the Hubei Provincial Government cooperated with the JD logistics to manage the disaster relief supply chains. By utilizing their supplier resources and logistics system, JD logistics has distributed over 2000 tons of relief supplies to Wuhan and its neighboring areas in the first twenty days since Wuhan's lockdown. Without the support of JD logistics, it would be extremely difficult for the local hospitals and community to acquire the whole and complete medical supplies and living necessities that were needed, and deliver them to the demand point in such a timely and cost-effective manner.

Using the e-commerce platform's delivery system is beneficial for the integration of humanitarian logistics activities. While there may involve thousands of converging supply chains in disaster relief operations, lack of coordination is the chief cause of material, manpower, and information convergence. In contrast, given that the e-commerce platforms provide systematic logistics services through bridging the real-time demand with their partner suppliers' capacity, sharing their logistics services can reduce the overall logistics cost and improve the efficiency of rescue and relief operations.

4.4 Retailer network

Demand is highly uncertain in the humanitarian context. Information concerning the disaster timing and location, available resources, functional infrastructure, type and size of beneficiaries' needs, is always limited, rough, inaccurate, and hysteretic. Prepositioned inventories held or managed by the government or humanitarian organizations may sometimes not satisfy the beneficiaries' demand in type and quantity, or cannot be transported to the disaster sites in time. In case of such failures, the local government and humanitarian organizations with the official background (e.g., the Red Cross) should make the full use of social resources. Requisitioning local retailers' reserves of needed commodities or signing agreements with local convenience or pharmaceutical chain stores to stockpile relief inventories are both honored to ensure the efficiency of disaster relief supply chains. In 2015, the Chinese Ministry of Civil Affairs has published the *Guidance on Strengthening the Construction of Relief Material Reserve System for Natural Disasters* wherein the social reserves are recommended as necessary supplements to the governmental reserves. Most local governments in the high-risk areas of natural disasters have established partnerships of stockpiling relief supplies with private sectors. Supported by its vast retailing network in Sichuan Province, for instance, the distribution centers of Chengdu Hongqi Chain Store were on standby 24/7 to guarantee the supply and allocation of emergency materials during the yearly flood season.

Other than supporting the last-mile delivery of relief supplies, the retailer network is also distributed as the outlet for the public to acquire relief and living commodities, particularly under the quarantine and distancing policy in prevention for the pandemic.

4.5 Community education

Education is the basis of reducing the impact of disasters on human security and sustainable development [16]. In the face of intensifying threats of disasters and crisis, education is no longer a schooling thing, but the whole community should be engaged in continuous learning and practicing in response to disasters. "Community-Based Disaster Risk Management" needs to establish with the linkages of disaster education to the community and household [17].

The strategies towards disaster education and risk reduction differ with the vulnerability of the community [17]. Strengthening the community capacity in terms of the individuals, families and communities is essential in the process of disaster education based on the local context [18], which concentrates on understanding the traditional knowledge, resources and practices of the local communities, and the corresponding skills, support, and structures for disaster preparedness and mitigation [16]. For instance, the social networks of the urban and rural communities are quite different. People live in the pretty intense urban areas usually have poorer communication and interactions with the community, whereas the rural communities have scattered population to connect in the process of disaster education.

Japan is widely acknowledged that has world-class disaster education in school. However, the long commuting distances have changed today's Japanese life style that few individuals and families can afford the time spent on establishing and maintaining relationships with the communities. The disaster education focused on the Japanese communities needs to emphasize the connection between self-help and collaboration with the community and pubic [17]. Community-based approaches to improve the resilience of the community and the people in response to disasters and crisis are also encouraged in the United States. The Los Angeles County Community Disaster Resilience (LACCDR) Project is structured on four levers: education, engagement, self-sufficiency, and partnership, in which disaster education lays the foundation of the other three levers by building the toolkit for the dissemination of information about preparedness, risks and resources [19].

In some other developing countries like China and India, disaster education at the community level must be addressed with other issues, such as the enhancement of school-based disaster education, the awareness of disaster risks and coping capability of less-educated population, the effective, sustainable and local-based cooperation of community issues and disaster education [17, 20].

Community-based disaster education through the understanding of the social networks in the local context is essential in building the resilience of disaster-prone areas. It should profoundly involve other stakeholders existing within the society, such as governmental agencies, educational institutions, social organizations, community leaders and groups, and policy-making bodies, to push the construction and development of disaster education systems at the community level in a large cooperative effort.

5. Conclusion

Disaster relief supplies, including consumable products and heavy equipment, play a vital role in natural disaster rescue and relief. To prepare sufficient relief supplies in terms of the availability in quality and quantity is an inevitable duty for all responsible government. DRS management should be initiated and supported by the government, yet the related cost issues are usually neglected. Due to the highly uncertain locations and timing of disaster events, DRS management is under-professional. Therefore, these supplies are usually well prepositioned but without

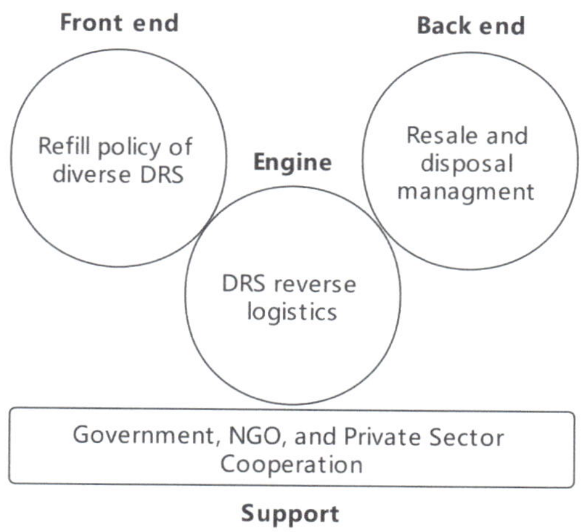

Figure 1.
A "green" view on DRS management.

proper consumption, which may cause a huge waste in practice both economically and environmentally. **Figure 1** presents our green view on the management of DRS.

To bring the reverse logistics strategies in conventional business practice into the disaster relief supply management not only benefits in cost reduction and environmental protection, but also improve the daily management of the DRS. The DRS reverse logistics, comparing with the conventional business reverse logistics, has many implementation advantages. The most critical point is the social trust and efficient marketing network provided by government coordination and international cooperation. Secondly, the stable quality level of these supplies and relatively integrated inventories bring the economies of scale in operations. After all, DRS reverse logistics would be expected to greatly energize the disaster rescue and relief activities.

Author details

Yusen Ye[1] and Hong Yan[2*]

1 Institute for Disaster Management and Reconstruction, Sichuan University, Chengdu, Sichuan, China

2 CPCE, The Hong Kong Polytechnic University, Kowloon, Hong Kong SAR, China

*Address all correspondence to: hong.yan@connect.polyu.hk

References

[1] Fisher, D., Hagon, K., Lattimer, C., O'Callaghan, S., Swithern, S., Walmsley, L. (2018). World Disasters Report 2018: Leaving No One Behind. Fisher, D., Swithern, S., Walmsley, L., Eds.

[2] Ye, Y., Jiao, W., Yan, H. (2020). Managing relief inventories responding to natural disasters: Gaps between practice and literature. Production and Operations Management, 29(4), 807-832.

[3] Luis, E., Dolinskaya, I. S., Smilowitz, K. R. (2012). Disaster relief routing: Integrating research and practice. Socio-economic planning sciences, 46(1), 88-97.

[4] Holguín-Veras, J., Pérez, N., Jaller, M., Van Wassenhove, L. N., Aros-Vera, F. (2013). On the appropriate objective function for post-disaster humanitarian logistics models. Journal of Operations Management, 31(5), 262-280.

[5] Lane, C., Glassman, A. (2008). Smooth and predictable aid for health: a role for innovative financing?. Brookings Global Health Financing Initiative Working Paper Paper, (1).

[6] FEMA 2018. FEMA Strategic Plan, 2018-2022. Published by Federal Emergency Management Agency.

[7] Van Wassenhove, L. N. (2006). Humanitarian aid logistics: supply chain management in high gear. Journal of the Operational Research Society, 57(5), 475-489.

[8] Papargyropoulou, E., Lozano, R., Steinberger, J. K., Wright, N., bin Ujang, Z. (2014). The food waste hierarchy as a framework for the management of food surplus and food waste. Journal of cleaner production, 76, 106-115.

[9] Garrone, P., Melacini, M., Perego, A., Sert, S. (2016). Reducing food waste in food manufacturing companies. Journal of Cleaner Production, 137, 1076-1085.

[10] Cluster, L. (2009). Logistics operational guide.

[11] Agrawal, V. V., Atasu, A., Van Wassenhove, L. N. (2019). OM Forum—New opportunities for operations management research in sustainability. Manufacturing & Service Operations Management, 21(1), 1-12.

[12] Van Wassenhove, L. N. (2019). Sustainable innovation: Pushing the boundaries of traditional operations management. Production and Operations Management, 28(12), 2930-2945.

[13] Beamon, B. M., Balcik, B. (2008). Performance measurement in humanitarian relief chains. *International Journal of Public Sector Management*.

[14] Whybark, D. C. (2007). Issues in managing disaster relief inventories. International journal of production economics, 108(1-2), 228-235.

[15] Yu, Y., Wang, X., Zhong, R. Y., Huang, G. Q. (2017). E-commerce logistics in supply chain management. Industrial Management & Data Systems.

[16] Shaw, R., Shiwaku, K., Takeuchi, Y. (Eds.). (2011). Disaster education. Emerald Group Publishing.

[17] Takeuchi, Y., Mulyasari, F., Shaw, R. (2011). Chapter 4 Roles of family and community in disaster education. Community, Environment and Disaster Risk Management, 7, 77-94.

[18] Victoria, L. (2009). Community capacity and disaster resilience. Disaster management: global challenges and local solutions. University Press, Hyderabad, India, 338-351.

[19] Eisenman, D., Chandra, A., Fogleman, S., Magana, A., Hendricks, A., Wells, K., Williams, M., Tang, J., Plough, A. (2014). The Los Angeles county community disaster resilience project—A community-level, public health initiative to build community disaster resilience. International journal of environmental research and public health, 11(8), 8475-8490.

[20] Zhu, T. T., Zhang, Y. J. (2017). An investigation of disaster education in elementary and secondary schools: evidence from China. Natural hazards, 89(3), 1009-1029.

Chapter 12

The Risk of Tsunamis in Mexico

Jaime Santos-Reyes

Abstract

The paper reviews the risk of tsunamis in Mexico. It is highlighted that the Pacific coast of the country forms part of the so called "Ring of fire". Overall, the risk of tsunami that has the potentiality to affect communities along the Pacific coast of the country are twofold: a). Local tsunami; i.e., those triggered by earthquakes originating from the "Cocos", "Rivera" and the "North American" plates (high risk); and b) the remote tsunamis, those generated elsewhere (e.g, Alaska, Japan, Chile) (low risk). Further, a preliminary model for a "tsunami early warning" system for the case of Mexico is put forward.

Keywords: tsunami, earthquake, Mexico, tsunami early warning

1. Introduction

A *tsunami* has been defined as "a series of travelling waves of extremely long length and period, usually generated by disturbances associated with earthquake occurring below or near the ocean floor ... Volcanic eruptions, submarine landslides, and coastal rock falls can also generate tsunamis, as can a large meteorite impacting the ocean" [1]. Also, tsunamis may be regarded as low frequency events but with high impacts in terms of human/infrastructure/economic losses. Their power of destruction has been more than evident in recent years [2–11]. It is believed that from the time period between 1998 and 2017, the losses inflicted by tsunami disasters were a total of US$280 billion and 251,770 causalities, in damages [7]. Moreover, the authors argue that the impact from this period has been 100 times higher than during the time period 1978–1997.

Following the 2004 tsunami in the Indian Ocean, there has been a large amount of literature published on several topics associated with tsunami science. For example, research has been conducted on the physics of tsunami waves [12], tsunami's impact and characteristics [1–3, 11, 13], tsunami early warning systems [14, 15], tsunami risk assessment [10, 11, 16], geology's perspective [17–19], to mention a few.

Recent tsunamis have highlighted the need for an effective early warning system. An early warning is defined as "the provision of timely and effective information, through identified institutions, that allows individuals exposed to a hazard to take action to avoid or reduce their risk and prepare for effective response" [20]. Moreover, the United Nations Inter-Agency Secretariat of the International Strategy for Disaster Reduction (UN/ISDR) argues that an "effective early warning system" should include the following four key elements: "the knowledge of risks", "the technical monitoring and warning service", "dissemination & communication of meaningful warnings to those at risk", and "the public awareness and preparedness to react to warnings" [20, 21].

The objective of the paper is to highlight the tsunami risk in Mexico. The data presented in the paper are based on previous studies on tsunamis in the country [15, 22]. Further, a preliminary "tsunami early warning" system which aims at

IntechOpen

integrating, for example, the four key elements proposed by the UNISDR [20] for the case of Mexico is presented.

2. The risk of tsunamis in Mexico

The "Pacific ring of fire" belt covers a vast area of highly active tectonic plate boundaries where most of the earthquakes originate and active volcanoes (**Figure 1**). It is believed that three quarters of all the volcanoes in the world are in the ring [23].

Further, the "Ring of fire" runs through several countries, such as Canada, USA, Russia, Chile, Peru, Guatemala, New Zealand, Japan, Indonesia, Philippines, Mexico.

Regarding the tsunami risk in Mexico, studies based on tsunami historical data showed that there are two zones of tsunami threat: local (i.e., generation of tsunamis) and remote (i.e., arrival of tsunamis) (**Figure 2**) [15, 22]. The authors defined these two zones by considering the nature of the faulting and tectonic plate interaction. In the subsequent subsection each of these will be addressed.

2.1 Local tsunami risk

According to [15, 22] at the west of the "Rivera plate" and along the "Middle America trench," the "Cocos plate" subduction beneath the "North American plate" at rates of 2.5 to 7.7 cm/year (**Figure 2**). Given the fact, that large earthquakes occur in this region; therefore, the zone has been regarded as a generator of tsunamis (**Table 1** and **Figure 3**).

According to historical data, the generated tsunamis that produced the highest wave heights were those that occurred in 1925 (7–11 m), 1932 (9–10 m), 1995 (2.9–5.10 m), 1985 (1–3 m). For example, the 1985 earthquake of M8.0 of magnitude generated a tsunami that affected several communities in this zone. It is believed that a key infrastructure port was affected with waves of 2.5 m and flooded the area about 500 m inland [15]. Also, several tourist resorts were affected by the tsunami; for example, waves for up to 2.5 m high were observed in Playa Azul [15].

Interestingly, a day after the main earthquake, a M7.5 aftershock hit the zone; it is thought the generated tsunami affected a local fishing community with waves ranging from 2 to 3 m high [15].

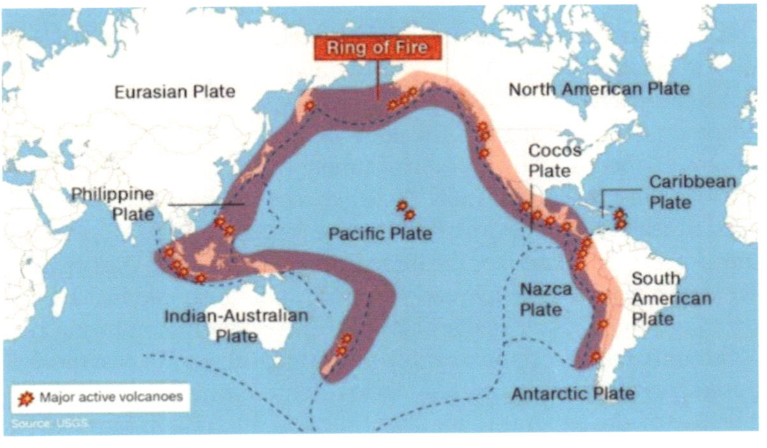

Figure 1.
The "Ring of fire" [23].

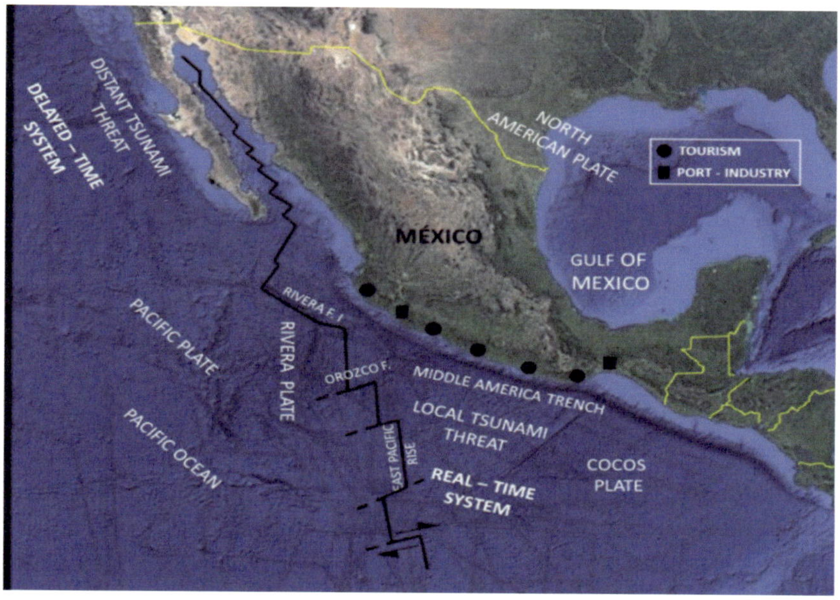

Figure 2.
Mexico's local & remote tsunami threat [15, 22].

Year	Region	Magnitude	Tsunami (places hit, Mexico)	Max. height waves (m)
1732	Guerrero	—	Acapulco	4.0
1754	Guerrero	—	Acapulco	5.0
1787	Guerrero	>8.0	Acapulco	3–8
1787	Oaxaca	—	Juquila Pochutla	4.0 4.0
1820	Guerrero	7.6	Acapulco	4.0
1852	B. C.	—	Río Colorado	3.0
1907	Guerrero	7.6	Acapulco	2.0
1925	Guerrero	7.0	Zihuatanejo	7.0–11.0
1932	Jalisco	8.2	Manzanillo San Pedrito	2.0 3.0
1932	Jalisco	7.8	Manzanillo	1.0
1932	Jalisco	6.9	Cuyutlán	9.0–10.0
1948	Nayarit	6.9	Islas Marias	2.0–5.0
1957	Guerrero	7.8	Acapulco	2.6
1973	Colima	7.6	Manzanillo	1.1
1978	Oaxaca	7.6	Puerto Escondido	1.5
1979	Guerrero		Acapulco	1.3
1985	Michoacán	8.1	Lázaro Cardenas Ixtapa Zihuatanejo Playa Azul Acapulco Manzanillo	2.5 3.0 2.5 1.1 1.0
1985	Michoacan	7.8	Acapulco Zihuatanejo	1.2 2.5

Year	Region	Magnitude	Tsunami (places hit, Mexico)	Max. height waves (m)
1995	Colima	8.1	Boca de Iguanas	5.10
			Barra de Navidad	5.10
			San Mateo	4.90
			Melaque	4.50
			Cuastecomate	4.40
			El Tecuán	3.80
			Punta Careyes	3.50
			Chamela	3.20
			Pérula	3.40
			Punta Chalacatepec	2.90
2003	Colima	7.8	Manzanillo	1.22
2017	Chiapas	8.1	Salina Cruz	1.10

Table 1.
Local tsunamis-only those with wave height > 1.0 m are shown [22].

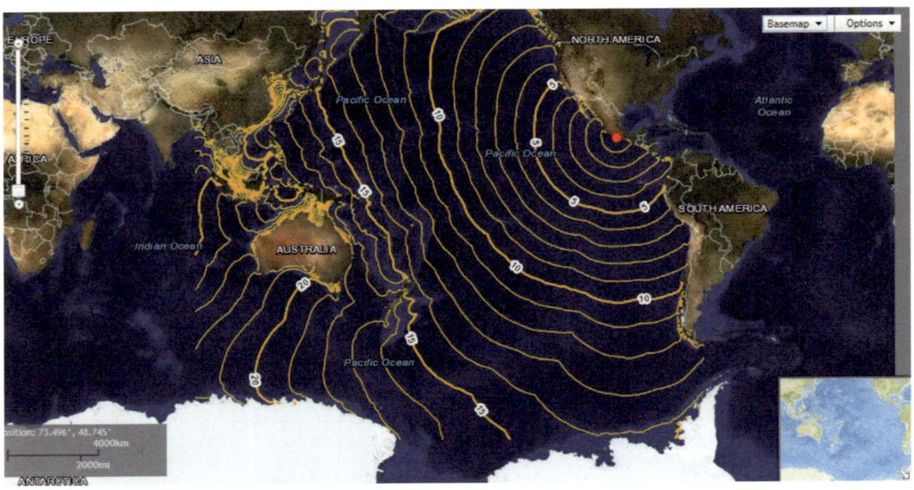

Figure 3.
Local tsunamis in the pacific coast of Mexico [24].

More recently, it has been found that instrumentally based assessments of "tsunamigenic" possibility of subduction zones in the Pacific coast have underestimated the frequency and magnitude of great earthquakes and tsunamis [25]. The authors argue that geological evidence shows that in fact great tsunamis (and earthquakes) have occurred in the subduction zone in the past, i.e., the stretch of the coasts of Guerrero and Oaxaca, the southern region of Mexico.

For example, it has been found evidence of two sand tsunami deposits, 1.5 km inland of the coast [25]. Further, it is believed that an earthquake of M8.6 of magnitude occurred in 1787 and produced a giant tsunami that flooded up to 6 km inland. The second tsunami (less documented) occurred in the year 1537. More importantly, the authors conclude that great tsunamis have occurred in the Pacific coast of the country.

On the other hand, it should be highlighted that another geographical region that is not mentioned in the official reports (e.g., in Ref. [22]) in relation to the potential tsunami source is that related to those originating in the Caribbean Sea (**Figure 4**). It is believed that geological events such as volcanoes and earthquakes

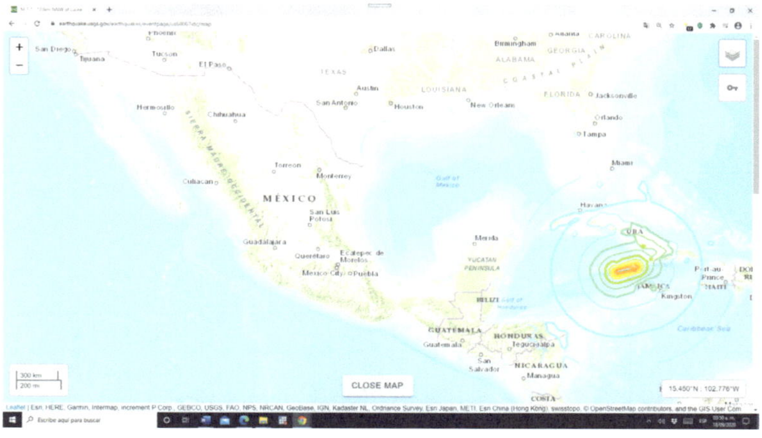

Figure 4.
The 123 km of Lucea, Jamaica earthquake in 2020 [26].

are common and therefore the region is geological active [27]. Further, the authors argue that historical data has shown that there has been the occurrence of "tele-tsunamis," tectonic tsunamis, landslide tsunamis, and volcanic tsunamis in the region [27], p. 60.

That is, there has been twenty-seven "verified tsunamis" and "nine are considered to be very likely true tsunamis" of a total of 97 reported waves that might be tsunamis in the Caribbean region [27]. Moreover, it is believed that one of the deadliest and most recent tsunamis that hit communities in Dominican Republic, Haiti, and Puerto Rico occurred in 1946; the tsunami killed 1790 people [27], p. 84.

More recently, this threat became more apparent with the occurrence of a strong earthquake in the region (**Figure 4**). That is, on 28 January 2020, an earthquake of M7.7 of magnitude (with a depth of 1o km) hit between the Cayman Islands and Jamaica and Cuba [26]. It is believed that the tremors were felt as far away as Miami, US. However, no causalities have been reported.

The earthquake prompted the issuing of tsunami warnings by the Pacific Tsunami Warning Center (PTWC) [28]. The PTWC's warning was: "hazardous tsunami waves are possible; it is thought the warning was for communities living along the coasts located within 300 km from the epicentre; i.e., those include coasts of the following countries: the Caiman Islands, Jamaica, Belize, Cuba, Honduras, and Mexico (i.e., the Caribbean coast of the Yucatan Peninsula, **Figure 4**). The tsunami warning was lifted off after a few hours.

Overall, it may be argued that a potential threat of tsunamis come from the Caribbean Sea, although it may be regarded as extremely low (i.e., there has not been any data of tsunamis hitting the Caribbean coast of the Yucatan Peninsula); however, tsunamis are unpredictable and communities, governments should always be prepared for the unthinkable, because as usual, this is what happens (see Section 3 for further details about this).

2.2 Remote tsunami risk

It is believed that on the Northwest of the "Rivera plate" (**Figure 2**), along the Gulf of California where the Pacific Plate slides north with respect to the North American plate, generation of tsunamis in this zone is unlikely [15, 22]. This is

Date	Region	Magnitude	Tsunami (places hit, Mexico)	Max. height waves (m)
1952	Kamchatka, USSR	8.3	La Paz, BCS	0.5
			Salina Cruz	1.2
1957	Aleutian Islands	8.3	Ensenada, B.C.	1.0
1960	Chile	8.5	Ensenada, B.C.	2.5
			La Paz, B.C.S.	1.5
			Mazatlán	1.1
			Acapulco	1.9
			Salina Cruz	1.6
1960	Peru	6.8	Acapulco	0.10
1963	Kuril, Islands, USSR	8.1	Acapulco	<1.0
			Salina Cruz	
			Mazatlan	
			La Paz, B.C.S.	
1964	Alaska	8.4	Ensenada, B.C.	2.4
			Manzanillo	1.2
			Acapulco	1.1
			Salina Cruz	0.8
1968	Japan	8.0	Ensenada, B.C.	<1.0
			Manzanillo	
			Acapulco	
1975	Hawaii	7.2	Ensenada, B.C.	<1.0
			Manzanillo	
			Puerto Vallarta	
			Acapulco	
1976	Kermadec Islands	7.3	San Lucas, B.C.S.	<1.0
			Puerto Vallarta	
			Manzanillo	
			Acapulco	
1995	Chile	7.8	Cabo San Lucas	<1.0
2004	Indonesia	9.0	Manzanillo	1.22
			Lazaro Cardenas	0.24
			Zihuatanejo	0.60
2010	Chile	8.8–9.0	Manzanillo	0.32
			Cabo San Lucas	0.36
			Acapulco	0.62
2011	Japan	9.0	Ensenada, B.C.	0.70
			Huatulco	0.70
			Puerto Angel	0.29
			Acapulco	0.72
2018	Indonesia	7.5	—	—
2018	Indonesia	AK Vulcano tsunami	—	—

Table 2.
Remote tsunamis- historical data taken from [22] except for the last two tsunamis that occurred in 2018.

consistent with historical data (**Table 2**); it can be seen that data on "small" and "moderate" tsunamis generated by remote sources; for example, the two most recent 2010 Chile and the 2011 tsunamis (**Figure 5**) where the maximum wave heights registered were < 1.0 m. However, it is worth mentioning that the historical data showed that there were two tsunamis that registered the height of waves up to 2.4 and 2.5 m; that is, those generated in Chile (1960) and Alaska (1964), respectively (**Table 2**).

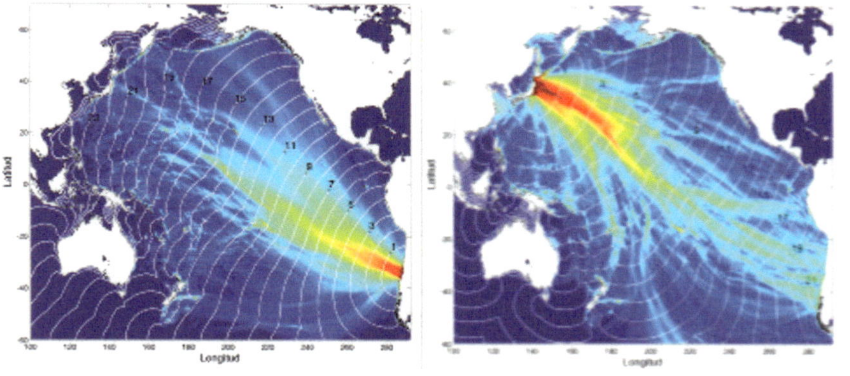

Figure 5.
The 2010 Chile tsunami (left) and the 2011 tsunami in Japan (right) [29].

3. A Mexican tsunami early warning system

As mentioned in previous sections, tsunamis (and earthquakes) are unpredictable and can happen any time. Therefore, there is a need for an effective tsunami early warning system (TEWS). A system which should include not only the technical aspect but also the human issue. This section presents a preliminary model for such a system.

In particular, it considers the Pacific and the Caribbean coasts of Mexico (Section 2). However, only those aspects associated with the "structural-organisation" of the proposed model will be discussed in some detail (i.e., the five interrelated subsystems associated with systems 1–5 and its channels of communication as shown in **Figure 6**). The proposed model is based on previous research on issues related to safety and disaster management systems [30–32].

In the context of this case study, the overall function of systems 2–5 (MTEW-SMU) is to establish the key tsunami safety policies aiming at maintaining tsunami risk within an acceptable range; this implies allocating the necessary resources, for example, to build response capabilities at national and community levels.

System 1, on the other hand, embraces the following three subsystems: TNZO (Tsunami Northern Zone Operations), TSZO (Tsunami Southern Zone Operations), and TCZO (Tsunami Caribbean Zone Operations) with their associated management units (TNZ-SMU, TSZ-SMU & TCZ-SMU). These three operations of system 1 were considered given the fact that the risk of tsunamis comes from local and remote tsunami sources as mentioned in Section 2.

Further, it is important to highlight that one of the key functions within the MTEW-SMU is that related to System 2, which is associated with what it is called here MTEW-CC (Mexican Tsunami Early Warning-Coordination Centre); its key function is the monitoring, detection of a tsunami through the following coordination centres: TSZ-CC (Tsunami Southern Zone-Coordination Centre), TNZ-CC (Tsunami Northern Zone Coordination Centre), and TCZ-CC (Tsunami Caribbean Zone Coordination Centre), as shown in **Figure 6**. The process of the flow of key information and decision making process is briefly described in **Table 3**; **Table 4**, on the other hand, presents some of the key actors involved in the existing system when compared with the features of the model.

In general, communities living in active seismic areas and along coastal regions are vulnerable to tsunamis. These natural hazards are not that common and unpredictable, but powerful and with devastating consequences to those communities in

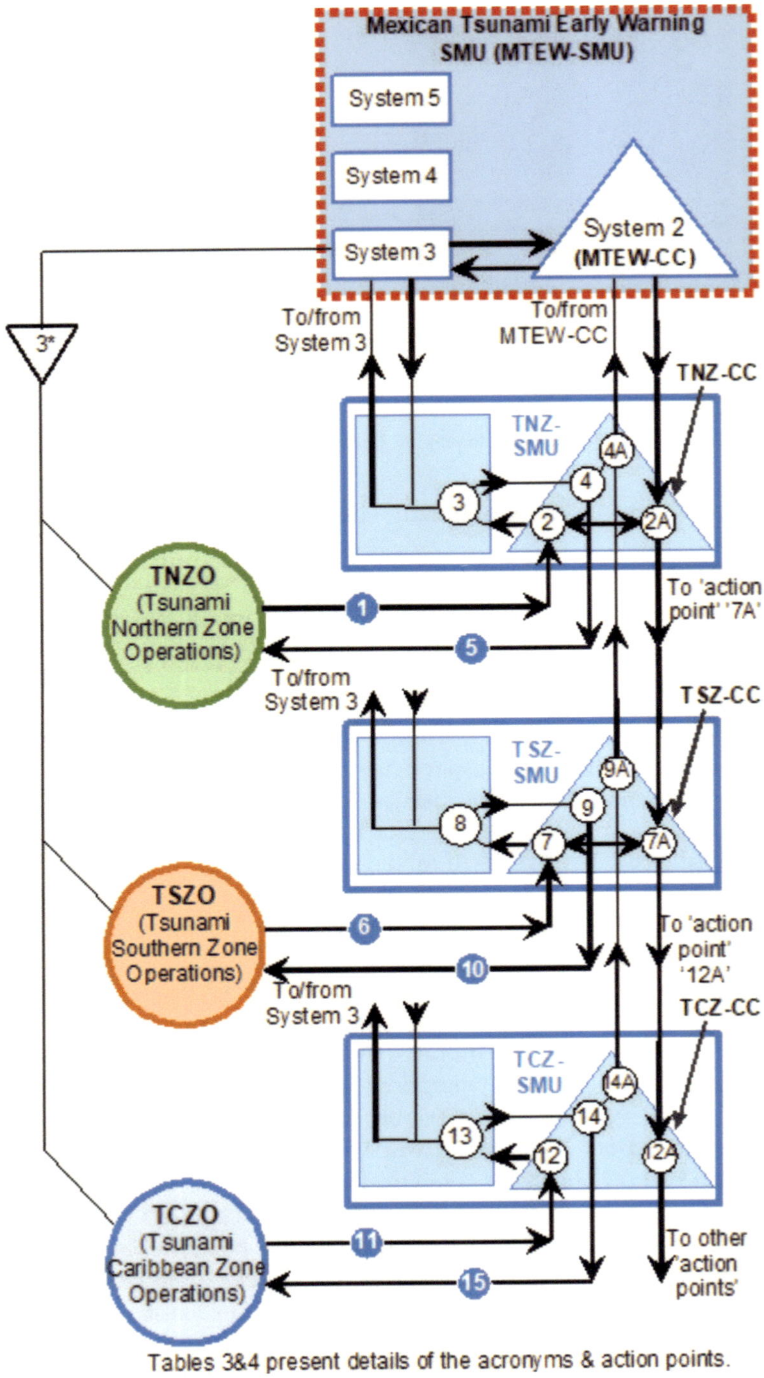

Figure 6.
A Mexican tsunami early warning system (MTEWS).

their path. It is believed that tsunamis are the deadliest in terms of the proportion of people being killed [34].

Following the 2004 tsunami in the Indian Ocean, the need for a tsunami warning system (TWS) was more than evident; however, it may be argued that the existing TWS may be deficient in dealing with the mitigation of impacts of such events; moreover, there are still regions worldwide without such systems.

"Action points"	Description
"1", "6", "11"	Flow of data on key variables monitored by MTEW-CC through TNZ-CC, TSZ-CC, TCZ-CC (e.g., earthquakes, pressure sensors, tide gauges, etc.). It should also be mentioned that this information is provided by the SSN (National Seismological Service), USGS, the PTWC (Pacific Tsunami Warning Centre), see **Table 4**.
"2&2A", "7&7A"	If a strong earthquake occurs, for example, within TNZO (Tsunami Northern Zone Operations), then in "2", the tsunami risk is assessed, if the key variable not withing the acceptable criteria (e.g., a tsunami), then it issues the tsunami warning to "2A", which in turn issues the warning to the TSZ-CC, even if the risk is low (Section 2), through the "action point" "7A". In the model, the TCZ-CC also receives the warning, although in the context of this scenario, it is not necessary to warn communities within TCZO (Tsunami Caribbean Zone Operations) to take some protective actions for obvious reasons. Nevertheless, the key decision-makers within this zone are on alert.
"3&4", "8&9", "13&14"	"Actions points" "3&4" plan and devise measures to respond to the tsunami emergency, e.g., design of risk maps, plans to conduct drills, evacuation plans; etc. All of these aiming at better prepare the vulnerable communities within TNZO. "Action point" "3" also issues the tsunami warning to MTEW-SMU (i.e., to System 3). In the same vein, "action points" "8&9" and "13&14" perform similar functions into their respective coordination centres (i.e., TSZ-CC & TCZ-CC), see **Figure 6** and **Table 4**.
"4A", "9A", "14″	Following the scenario herein, "4A" communicates the protective measures taken (e.g., evacuation) to the MTEW-CC, which in turn may devise further actions given its synergistic view of the total system through system 3, as shown in **Figure 6**. The same rationale applies to "9A" and "14" within their respective coordination centres.
"5", "6", "15″	"Action point" "5" issues the tsunami warning to the affected communities within this zone (e.g. B.C, B.C.S., Sinaloa, Manzanillo, etc.). Further, it implements all the protective measures to mitigate the impact of the tsunami in the coastal areas, e.g., evacuation to safe areas, etc. Moreover, it also implements plans to relocate the affected people to safe areas if necessary. Similarly, as in "5", "6" issues the tsunami warning to the affected communities within this zone (e.g. Acapulco, Oaxaca, Manzanillo, Zihuatanejo, etc.). "Action point" "15", issues the warning to the communities vulnerable to tsunamis within this zone (e.g. Cancun, etc.), see **Figure 6**.

Table 3.
Description of the key action points of the model in ***Figure 6****.*

Recent tsunami disasters have highlighted some of these deficiencies; for example, in the case of the 2010 tsunami in Chile, the entity in charge of issuing a tsunami warning failed to do so [5], p. 30 (see "action point" "2"& "7" in **Figure 6** and **Table 3**). The failure to perform this action contributed to fatalities in the coastal communities. More recently, the 28 September Sulawesi tsunami and the 24 December Anak Krakatau (AK) volcano tsunami, both in Indonesia, illustrate deficiencies in TWS too. In the former case, the tsunami warning was issued but the warning was lifted over thirty minutes [4]. However, the city of Palu, located in a narrow bay, was hit hard with waves reaching six metres of height; why were not they warned? the head of the BMKG (Indonesia Agency for Meteorology, Climatology and Geophysics) argued that "we have no observation data at Palu...", "If we had a tide gauge or proper data in Palu, of course it would have been better" [4]. The tsunami (and earthquake) killed over 2000 people [2]. Finally, regarding the AK volcano tsunami, it is thought that there was not a tsunami warning system for the case of volcano-induced tsunamis; however, the tsunami killed 437 people [3].

It may be argued that a TWS should not be only concerned with the technical infrastructure systems (e.g., tidal gauge, network of buoys, etc.), but also the organisational and human components. Further, it may be argued that the most

Some of the key features of the model		**Examples of what perform some of the functions of the existing system (left & Figure 6)**	
System	Key Components	SMU ("square boxes")	Operations ('circles')
Systems 2–5	MTEW-SMU ("Mexican Tsunami Early Warning-SMU")	1. Secretariat of the Navy (SEMAR) manages the Tsunami Warning Centre (CAT); the monitoring, detection, and forecasting centre [33]. 2. Receives information from the SSN (National Seismological Service), 3. Receives the input from the PTWC. 4. Receives the input from the USGS. 5. Other (e.g. CICESE, etc. [15, 22])	-
	MTEW-CC ("Mexican Tsunami Early Warning Coordination Centre")	Warning coordination centres within the CAT.	-
System 1	TNZ-SMU ("Tsunami Northern Zone- SMU")	Same as with 1,2 & 5 above, and local/ regional decision-makers, e.g., civil protection, etc.	—
	TNZO ("Tsunami Northern Zone Operations")		Local communities living in the zone, including tourists & those working in touristic resorts, such as 'Los Cabos', etc.
	TNZ-CC ("Tsunami Northern Zone- Coordination Centre)	Same as with 1,2 & 5 above, and local/ regional decision-makers, e.g., civil protection, etc.	-
System 1	TSZ-SMU ("Tsunami Southern Zone- SMU")	Same as with 1,2 & 5 above, and local/ regional decision-makers, e.g., civil protection, etc.	-
	TSZO ("Tsunami Southern Zone Operations")		Local communities living in the zone, including tourists & those working in touristic resorts, such as 'Puerto Vallarta', 'Acapulco', 'Huatulco', etc.
	TNZ-CC ("Tsunami Southern Zone- Coordination Centre)	Same as with 1,2 & 5 above, and local/ regional decision-makers, e.g., civil protection, etc.	-

Some of the key features of the model		Examples of what perform some of the functions of the existing system (left & Figure 6)	
System 1	TSZ-SMU ("Tsunami Caribbean Zone- SMU")	Same as with 1,2 & 5 above, and local/ regional decision-makers, e.g., civil protection, etc.	-
	TCZO ("Tsunami Caribbean Zone Operations")		Local communities living in the zone, including tourists & those working in touristic resorts, such as 'Cancun', 'Playa del Carmen'.
	TNZ-CC ("Tsunami Caribbean Zone-Coordination Centre)	Same as with 1,2 & 5 above, and local/ regional decision-makers, e.g., civil protection, etc.	-

Table 4.
*Examples of the key players that perform some of the functions of the system in place when compared with the model (**Figure 6**).*

difficult aspect is the human factor; there is a need to better understand human behaviour during these events, so that make these communities less vulnerable and resilient to tsunamis. In other words, there is a need for an effective tsunami early warning system able to consider all these components in a coherent manner, such as the system being proposed herein and elsewhere. Further, these systems should be "people-centred" [21, 35].

4. Conclusions

The paper has presented the risk of tsunamis in Mexico. The approach has been a review of existing literature on historical data of tsunami occurrence in Mexico. The literature survey showed that the tsunami threat comes from local and remote zones. Overall, the review showed that the highest tsunami risk comes from tsunamis induced by earthquakes occurring in the Southern zone of the country (i.e., local zone). The paper has also put forward a preliminary model of a TEWS (Tsunami Early Warning System) for the case of Mexico. However, it needs further research to design the whole networks of the flows of information not only for the case of tsunamis, but also for the case of earthquake early warning "people-centred" systems.

Acknowledgements

This research was supported by the following grants: SIP-IPN-20201790; CONACYT-No:248219.

Conflict of interest

The author declares that he has no competing interests.

Author details

Jaime Santos-Reyes
Grupo de investigación SARACS, SEPI-ESIME, ZAC., Instituto Politécnico Nacional, Mexico

*Address all correspondence to: jrsantosr@hotmail.com

References

[1] UNESCO/IOC (United Nations Educational, Scientific and Cultural Organization/Intergovernmental Oceanographic Commission)., 2013. Tsunami Glossary. Revised Edition 2013. IOC Technical series, 85. UNESCO, Paris (IOC/2008/TS/85rev).

[2] Widiyanto, W., Santoso, P.B., Hsiao, S.C., Imananta, R.T., 2019. Post-event field survey of 28 Septemeber Sulawesi earthquake and tsunami. Nat. Hazards Earth Syst. Sci., 19, 2781-2794.

[3] Muhari, A., Heidarzadeh, M., Susmoro, H., Nugroho, H.D., Kriswati, E., Supartoyo, Wijanarto, A.B., Imamura, F., Arikawa, T., 2019. The December 2018 Anak Krakatau Volcano Tsunami as inferred from post-tsunami field survey and spectral analysis. Pure Appl. Geophys. 176, 5219-5233.

[4] BBC. 2018. Indonesia earthquake and the tsunami: How warning system failed the victims. Available at: https://www.bbc.com/news/world-asia-45663054 (Accessed date: 24/12/2019).

[5] Valenzuela, I.B., Camus, P.M., 2012. Chile pre y post catástrofe: algunas claves para aproximarse a los desafíos de reconstrucción. In: Brain y Mora (ed). Emergencia y reconstruccion: El antes y después del terremoto y tsunami del 27 F en Chile. Fundacion Mapfre y Pontificia Universidad Católica de Chile. Chile.

[6] Fakhrul-Razi, A., Ridwan-Wong. M.M., Mat-Said, A. 2020. Consequences of the 2004 Indian Ocean Tsunami in Malaysia. Safety Science, 121, 619-631.

[7] Imamura, F., Penmellen-Boret, S., Suppasri, A., Muhari, A. 2019. Recent occurrences of serious tsunami damage and the future challengues of tsunami disaster risk reduction. Progress in Disaster Science, 1, 100009.

[8] Moreno, J., Lara, A., Torres, M. 2019. Community resilience in response to the 2010 tsunami in Chile: The survival of a small-scale fishing community. International Journal of Disaster Risk Reduction, 33, 376-384.

[9] Paula Dunbar, Heather McCullough, George Mungov , Jesse Varner & Kelly Stroker. (2011) 2011 Tohoku earthquake and tsunami data available from the National Oceanic and Atmospheric Administration/National Geophysical Data Center, Geomatics, Natural Hazards and Risk, 2:4, 305-323, DOI: 10.1080/19475705.2011.632443

[10] Leon-Canales, J., Vicuña del Río, M., Gubler, A. 2019. Increasing tsunami risk through intensive urban densification in metropolitan areas: A longitudinal analysis in Viña del Mar, Chile. International Journal of Disaster Risk Reduction, 41, 101312.

[11] Li, Z., Yu, H., Chen, X., Zhang, G., Ma, D. 2019. Tsunami-induced traffic evacuation strategy optimization. Transportation Research Part D, vol.77, 535-559.

[12] Robke, B.R., Vott, A. 2017. The tsunami phenomenon. Progress in Oceanography, 159, 296-322.

[13] Goff, J., Terry, J.P., Chagué-Goff, C., Goto, K. 2014. What is a mega-tsunami? Marine Geology, 358, 12-17.

[14] Chaturvedi, S.K. 2019. A case study of tsunami detection system and ocean wave imaging mechanism using radar. Journal of Ocean Engineering and Science, 4, 203-210.

[15] Ferraras, S.F., Sanchez, A.J., 1991. The tsunami thereat on the Mexican west coast: A historical analysis and recommendations for hazard mitigation. Natural Hazards, 4, 301-316.

[16] Jaimes, M.A., Reinoso, E., Ordaz, M., Huerta, B., Silva, R., Mendoza, E., Rodríguez, J.C. 2016. A new approach to probabilistic earthquake-induced tsunami risk assessment. Ocean & Coastal Management, 119, 68-75.

[17] Trejo-Gómez, E., Ortiz, M., Nuñez-Cornú, F.J. 2015. Source model of the Octobre 9, 1995 Jalisco-Colima Tsunami as constrained by field survey reports, and on the numerical simulation of the tsunami. Geofísica International, 54-2, 149-159.

[18] Roy, P.D., Jonathan, M.P., Consuelo-Macias, M., Sanchez, J.L., Lozano, R., Srinivasalu, S. 2012. Geological characteristics of 2011 Japan tsunami sediments deposited along the coast of southwestern Mexico. Chemie der Erde, 72,91-95.

[19] Ramirez-Herrera, M.T., Bogalo, M.F., Cerny, J., Goguitchaichvili, A., Corona, N., Machain, M.L., Edwards, A.C., Sosa, S. 2016. Historic and ancient tsunamis uncovered on the Jalisco-Colima Pacific coast, the Mexican subduction zone. Geomorphology, 259,90-104.

[20] United Nations International Strategy for Disaster Reduction (UN/ISDR), 2006. Global survey of early warning systems. UNISDR, Geneva, Switzerland.

[21] Bashir, R., 2006. Global early warning systems for natural hazards: systematic and people centred. Philosophical Transactions of the Royal Society A 364, 2167-2182.

[22] Ferraras, S.F., Dominguez-Mora, R., Gutierrez-Martinez, C.A., 2014. Tsunamis. Centro Nacional de Prevencion de Desastres (CENAPRED), Secretaria de Gobernación, Mexico.

[23] USSG. Ring of fire map. Available at: http://www.geologyin.com/2018/01/the-ring-of-fire.html#kMGfZ4P00Ue21LoI.99 (Accessed date: 24/12/2019).

[24] NOAA, 2019. Estimated tsunami travel times to coastal locations-Acapulco Mexico. National Centers for Environmental Information (NOAA). Available at: https://maps.ngdc.noaa.gov/viewers/ttt_coastal_locations/ (Accessed date: 24/12/2019).

[25] Ramirez-Herrera, M.T., Corona, N., Cerny, J., Castillo-Aja, R., Melgar, D., Lagos, M., Goguitchaichvili, A., Machain, M.L., Vasquez-Caamal, M.L., Ortuño, M., Caballero, M., Solano-Hernandez, E.A., Ruiz-Fernandez, A.C, 2020. Sand deposits reveal great earthquakes and tsunamis at Mexican Pacific Coast. Scientific Report, 10, 1145. DOI: 10.1038/s41598-020-68237-2

[26] USGS, 2020. M7.7-123 km NNW of Lucea, Jamaica. Available at: https://earthquake.usgs.gov/earthquakes/eventpage/us60007idc/executive (Accessed date: 20/04/2020).

[27] Lander, J.F., Whiteside, L.S., Lockridge, P.A., 2002. A brief history of tsunamis in the Caribbean Sea. Science of Tsunami Hazards, vol. 20(1), 57-94.

[28] PTWC. 2019. Pacific Tsunami Warning Centre. Available at: https://ptwc.weather.gov/ (Accessed date: 24/12/2019).

[29] Ortiz-Huerta, L.G., Ortiz-Figueroa, M., 2014. ¿Cómo me puedo preparar ante un Tsunami? Guía para el Maestro. SEGOB, CENAPRED, CICESE, Mexico.

[30] Santos-Reyes, J., Padilla-Perez, D., Beard, A.N., 2019. Transport infrastructure interdependency: Metro's failure propagation in the road transport system in Mexico City. Sustainability, vol. 11, 4757, 1-24. DOI:10.3390/su11174757.

[31] Santos-Reyes, J., & Beard, A.N., 2017. An analysis of the

1996 Channel tunnel fire. Journal of Rail and Rapid Transit: Proc. IMechE, Part F. vol. 231(8), 850-876. DOI:10.1177/0954409716647093.

[32] Santos-Reyes, J., & Beard, A.N., 2011. Applying the SDMS model to the analysis of the Tabasco flood disaster in Mexico. Human and Ecological Risk Assessment: An International Journal, vol. 17(3), 646-677. DOI:10.1080/10807039.2011.571099.

[33] CAT, 2020. CAT-Tsunami Warning Centre. Available at: https://digaohm.semar.gob.mx/cat/centroAlertasTsunamis.html (Accessed date: 22/05/2020).

[34] CRED, 2016. Tsunami disaster risk-Past impacts and projections. Centre for Research on the Epidemiology of Disasters (CRED), UNISDR.

[35] Santos-Reyes, 2019. How useful are earthquake early warnings? The case of the 2017 earthquakes in Mexico City. International Journal of Disaster Risk Reduction, vol. 40, 101148, 1-11.

Section 2

Simulation and Modeling

Chapter 13

Natural Hazards - Impacts, Adjustments and Resilience

Bhushan Mohan Raisinghani

Abstract

Reinforced concrete is a global material, the utilization of which has no limits. India is a country that uses mostly RC framed structures as the routine building construction type. The building is made of inter-connecting elements in horizontal and vertical directions. To showcase the effectiveness of high grade of concrete and confining reinforcement much research has been carried out till date from 1980s. However, in design of structures we do not consider the effect of confining reinforcement in resisting stress in any member element. Various tools have been developed to find the capacity of member at element level to resist forces. For performance-based design of buildings, it is necessary to evaluate the performance at individual local level and at global levels. In this study, the effect of available tools (for section analysis) and design codes for member limit calculation is demonstrated and structure is evaluated for the threshold limits given in ASCE-41. It is observed that the code designed members are sufficient to resist lateral earthquake forces effectively for the estimated hazards if proper design tools are employed.

Keywords: disaster mitigation, resilient structures, performance-based design, design tools, displacement-based design, seismic performance evaluation

1. Introduction

Indian subcontinent is a region experiencing seismic activities since ancient times. According to BIS seismic zoning map, over 65% of the country is prone to earthquakes of intensity MSK VII or more, putting 38 cities in high risk zones. In 2011, there were 80 earthquakes in India, with magnitude ranges between 3.5 M to 6.5 M on Richter scale. The year 2012 was an equally eventful year with 19 earthquakes by 5th March. Great earthquake having magnitude of 8.0 M or higher occurs somewhere in the world every year. Seismic engineering of structures is in discussion since decades while the aspects of risk mitigation and hazard assessment are relatively new in this field which are in concern with our preparedness for future events. The aim of a structural design engineer is always the safety as against the sole aim of economy. However, architectural needs and budget compromises the higher target of an engineer to bring resilience as an important consideration in design of buildings in seismic prone regions. The code provisions give legal benefit to developers while the resilience aspects highlight the need to bring better scientific methods to safeguard the community during strong earthquakes. With the advancement in knowledge and promotion of performance-based design procedures, it has become possible to safeguard our interests against the fury of nature.

IntechOpen

The Bhuj earthquake (2001), had caused severe damage to property and more than 20,000 people were killed in Kuchchh area [1]. The reported PGA of the earthquake was 0.38 g having magnitude 7.7 Mw. Around 70 multistorey buildings collapsed in Ahmedabad, with reported PGA of 0.1 g for which the damage observed was higher than expected. Hence, the influence of present process of design of buildings on the performance of RC structures needs further check. There is another significant parameter for design consideration - the accurate hazard estimation. The observed damages better indicate hazard than the PGA values especially for mid-high rise buildings. The attenuation in Ahmedabad city is found to be around 1.8–2.0 [2], not considered in design. This is the serious limitation of force-based design wherein inaccurate estimate of hazard will give a sense of design to the engineer but is a gross-error in seismic design force calculations.

The expression of risk gives a good overview of each part that contributes to either the safety or to the weakness. In the equation, the term Hazard (H) represents the severity of earthquake expected in the region considered for risk estimation (refer Eq. (1)).

$$\text{Risk} = \text{Hazard (H)} \times \text{Vulnerability (V)} \times \text{Exposure (E)/Asset} \tag{1}$$

If the estimation of hazard is an under-estimate, then the risk increases substantially, though the design may be done with utmost care. Seismic micro-zonation of Ahmedabad gives understanding of hazard to design engineers.

The building typology and design methods followed in a region suggest vulnerability aspects. Without disturbing the cultural heritage of local region, the capacity of existing buildings can be increased to reduce the quantum of vulnerable stock to various intensities of earthquake likely to hit there [3]. The role of developer, owner and administration plays a vital role in upgradation of vulnerable facilities and investing efforts towards the mitigation of earthquake hazard by identifying the vulnerability (V) component in a region. The density of buildings in an area, important structures, and the type of facility under threat is covered by the risk as the exposure or asset (E) parameter to estimate the threat susceptibility. The impact of earthquakes in India show the concern due to recent earthquakes (2001;2005) on the number of people succumbed due to lesser engineering (refer **Table 1**).

Hence, if we are better able to know our region, dedicated in identifying weak structures for upgradation and protect the important assets by identifying their importance, we can control the impact of earthquakes on our habitat. This study focuses on adopting resilience in mid-rise buildings using design tools.

Event	Year	Magnitude	PGA	Intensity	Casualty
India-Burma	1988	7.2	0.34 g	VIII	709
Garhwal	1991	7.1	0.3 g	VIII	768
Uttarkashi	1991	7	0.29 g	IX	>2000
Koyna	1967	6.5	0.4 g	VIII	1500
Chamoli	1999	6.6	0.34 g	VIII	103
Bhuj	2001	7.7	0.38 g	VIII	20,000
Kashmir	2005	7.6	0.23 g	VIII	>80,000
Sikkim	2011	6.9	0.35 g	VI	111
Nepal	2015	7.9	—	IX	>8000

Table 1.
List of severe earthquakes in India (7 Mw): Magnitude-PGA-intensity.

2. Methodology

2.1 Problem description

A fifteen storey reinforced concrete building located in Ahmedabad city of Gujarat state is to be designed using Indian Standards (IS 456; IS 1893). The performance of this force-based design building will be evaluated using displacement-based method for performance-based design (PBD) using the procedures mentioned in **Figure 1**. The plan and section details are given in **Figure 2**.

2.2 Literature review

The building design code for earthquake was first developed in Japan to consider 10% of dead weight as the lateral load. The code procedure developed based on understanding of severe damages in strong earthquakes around the globe [4]. Seismic risk mitigation is dependent on understanding exposure and vulnerability apart from seismic hazard for which effective design provisions are required for built environment to sustain the next event [5]. The present codes include the cyclic behavior through simpler procedures for performance-based design of building structures with reduced vulnerability [6–9]. TEC-2007 has been upgraded to

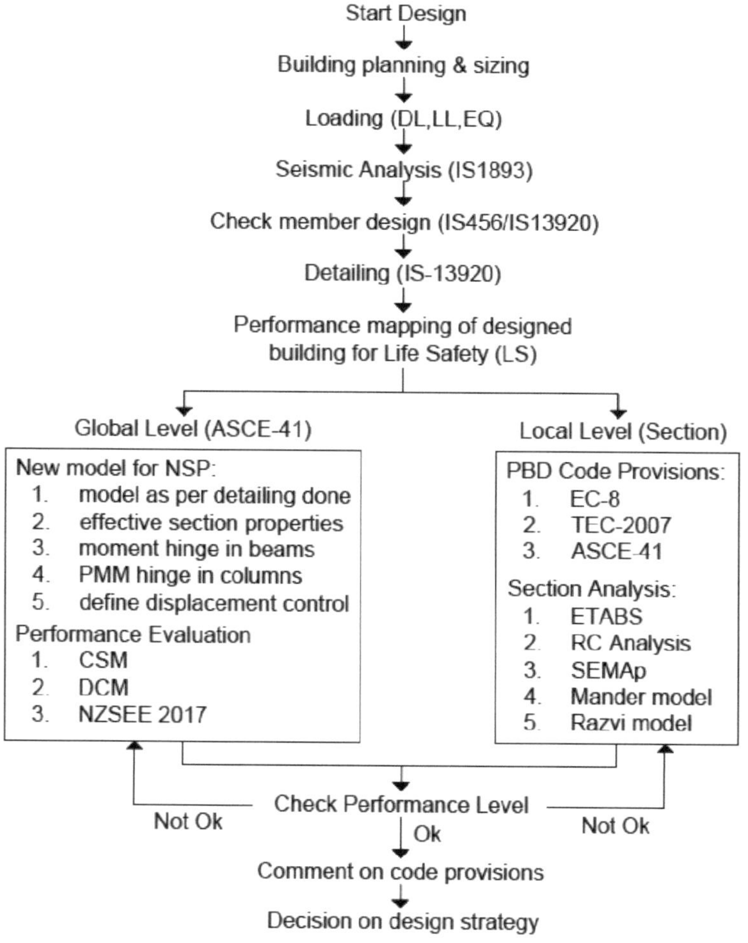

Figure 1.
Scheme for performance-based design method for standardization.

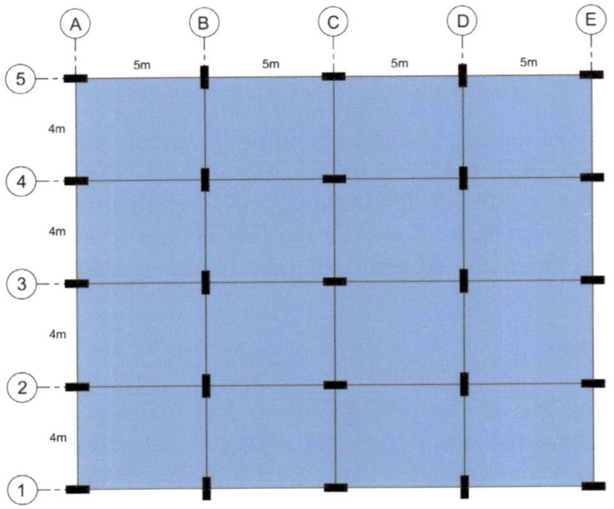

Figure 2.
Plan and section sizes of building (20 m × 16 m).

TBEC-2018 for design of new buildings with taller stature [10]. Also, the design of buildings with FIB Model Code 2010 is considered to upgrade the force-based provisions of EC-8 to displacement-based procedure [11]. Various studies have suggested competence of EC-8 towards better design of buildings compared to Indian and ACI codes [12]. SIRCO tool is developed to simulate and estimate seismic risk in Portugal to plan, prevent and respond to future earthquakes [13]. Engineering parameters are identified to rate the world-wide seismic code quality index and formally 166 countries have some form of seismic code with 510 revisions since 1900 [14]. Optimum solution technique is proposed to reduce computational efforts towards performance-based design using displacement procedure [15].

The above literatures do not focus towards mapping of performance of building and its elements using design codes and design tools to move towards standardization in performance-based design. The effectiveness of code provisions for building design and detailing are put under the lens, in-order to plan structures for the future H-V-E model in seismic prone regions.

3. Performance evaluation of building

The aim of design engineers is always towards the optimum use of materials with proper strength regulations. Seismic engineering requires separate framework to keep the structures functioning or regulate damages under multiple earthquake events. The providence of ductility and sizing hierarchy through strong column-weak beam philosophy is a different routine for design of structures. The importance of appropriate seismic hazard (PGA/PGV/Mw), modeling parameters, methodology and mapping of required performance thresholds are part of performance-based design procedure. The performance of building is evaluated following ASCE-41. The performance target for the building is Life Safety (LS) under MCE level earthquake (7.0 M in Bhuj, located at 250 kms from building).

3.1 Estimation of hazard for seismic demand

The building is in a city where risk reduction measures are employed after observing severe damages in Bhuj earthquake in 2001. The hazard actions are handled by Institute of Seismological Research (ISR) and mitigation measures are suggested by Gujarat State Disaster Management Authority (GSDMA). The understanding of local seismic hazard of Ahmedabad city is done and there seems to be under-estimation of hazard in BIS zoning map (refer **Figure 3**). Hence, the three hazard levels suggested for Ahmedabad city will be used to check the performance of building using capacity-spectrum method (CSM, FEMA-440) and displacement-coefficient method (DCM, ASCE-41). The equivalent static analysis results are shown in **Table 2**.

Building information:

No. of stories: 15	Full height: 45 m
Location: Ahmedabad city	Distance from fault: 250 kms
Time period (X,Y): (0.9,1.0)s	Base shear: refer **Table 2**

Hazard information:

BIS hazard zone: III (0.16 g)	Attenuation: 1.5–2 (ISR)
PSHA maps: 0.22 g (NDMA)	Seismic micro-zonation: 0.18 g (ISR)

3.2 Performance evaluation using seismic design codes and ASCE-41

Bureau of Indian Standards (BIS) is the governing body that regulates the professional minimum standards to be followed in India. The codes that are applicable for seismic design of buildings are IS 1893 (P1)-2016 and IS 13920–2016. The building is designed using these provisions and the capacity of building at structure level is noted for comparison with design standards of US (ACI-318) and Europe (EC-8). This can be a significant evaluation criterion in terms of performance evaluation, as the design and evaluation standards are different due to lack of performance evaluation guidelines in India.

The nonlinear static analysis (NSA) is a simpler means to evaluate performance and behavior of buildings under multiple excitations, (4 Mw – 7 Mw). The capacity spectrum method (CSM) is much debated as the demand-capacity are compared in one graph where hazard may be a partial information. Displacement

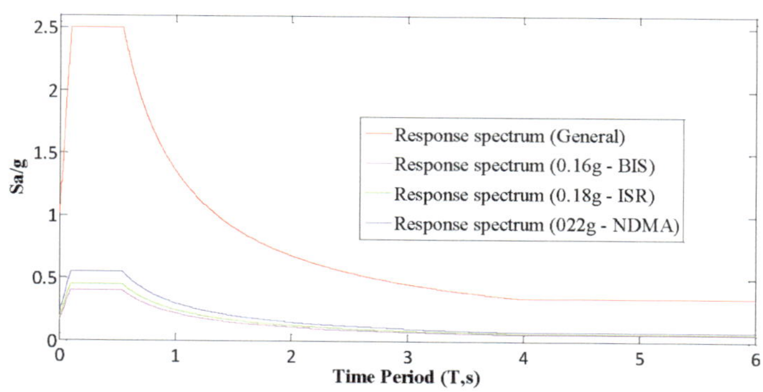

Figure 3.
Response spectrum for different hazard levels as per IS 1893–2016.

Time period (s)	Weight (W)	Acceleration (A_h)		Base Shear (V_B)
X dir. = 0.9	31461.54 kN	BIS	0.058	1636 kN
Y dir. = 1.0	DL = 100%	ISR	0.065	2054 kN (+25%)
(with infill)	LL = 25%	NDMA	0.079	2485 kN (+52%)

Table 2.
Equivalent static analysis as per IS-1893-2016.

coefficient method (DCM) is a better means to employ performance-based design procedure (PBD) as displacement gives better control. The target displacement (δ_t) is estimated to obtain nonlinear displacement limit at structure level (refer Eq.(3)).

$$\delta_t = C_0 C_1 C_2 S_a \frac{T_e^2}{4\pi^2} g \tag{2}$$

δ_t = (306.50, 348.60, 420.01) mm for three hazard levels

The building passes Life Safety (LS) performance under MCE level earthquake in case of 0.16 g and 0.18 g seismic hazard in Ahmedabad city for IS code-design case (refer **Figure 4**). Also, the building designed with IS code had performance of Collapse Prevention (CP) level while the building design with ACI-318 and EC-8 will collapse in 0.22 g hazard level. However, the location of failure hinge shown in **Figure 5**, describes the better behavior of structure designed using US and EN code provisions for buildings as only the base columns are failing. The columns at 4th and 7th level are failing in IS design case, hence the energy dissipation in the graph is poor. Important to note the strength of building designed as per IS code is highest and ductility achieved is more than the other design codes though the location of failure is not acceptable.

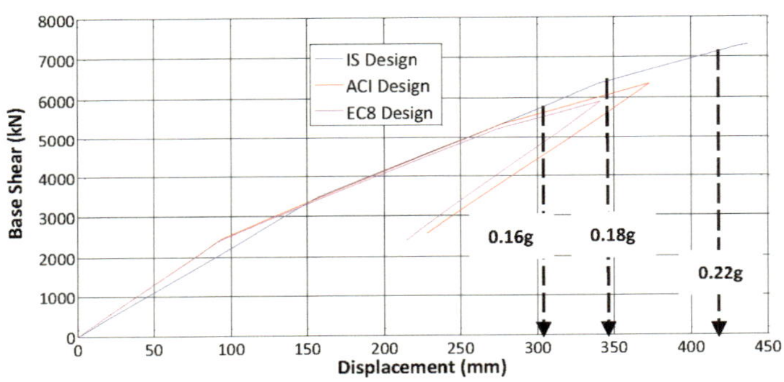

Figure 4.
Comparison of capacity curve of building as per design codes.

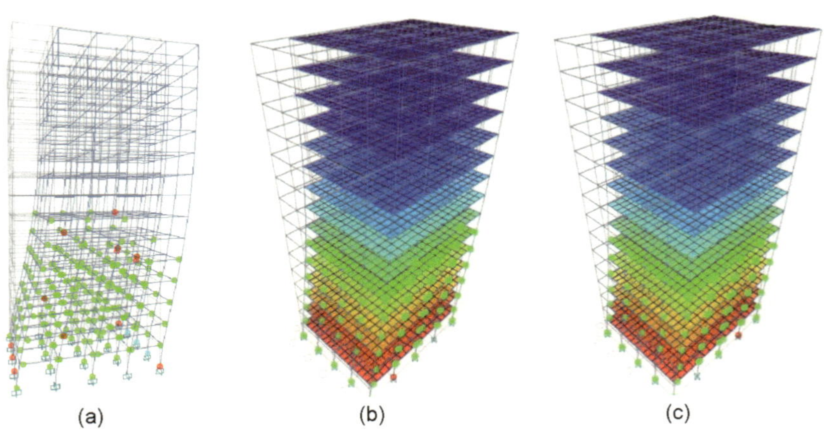

Figure 5.
Location of failure hinge at the end of POA. (a) IS design. (b) ACI design. (c) EC8 design.

3.3 Element wise performance in the building using ETABS v15.0

The performance of building at structure level is shown in previous section, which gives the global picture of stability in the building and the performance level achieved at the end of pushover analysis. However, the deformation in each element of building (beam and column) during each step of nonlinear analysis gives the local performance level for that element and the concerned storey of building as the weak link may be established. The elements are modeled in ETABS using confined concrete model [16]. The performance of elements that failed in the structure at end of POA are shown in **Figure 6**. Performance of elements designed as per ACI and EN code is found to be higher in terms of post yield deformation and hence better performance will be available at local level.

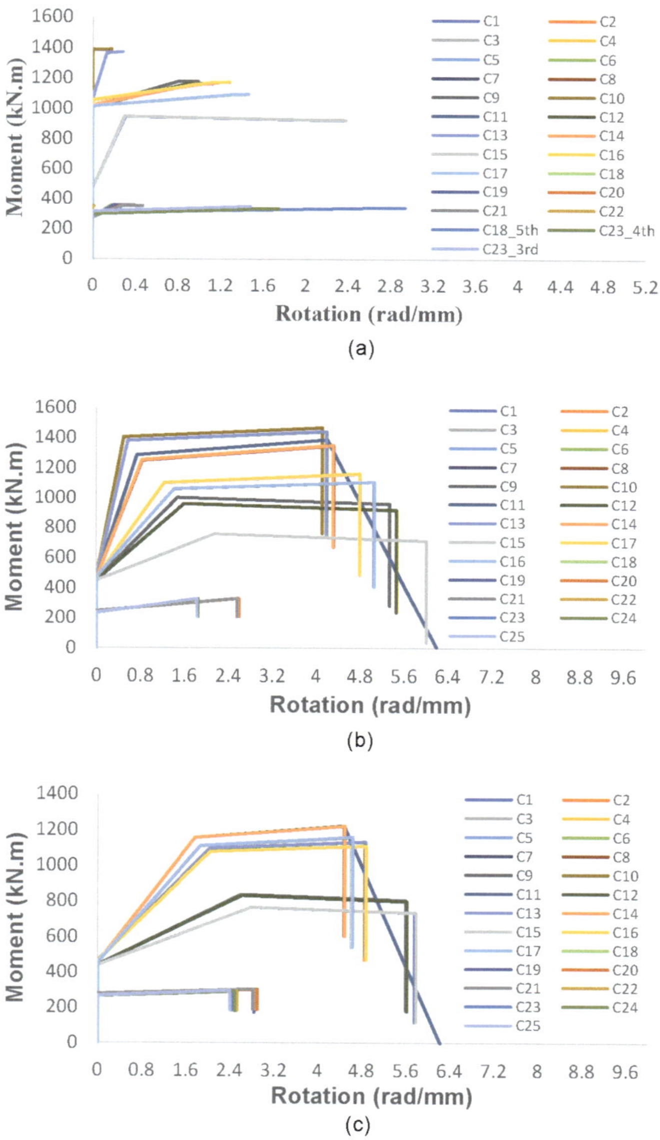

Figure 6.
Results of $M - \theta$ of all columns of building (ETABS 2015) at ground floor. (a) Ground storey column hinge results for IS code designed building. (b) Ground storey column hinge results for ACI code designed building. (c) Ground storey column hinge results for EC8 code designed building.

4. Use of design tools for element level performance estimation

The building is designed using ETABS software which gives the structure and element level performance of building. The performance of building elements can be evaluated using prevalent concrete models of confined concrete and high strength concrete. RC Analysis-Columns [17], web-portal is developed by Virtual Laboratory for Earthquake Engineering (VLEE, UTPL). Also, the SEMAp software developed by Ozmen (2007) gives the element capacity based on four concrete models [18]. The element level performance is suggested by Euro-code and Turkish-code for design of buildings for earthquakes [8, 9]. Hence, evaluating the performance of columns (i.e. failed element at the end of POA) using ETABS, EC-8, RC-Analysis (RCA) and SEMAp is interesting segment for assessing the performance.

4.1 Understanding the element deformations

The change in geometry of elements is picturized in form of curvature (φ), rotation (θ) and displacement (δ). The ductility in these elements will give them ability to absorb seismic forces and display higher performance before failure. The curvature ductility (φ_u/φ_y) is more meaningful than rotational ductility (θ_u/θ_y) as the calculation of yield rotation is not singular value [19].

The testing of beams and columns under transverse loading will be the basis for element level performance (refer **Figure 7c**). The building elements follow double curvature and under lateral loading both ends undergo plastic deformations

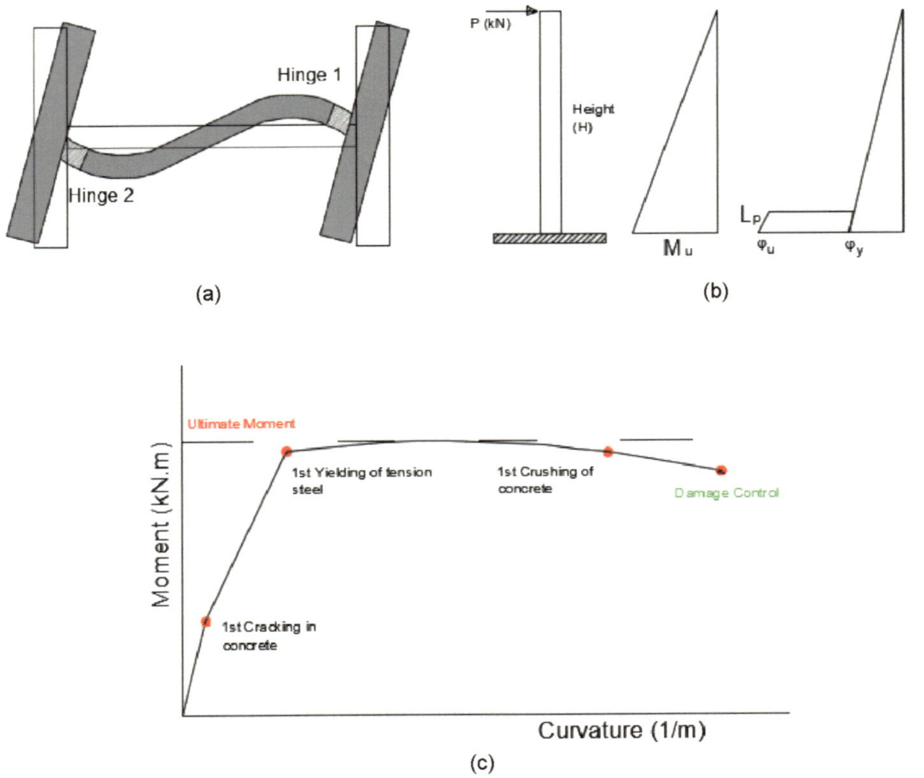

Figure 7.
Element level analysis framework under transverse loading. (a) Double curvature in beams/columns. (b) Plastic analysis of column. (c) Testing of elements under transverse loading.

DOI: http://dx.doi.org/10.5772/intechopen.94303

forming two hinges (refer **Figure 7a**). The curvature can be found using strain rates and member ($M - \varphi$) curve is used for managing performance at element level. The number of lateral load resisting beams and columns that can be accepted to undergo damage is the decision of either the regulatory bodies or the stakeholders which will influence the target performance to be achieved by design engineer. A sample performance level prescribed by TEC-2007 for Life Safety (LS) gives the basis for defining and making a building with mitigated risk (refer **Figure 8**).

However, such limits are not prescribed by IS-1893 and makes it more critical for engineers to select and debate performance levels with their clients or stakeholders. Similarly, ASCE-41 suggests the rotation limits for Life Safety (LS) performance level which is 75% use of post-elastic rotation of the member (refer **Figure 9**).

The ultimate rotation limit is suggested by EC-8 for Near Collapse (NC) performance i.e. 100% use of post-yield deformation and Life Safety (LS) is 75% of ultimate rotation limit (refer Eqs. (4)–(7)).

Ultimate chord rotation capacity for Near Collapse (NC) performance level:

$$\theta_{um} = \theta_y + \left(\varphi_u - \varphi_y\right)L_{pl}\left(1 - \frac{0.5L_{pl}}{L_V}\right) \tag{3}$$

Chord rotation at yield:

$$\theta_y = \varphi_y.L_V/3 \tag{4}$$

Length of plastic hinge:

$$L_{pl} = 0.025L_V + 0.125h + 0.02d_b f_y \tag{5}$$

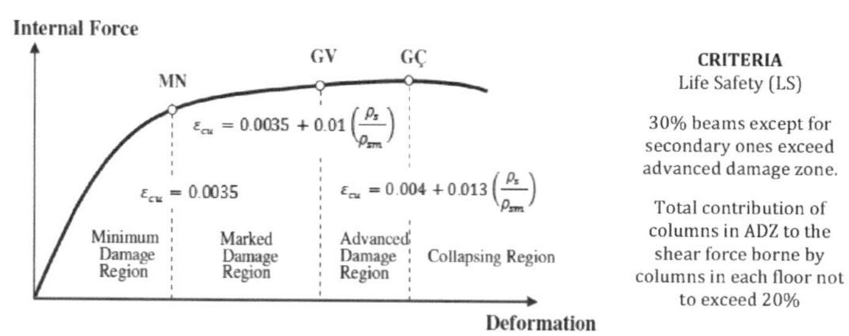

Figure 8.
Definition of criteria for life safety (LS) performance limit (TEC 2007).

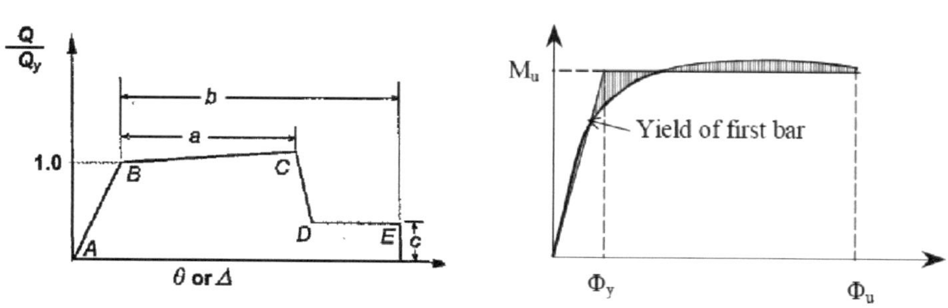

Figure 9.
Definition of criteria for life safety (LS) performance limit (ASCE-41).

Ultimate compression strain in concrete (EC8):

$$\varepsilon_{cu} = 0.004 + 1.4\rho_s f_{yh}\varepsilon_{sm}/f'_{cc} \quad (6)$$

f'_{cc} = confined compressive strength of concrete
f_l = confining pressure
A_{sh} = area of confinement reinforcement

4.2 Tools for section analysis of building elements

The section analysis of confined concrete is dependent on the type of detailing done for beams and columns which undergo in-elastic deformations. ETABS software uses *Mander*-model for confined concrete to estimate increased strength due to lateral reinforcements. However, without complete detailing of the column reinforcement it is not suitable to directly assume the performance. Hence, it is suitable to use a separate tool to generate $(\boldsymbol{M} - \boldsymbol{\varphi})$ and $(\boldsymbol{\sigma} - \boldsymbol{\varepsilon})$ plots and the results of which can be uploaded in ETABS software to give more accuracy in terms of overall performance of building. The two such tools are RCA and SEMAp, which are used to compare the performance limits (refer **Figures 10–12**).

4.3 Mapping of section analysis using design tools and design codes

The mapping of threshold limits using design tools and codes gives an envelope for comparison. The ultimate rotation (θ_u) is sum of yield rotation (θ_y) and plastic rotation (θ_p). The length of plastic hinge is taken as 0.5D and absolute concrete strain (ε_{cu}) is calculated [20, 21]. It is observed that for IS designed building, the

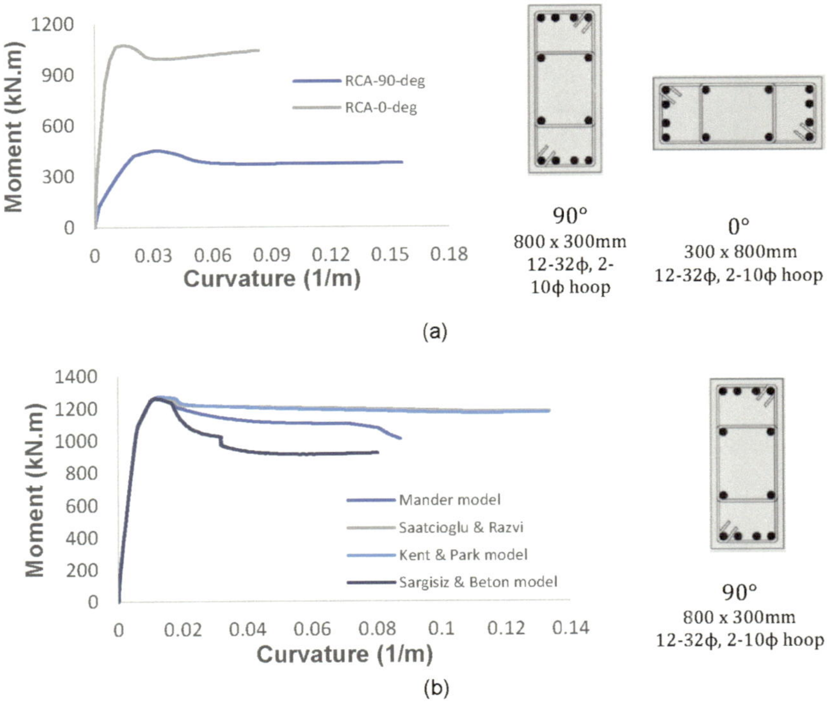

Figure 10.
Section analysis results for RC column designed as per IS code. (a) $M - \varphi$ curve as per RC analysis tool for column designed as per IS code. (b) $M - \varphi$ curve as per SEMAp tool for column designed as per IS code.

DOI: http://dx.doi.org/10.5772/intechopen.94303

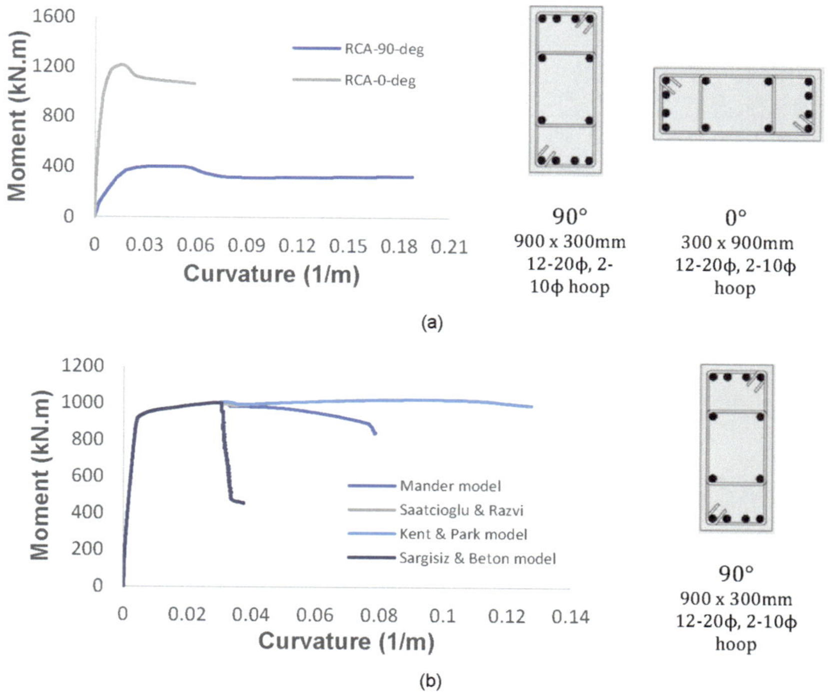

Figure 11.
Section analysis results for RC column designed as per ACI-318 code. (a) $M - \varphi$ curve as per RC analysis tool for column designed as per ACI-318 code. (b) $M - \varphi$ curve as per SEMAp tool for column designed as per ACI-318 code.

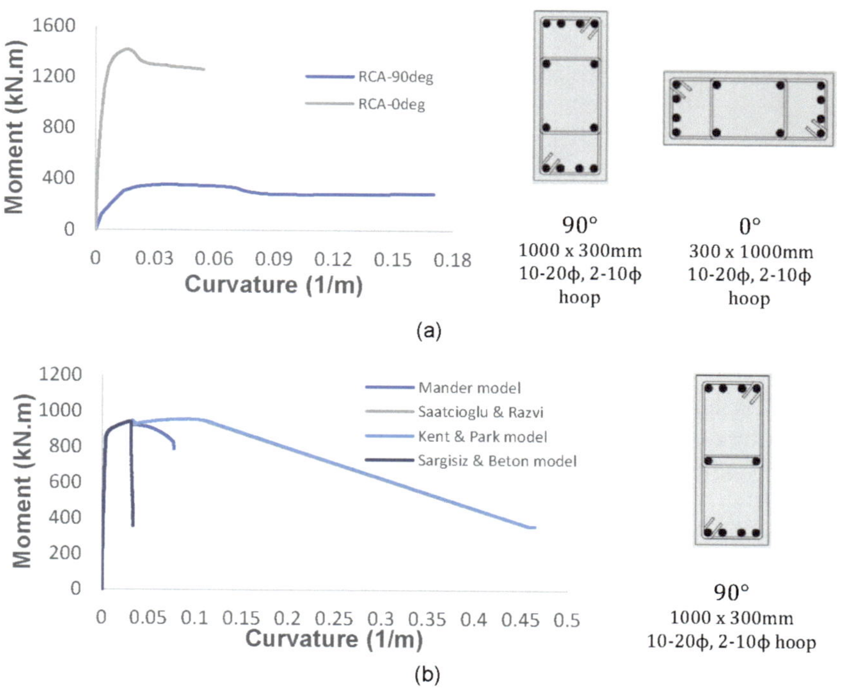

Figure 12.
Section analysis results for RC column designed as per EC8 design code. (a) $M - \varphi$ curve as per RC analysis tool for column designed as per EC8 design code. (b) $M - \varphi$ curve as per SEMAp tool for column designed as per EC8 design code.

[1]	[2]	[3]	[4]	[5]	[6]	[7]	[8]	[9]	[10]
Design case	Tool/Code	M_y (kN.m)	M_N (kN.m)	M_u (kN.m)	φ_y (1/m)	φ_N (1/m)	φ_u (1/m)	ε_{cu}	θ_u
IS Design 300 × 800 12-32ϕ	RCA	120.42	539.86	560.82	0.0021	0.02	0.22	0.024	0.032
	SEMAp	94.78	430.98	364.14	0.0025	0.025	0.22	0.035	0.036
	TEC'07	—	—	369.65	0.0025	0.025	0.139	0.018	0.029
	EC8	—	—	361.06	—	—	—	0.032	0.053
	ETABS	314.58	314.58	348.49	—	—	—	—	0.003
ACI Design 300 × 900 12-20ϕ	RCA	120.77	399.97	329.53	0.0028	0.03	0.187	0.025	0.038
	SEMAp	103.16	259.14	183.21	0.0035	0.028	0.245	0.030	0.044
	TEC'07	—	—	369.65	0.0035	0.028	0.139	0.018	0.031
	EC8	—	—	366.34	—	—	0.192	0.026	0.055
	ETABS	234.53	234.53	330.91	—	—	—	—	0.002
EC8 Design 300 × 1000 10-20ϕ	RCA	125.92	314.63	293.42	0.0026	0.043	0.133	0.011	0.035
	SEMAp	103.2	248.95	173.86	0.0033	0.027	0.218	0.028	0.042
	TEC'07	—	—	372.58	0.0033	0.027	0.103	0.018	0.025
	EC8	—	—	366.16	—	—	0.193	0.026	0.041
	ETABS	265.30	265.30	296.86	—	—	—	—	0.003

Table 3.
Section analysis using design tools and design codes (limits mapping).

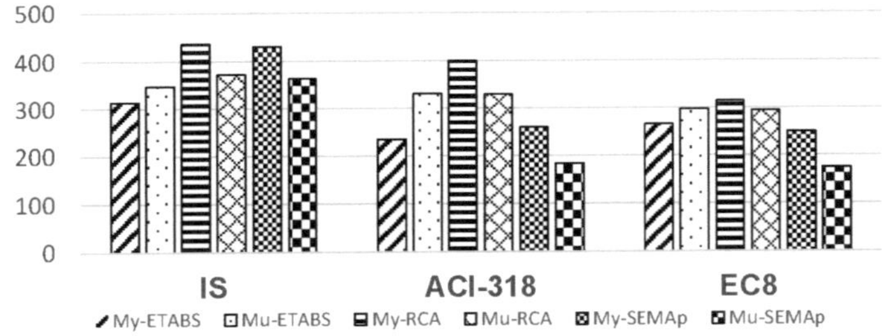

Figure 13.
Comparison of moment (M) at yield and ultimate state for column C19.

results of RCA and SEMAp are matching in terms of moments (**Table 3**; [5]). The results of ultimate strain limits for NC-PL of EC-8 and SEMAp are matching (**Table 3**; [9]). Similarity is there in ultimate rotation values of RCA and SEMAp (**Table 3**; [10]). The ultimate rotation values of columns designed using ACI-318 and EC-8 are found to be in similar range and are higher than rotation limit of IS designed building (**Table 3**; [10]). The variation in moments for all cases is shown in **Figure 13**.

5. Discussion of the problem and its results

The discussion on results is based on design output for three seismic design codes for buildings to mark the comparisons. The details of performance and failure

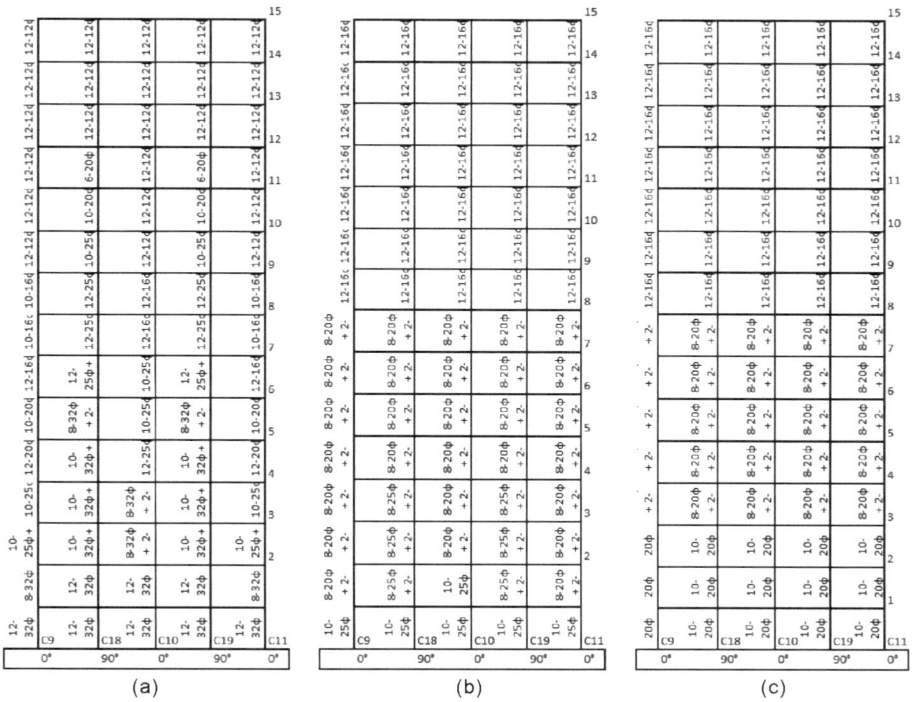

Figure 14.
Comparison of reinforcement in columns. (a) IS code design. (b) ACI-318 code design. (c) EC8 design.

pattern in the following part are discussed to encompass: hinge results, hinge locations, performance point and element level indexes.

5.1 Comparison of design codes on building performance

- The building designed as per Indian Standard codes lead to higher performance in terms of overall capacity of building owing to section sizes and higher rebar requirements. The distribution of steel is better in building designed using ACI-318 and EC-8 (refer **Figure 14**).

- The concept of formation of hinges in columns is not appreciated by seismic design engineers for which the beam to column capacity ratio is maintained in all seismic design codes viz. 1.2 in ACI and 1.4 in IS-13920. Hence, strong column-weak beam is the basic philosophy followed by all codes in different capacities (refer **Figure 15**).

5.2 Comparison based on nonlinear static analysis in ETABS

- It was found that the building designed using IS code resulted in the formation of hinges in columns and the failed columns were at the GF, 3rd, 4th and 5th storey (refer **Figure 5a**). The hinge locations for three designs are given in **Figure 16**. The performance point is compared in **Figure 17**.

- The capacity of building designed as per IS code is higher due to higher percentage of steel in the building elements:

  Case 1: IS-1893 V (kN): 7200 D (mm): 430

Case 2: ACI-318 V (kN): 6200 D (mm): 360

Case 3: EC-8 V (kN): 6000 D (mm): 340

5.3 Element level performance evaluation: Code & tool comparisons

- The yield strain (ε_y) is dependent on yield moment (M_y) and ultimate strain (ε_{cu}) is given by section properties as per EC-8 and TEC-2007. This helps the design engineer to make buildings with specific target.

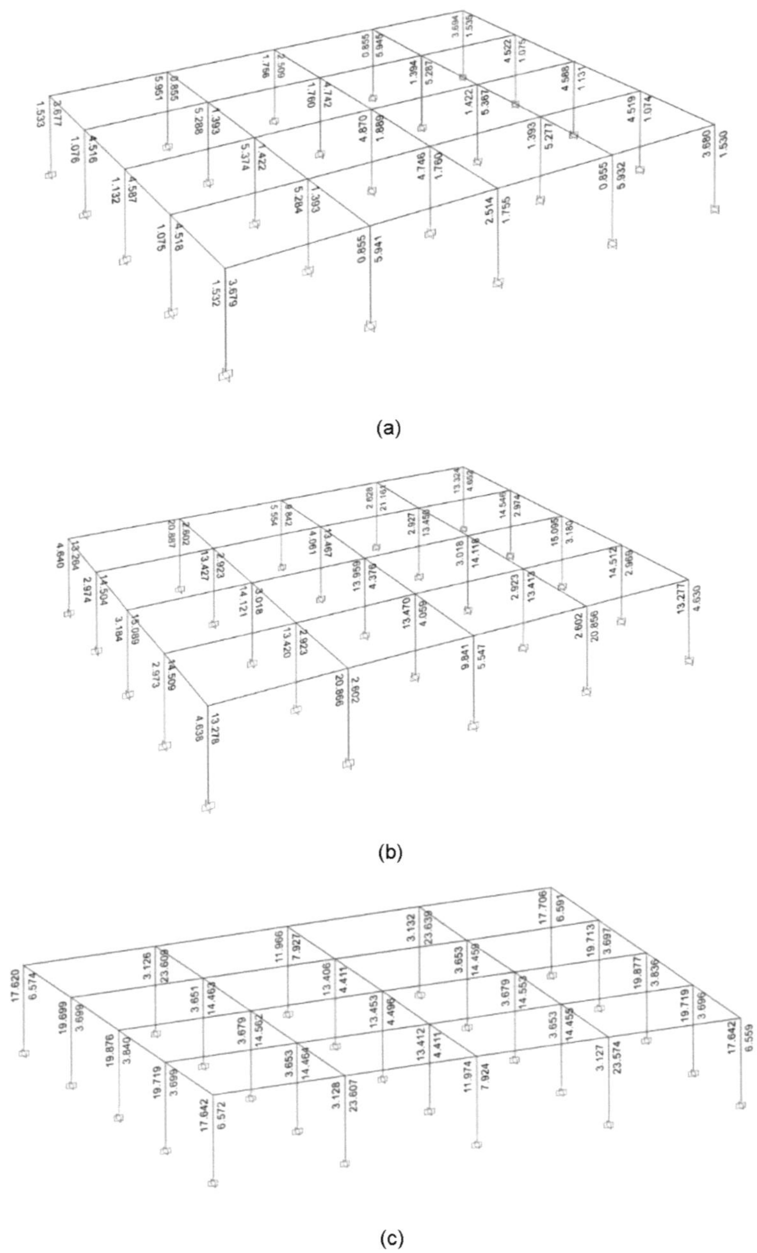

Figure 15.
Comparison of column to beam capacity ratio for GF elements. (a) Design as per IS-13920: Minimum 1.4 ratio in strong direction. (b) Design as per ACI-318: Minimum 2.6 in weak direction. (c) Design as per EC-8: Minimum 3.13 in weak direction.

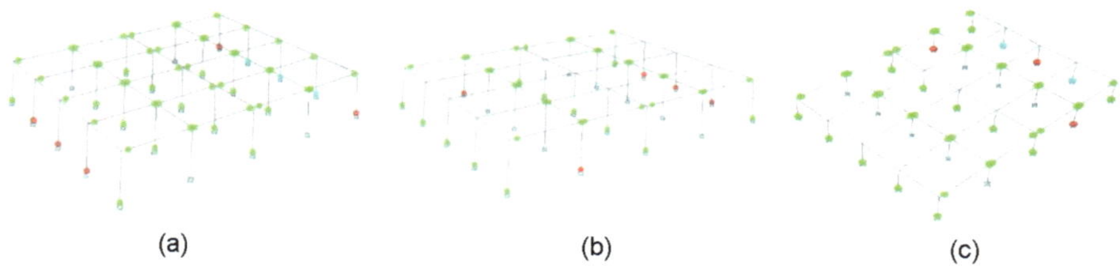

Figure 16.
Failure hinges in GF elements designed as per seismic codes. (a) IS design. (b) ACI design. (c) EC-8 design.

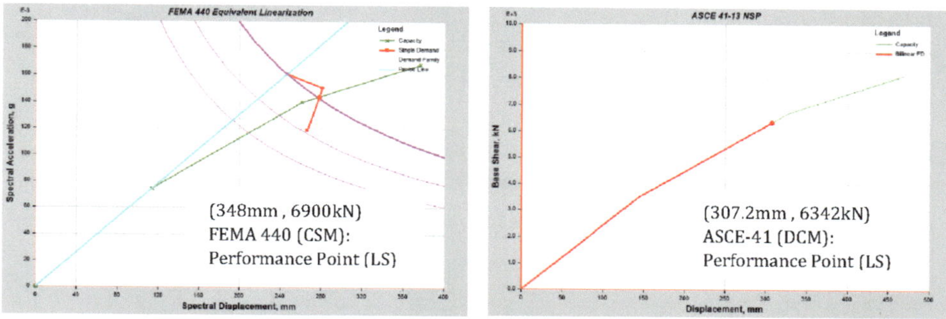

Figure 17.
Performance point of building designed as per IS codes.

- The moment estimates using ETABS, RCA and SEMAp show similarities. The ultimate moment estimation of ETABS and RCA are nearer while the estimate of yield moment of ETABS and SEMAp are nearer. The ductility values are different in RCA and SEMAp (refer **Table 4**).

- RCA and SEMAp tools have distinct advantages to explore. Both the platforms help to exercise performance-based design. These tools give engineers the domain to evaluate elements considering different concrete models and different expressions for plastic hinge length to give more practical output in terms of performance.

- The mapping of threshold limits for Life Safety (LS) performance for column C19 is done based on results of **Table 3**. The ultimate curvature values obtained from RCA and SEMAp are the same i.e. 0.22 (1/m). However, the difference in LS threshold for C19 is in the limiting moment (refer **Figure 18**).

- To map the threshold limits based on the ultimate strain (ε_{cu}), the moments vary significantly when seen from the window of TEC-2007, EC-8, RCA and SEMAp (refer **Figure 19**). Based on the strain limit values, SEMAp is able to map the criteria set by EC-8 for LS performance while RC analysis results show the EC-8 (LS) threshold to be at Near Collapse (NC) performance.

Tool	IS-1893	ACI-318	EC-8
RCA	6.67	6.23	3.09
SEMAp	7.32	8.75	8.07

Table 4.
Curvature ductility (φ_u/φ_y) for column C19.

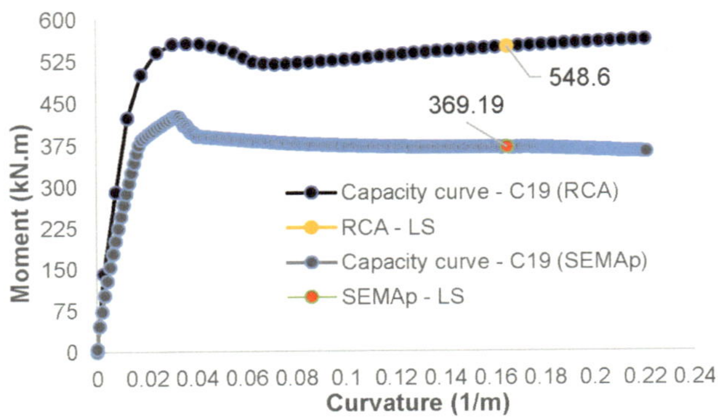

Figure 18.
Comparison of column C19 limits based on ultimate curvature.

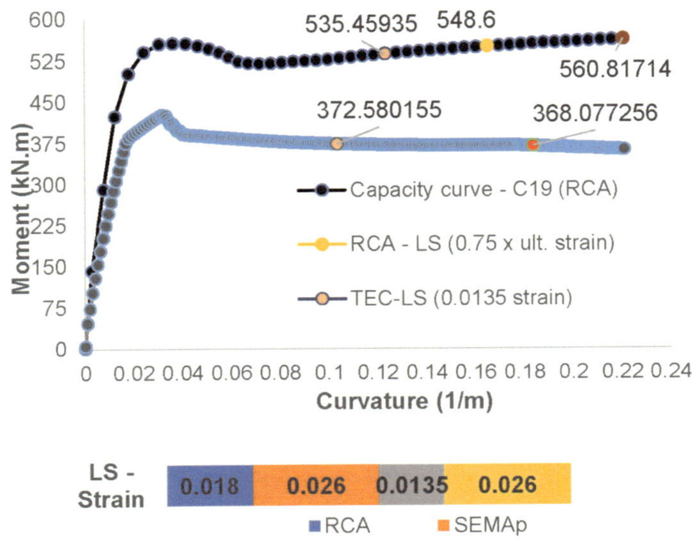

Figure 19.
Comparison of column C19 thresholds based on strain-limits.

5.4 Evaluation of effectiveness of code provisions for target performance

- The building designed using IS codes showed Life Safety (LS) performance level under 0.16 g and 0.18 g hazards while the building performance was Near Collapse (NC) for 0.22 g hazard.

- The building designed using ACI-318 and EC-8 behaved better in terms of failure mode, but it did not meet the LS performance target in three hazard levels. The effectiveness of ACI and EC-8 design provisions as the performance under defined hazard of 0.16 g was better in all respects.

- Controlling the element performance is necessary for improving the overall performance. The use of EC-8 provisions, which shows lowest capacity of building in current case, is suggested. ASCE-41 can be used for existing buildings while the provisions of EC-8 apply towards design of building and leads to less iterations for achieving LS-NC performance level.

- Similarly, TEC-2007 is systematic towards achieving target performance.

6. Conclusion

The building is designed as per current versions of seismic design codes (IS, US, EN) to understand the parameters that swing the performance. The effectiveness of design provisions was evaluated using the non-linear static analysis procedures outlined in ASCE-41 and tools available to support performance-based design. The following points can be summarized from the study:

- Though, pushover analysis and displacement procedures have inaccuracy for capturing higher modes, they significantly contribute towards understanding design problems and highlight limitations of code designed buildings.

- The building to be designed in Ahmedabad city shall consider higher hazard levels than that suggested by BIS.

- Code provisions need to be updated for Gujarat state as it is having five seismic zones (0.36 g – 0.1 g) and a hazard map is required to consider it in design of buildings. BIS provisions are based on intensity parameters, but the seismic-micro-zonation results are not considered in IS-1893.

- The building designed as per IS-1893 can sustain hazard of 0.22 g though it is designed for 0.16 g, due to higher reinforcement in elements (4.2% against 2% in ACI/EN for C19). However, the building did not satisfy the beam mechanism principle and hinges formed in mid stories.

- The building designed as per ACI and EC have better performance in terms of energy dissipation and control of failure. EC-8 has better potential to apply more control for achieving target performance.

- Design tools have significant effect on performance evaluation of building. RCA is based on ACI-08 while SEMAp covers TEC-07 & ACI-08 parameters.

- SEMAp software gives results for four concrete models which can be a feature to use in performance evaluation and hence it has a wider range as compared to RCA tool.

- RCA has feature of generating multiple curves for analysis of many sections and hence it can give faster results. Also, RCA tool has other important curves that can show better analysis insights.

- The results of SEMAp for four concrete models show significant variations that can lead to change in local and global performance of building. This feature is a support to the design and quick method to attempt PBD.

- The tools and software packages shall be checked for failure modes before applying them in design. Hence, correlation of results with physical testing shall be done.

- The design tools support the outputs for new design as they simulate results better for flexure failure. The limitations are discussed in literature [22].

Without knowledge of hazard, design of building is a fluke attempt and seismic-micro-zonation helps in mitigation plan. The vulnerability is judged by design principles and its effectiveness towards safety. The study was concentrated towards design of buildings which will reduce its vulnerability in MCE level earthquakes.

The exposure can be controlled only when design engineers with wholesome view of hazards, methods and safety are able to change the outcome from local level to global level which is the new norm of *fib*-MC2010 [11]. Optimization in design using analysis procedures for code provisions is needed [23]. Attempt shall be towards reduction in non-fatal injuries which form about 96% of the claims of government expenses post disaster [24]. NZSEE 2017 proposes index for existing buildings with design having hazard under-estimated to manage the next earthquake [25].

The focus in India is towards use of ASCE-41 performance limits while this scheme in design is not yet achieved. The performance of buildings has improved from past and newer methods are getting a place in design offices, yet the mechanism to control the performance of buildings shall be the way ahead in line with the disaster mitigation needs. Tools for displacement-based approaches need to be further developed. Experimental and computational procedures cannot be alienated.

Author details

Bhushan Mohan Raisinghani
Consulting Engineer (Seismic), Ahmedabad, Gujarat, India

*Address all correspondence to: raisinghanibhushan@gmail.com

References

[1] Mishra PK. The Kutch Earthquake 2001: Recollections, Lessons & Insights. National Institute of Disaster Management (NIDM): Report published; 2004. 258p 81-8347-000-9

[2] Sairam B, Rastogi BK, Patel V, Pancholi V. Site effects: Case study of the 2001 Bhuj earthquake damages in the Ahmedabad city, Gujarat. India. Bulletin of Seismological Society of America. 2018;**4**:2170-2182. DOI: 10.1785/0120170266.

[3] Porter K. A beginner's guide to fragility, vulnerability, and risk. University of Colorado Boulder, SPA Publications. 2020. 136 p. https://www.sparisk.com/pubs/Porter-beginnersguide.pdf. (also getting published in Springer, 2021)

[4] Celep Z, Ilki A. Earthquakes, existing buildings and seismic design codes in Turkey. Arabian Journal for Science and Engineering. 2012;**37**(2):365-380. DOI: https://doi.org/10.1007/s13369-012-0183-8

[5] Roeslin S, Ma M.T.Q, Garcia H J. Damage assessment on buildings following the 19th September 2017 Puebla, Mexico earthquake. Frontiers in Built Environment: Earthquake Reconnaissance – Building the Risk and Resilience Evidence Base. 2018;4–72:1–18. DOI: https://doi.org/10.3389/fbuil.2018. 00072.

[6] IS 1893-Part 1 (2016). Criteria for earthquake resistant design of structures: general provisions and buildings. Bureau of Indian Standards (BIS), New Delhi.

[7] ASCE 41. Seismic Rehabilitation of Existing Buildings. American Society of Civil Engineers, ASCE/SEI Standard. Reston: VA; 2017. p. 41

[8] CEN. Design of structures for earthquake resistance – Part 1: General rules, seismic actions and rules for buildings. In: EN 1998–1:2004, EC-8. Cmite Europeen de Normalisation: Brussels; 2004

[9] Turkish Earthquake Code (TEC-2007). Specifications for buildings to be built in seismic areas. Ministry of Public Works and Settlement, Ankara, Turkey. 2007.

[10] Aksoylu C, Mobrak A, Arslan MH, Erkan IH. A comparative study on ASCE7-16, TBEC-2018 and TEC-2007 for reinforced concrete buildings. Journal of Construction, Pontificia Universidad Católica de Chile. 2020; **19**(2):282-305

[11] Michael F. N. From performance and displacement-based assessment of existing buildings per EN1998–3 to design of new concrete structures in *fib* MC2010. In: Ansal A. (eds) Perspectives on European Earthquake Engineering and Seismology. Geotechnical, Geological and Earthquake Engineering, Springer Cham. 2014;34:227–266. DOI: https://doi.org/10.1007/978-3-319-07118-3_7.

[12] Hassan W, Anwar N, Norachan P, Majam F. The seismic performance evaluation of RC high-rise buildings designed to various building codes. In: Proceedings of IABSE conference – Engineering the developing world; 25-27th April 2018; Kuala Lumpur. Malaysia.

[13] Sa Luis, Esteban A. M, Durand P. A seismic risk simulator for Iberia. Bulletin of Seismological Society of America. 2016;106–3:1198. DOI: http://dx.doi.org/10. 1785/0120150195.

[14] Daniell JE, Wenzel F, Khazai B, Santiago JG, Schaefer A. A worldwide seismic code index, country-by-country global building practice factor and socio-economic vulnerability indices for

use in earthquake loss estimation. In: Proceedings of 2nd European Conference on Earthquake Engineering and Seismology (2ECEES); 25–29 August 2014; Istanbul, Paper no. 1400

[15] Panagiotis EM. Efficient optimum seismic design of reinforced concrete frames with nonlinear structural analysis procedures. Structural and Multidisciplinary Optimization. 2018; **58**(6):2565-2581. DOI: 10.1007/s00158-018-2036-x

[16] Mander JB, Priestley MJN, Park R. Theoretical stress-strain model for confined concrete. Journal of Structural Engineering, ASCE. 1988;114–8: 1804–1825.

[17] RC Analysis software. Program for Moment-Curvature-Shear analysis of RC sections [Internet]. 2008. Virtual Laboratory of Earthquake Engineering (VLEE), Universidad Tecnica Particular De Loja (UTPL). Available from: http://ingenieriasismica. utpl.edu.ec/?q=es/experimentos-hormigon.

[18] Ozmen HB, Inel M, Bilgin H. Modelling nonlinear behavior of reinforced concrete members. In: Proceedings of 6th National Conference on Earthquake Engineering (Oct'07); 16–20 October 2007. Istanbul: Turkey. pp. 207-2015

[19] Park R, Paulay T. Reinforced Concrete Structures. Inc.: John Wiley & Sons; 1975. 769 p. DOI: 10.1002/9780470172834

[20] Englekirk R. E. Seismic Design of Reinforced and Precast Concrete Buildings. John Wiley & Sons, Inc.; 2003. 848p p. ISBN: 978-0-471-08122-7.

[21] Priestley MJN, Paulay T. Seismic Design of Reinforced Concrete and Masonry Buildings. John Wiley & Sons Inc.; 1992. DOI: 10.1002/9780470172841

[22] Chasioti S. G, Megalooikonomou K. G, Pantazopoulou S. Shear strength assessment of reinforced concrete columns. In: Proceedings of 4th ECCOMAS Thematic Conference on Computational Methods in Structural Dynamics and Earthquake Engineering. 12-14th June 2013. 927–940 p. DOI: 10.7712/120113.4567.C1506.

[23] Panagiotis EM. Minimum cost performance-based seismic design of reinforced concrete frames with pushover and nonlinear response-history analysis. Structural Concrete. 2019;**21**(2):599-609. DOI: 10.1002/suco.201900211

[24] Porter K. Value of injuries in the Northridge earthquake. Earthquake Spectra 2006;22–2:555. DOI: http://dx.doi.org/10.1193/1.2194529.

[25] NZSEE (2017). Seismic Assessment of Existing Buildings. EPB Methodology. Available at: www.building.govt.nz. ISBN:978-0-478-43366-1.

Chapter 14

Performance-Based Design for Healthcare Facilities

Wilfrid Gbenankpon Djima, Abdullah Can Zulfikar and Cüneyt Tüzün

Abstract

Healthcare buildings are one of the most critical facilities in any country for their important roles just after an earthquake. In this context, working on the resistance of healthcare facilities against earthquake is of great importance for a probable future earthquake. However, in today's world, in either private or governmental agencies, buildings earthquake resistant design is not within the primary criteria such as social facilities and architectural details need for the residents. While the structural system in any building are often considered the most important in the performance, it represent approximately only 20% of the total building cost. Consequently, structural engineers should look the seismic performance in an extensive context, looking at all the systems of the building than just the damage to structural items and life-safety. So to response to this issue, a next generation of seismic performance-based design methodology and tools have been outlined in the FEMAP58 documents to allow engineers to query out the seismic performance of an entire building in terms of future life loss, facility repair cost and repair time and that we summarized and applied in this chapter for a six (6) story special moment frame healthcare building.

Keywords: Seismic Performance Assessment, Performance-Based Design, Earthquake Loss Assessment, Healthcare Facilities

1. Introduction

Many essential facilities such as hospital buildings are in high seismic zones throughout the world, and some of them were designed and built at a time without sufficient earthquake knowledge nor performed and are consequently susceptible to earthquakes.

The typical building design process is not performance-based. In the typical process, design professionals select, proportion, and detail building components to satisfy prescriptive criteria contained within the building code. Many of these criteria were developed with the intent to provide some level of seismic performance. However, the intended performance is often not obvious, and the actual ability of the resulting designs to provide the intended performance is seldom evaluated or understood.

Therefore, it has been noted in this period generation procedures some limitations in the: accuracy and reliability of available analytical procedures in predicting actual building response, the level of conservatism underlying the acceptance criteria, the inability to reliably and economically apply performance-based procedures to the design of new buildings and the need for alternative ways of communicating performance to stakeholders that is more meaningful and useful for decision- making

IntechOpen

purposes [1]. Other limitations in the performance based-design procedure were also the non-account of non-structural equipment's very important economically but also regarding their behavior during an earthquake. For example, 50% of the injuries and 3% of the deaths in the 1999 Kocaeli Mw7.4 earthquake were caused by non-structural elements and 30% of the losses were found to be furniture, white goods, electronic equipment and other valuable items [2, 3]. In addition, in the 1989 Loma Prieta and 1994 Northridge earthquakes, 10 large hospitals were evacuated or had to be closed due to damage caused by non-structural elements (plumbing) [4, 5].

So, to fulfill the promise of performance-based engineering, FEMA started the development of next-generation performance-based design procedures to address the above limitations. By result, it has been finalized the FEMAP58 [1, 6, 7] guideline to count not only the structural damage but also non-structural damage in the performance assessment. Specifically, others research also focused on the study of the non-structural seismic behavior and assessment and for hospital building [8, 9].

This paper provides practical guidance principally on implementing the seismic performance assessment methodology set forth in FEMA P-58-1 and the guidelines for Seismic performance assessment of buildings, [1, 10], to assess the seismic performance of individual buildings based on their unique site with structural, non-structural, and occupancy characteristics, expressed in terms of the probability of incurring casualties, repair and replacement costs, repair time. The FEMA-P58-2 Implementation Guide [2] contains examples illustrating the performance assessment process, including selected calculation and data generation procedures, by using the selected electronic materials provided in Volume 3 – Supporting Electronic Materials and Background Documentation [7, 11].

This study does a nonlinear static analysis for an existing typical six (6) story hospital building following the Turkish Building Earthquake Code [6] and the ASCE 41 [9] provisions as well as ACI-318 for reinforced concrete and masonry structure [7, 8], aiming to provide a more realistic estimate of the seismic demands and economic-effective assessment strategy. The PACT (Performance Assessment Calculation Tool) is used in the analysis of the sample hospital building [12–14].

Many financial institutions including lenders, investment funds, and insurers use Probable Maximum Loss (PML), Scenario Expected Loss (SEL), and Scenario Upper Loss (SUL) as preferred performance measures. These performance measures are quantitative statements of probable building repair cost, typically expressed as a percentage of building replacement value [1]. Some building owners, developers, and tenants have also relied on these performance measures to quantify seismic performance. In this regard, it is believed that this study will be a sample study for evaluation of seismic performance of a typical hospital building and its probable consequences.

2. Building description

The building is a six-stories healthcare building with a moment frame structure that has plan dimensions of 36,57 m by 54,86 m (see **Figures 1** and **2**). Floor-to-floor height is 4.6 m at the lower story and 4 m at other stories (see **Figure 3**). The structure has reinforced concrete special moment frames around the building perimeter. The floors and roof are two-way post tensioned flat slabs (0.2 m thick) supported by the perimeter moment frame and interior reinforced concrete columns on a 9.14 m by 9.14 m grid (see **Figures 4** and **5**). When entering the building information into the PACT direction 1 is arbitrarily aligned with the North–South (Y) axis and direction 2 is aligned with the East–West (X) building axis.

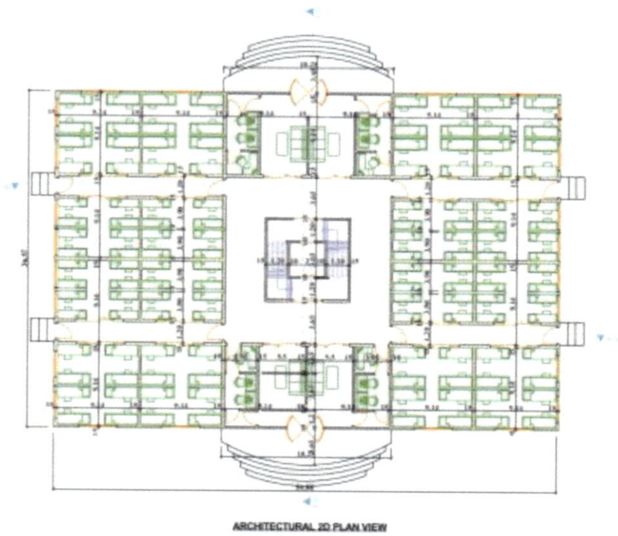

Figure 1.
Architectural 2D view of the healthcare building.

Figure 2.
Architectural Façade view of the healthcare building.

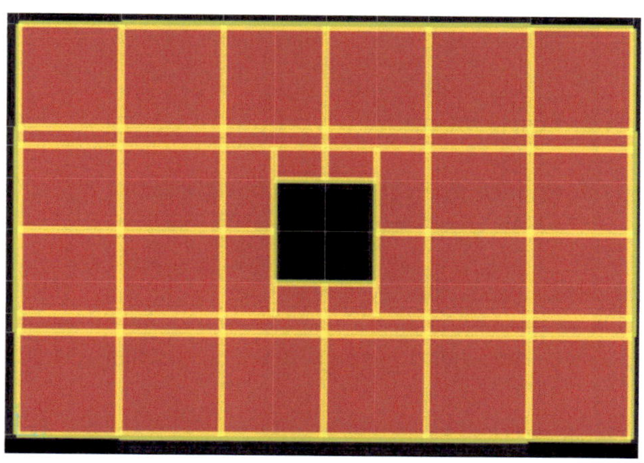

Figure 3.
Floor plan of building.

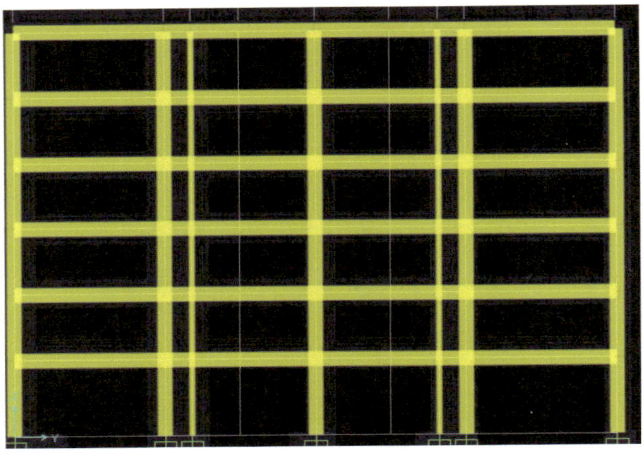

Figure 4.
Typical elevation E-W view.

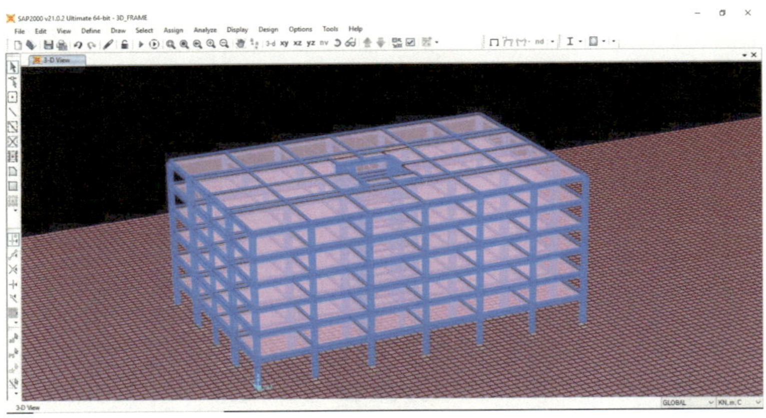

Figure 5.
3D view of the structural frame.

The building is designed and detailed according to the requirements of Turkish Building Codes 2018 [6, 8, 9]. Vulnerable non-structural building features include exterior glazing, gypsum board partitions, suspended acoustical ceilings, fire sprinkler system, traction elevator, concrete roof tiles on a perimeter mansard, hot and cold-water piping, and HVAC ducting.

3. Building performance model

The building performance model has been constructed in PACT by the following order:

- providing project information,
- building characteristics selecting fragility specifications and performance groups,
- identifying collapse fragility and collapse modes,
- and providing residual drift fragility

For this case study, information's input is as follow (see PACT input in **Figures 6** and **7**):

- Number of Stories: 6.
- Total Replacement Cost: Estimated as $\$2500/m^2 \times 12960\ m^2$ or \$32,400,000.
- Replacement Time: Estimated as 825 days.
- Core and Shell Replacement Cost: Estimated as $\$1000/m^2 \times 12960\ m^2$ or \$12,960,000.
- Maximum Workers per Square Foot: Default value of 0.001 is used.

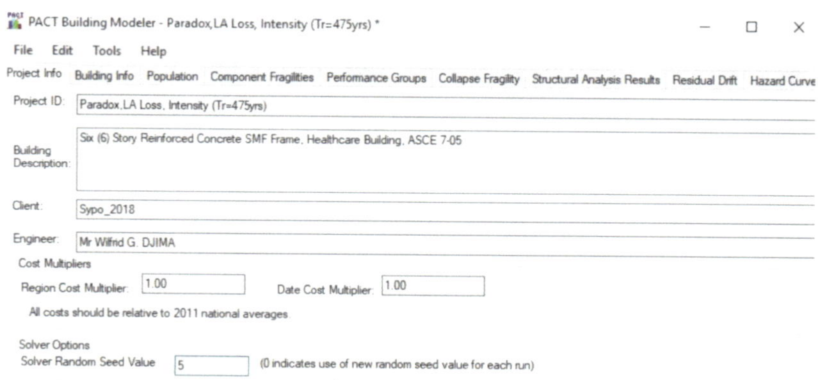

Figure 6.
PACT project information tab.

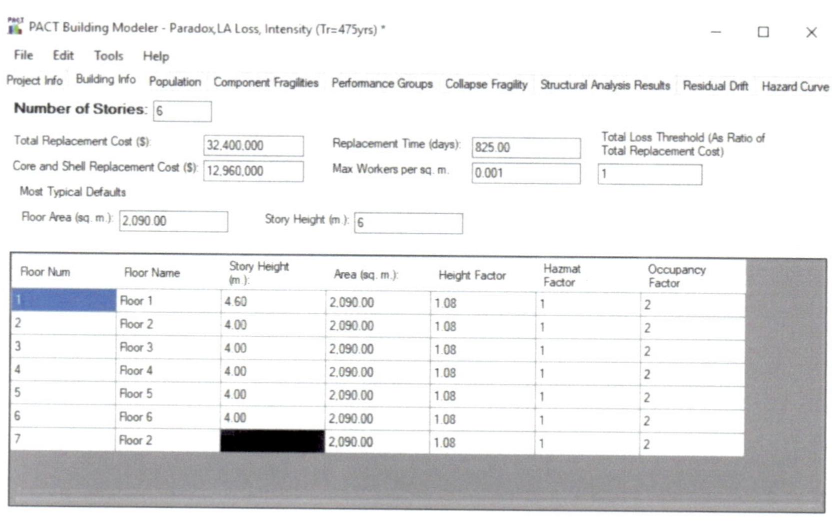

Figure 7.
PACT building information t tab.

- Total Loss Threshold (as Ratio of Total Replacement Cost): Default value of 1.0 is used.
- Floor Area: 2090 m^2
- Floor Height: 4 m Variation in floor height is input via the Floor Number drop down selector, which also permits input of non-typical floor areas.

Figures 8 and **9** show the PACT panel input for Population Modeling.

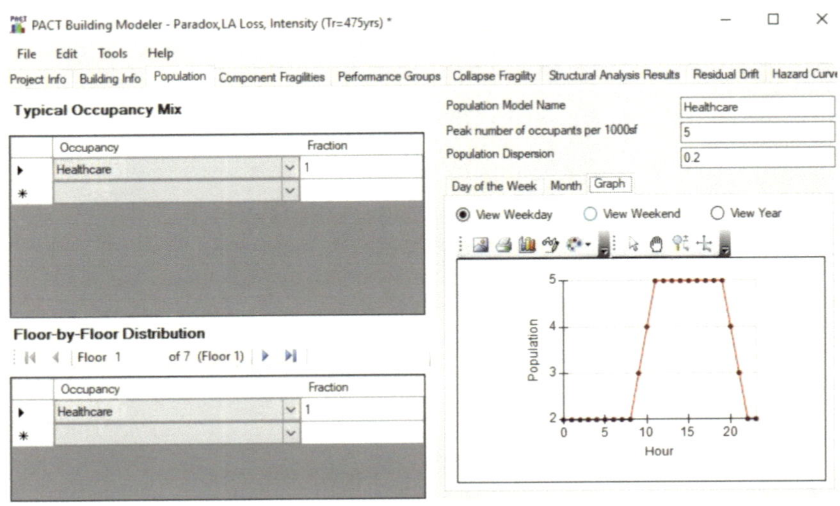

Figure 8.
PACT project population modeling tab graph.

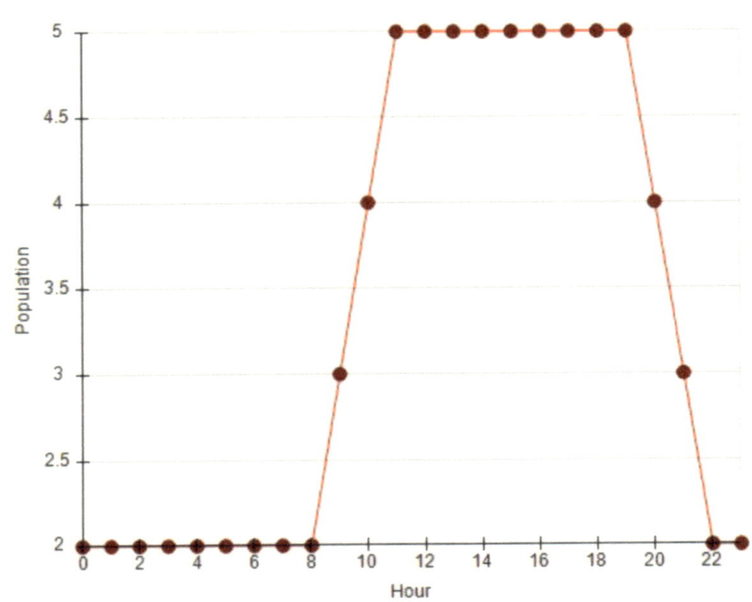

Figure 9.
PACT healthcare population.

4. Structural components

4.1 Structural component fragility specification

The structural components are inputs based on the basic building characteristics previously described. The selection process proceeds progressively from foundations through super structure. The following information summarizes the structural components included in the performance assessment model.

For each floor, the number of special moment frame beam-column joints vulnerable to story drift in each building direction are entered for each of the preselected specifications. **Table 1** and **Figure 10** summarize the defining performance groups in PACT with A, B, C, D, E the building axes in X (2) direction and 1, 2, 3, 4, 5, 6, 7 in Y (1) direction. Input of the post-tensioned slab/column joint information is similarly inserted at each floor; however, these fragilities are input as non-directional. There are 15 joints per floor **(Table 2)**.

4.2 Structural component performance group

The performance group definition process is repeated for each floor and for each direction (including non-directional) as shown in **Figures 11** and **12. Table 3** summarize the total number of performance group per floor.

Joint Location	Fragility Classification Number	Input Direction	Joint Location	Fragility Classification Number	Input Direction
A-1	B1041.002a	1	A-5	B1041.003b	2
A-2	B1041.002a	2	B-1	B1041.003b	1
A-6	B1041.002a	2	B-7	B1041.003b	1
A-7	B1041.002a	1	C-1	B1041.003b	1
E-1	B1041.002a	1	C-7	B1041.003b	1
E-2	B1041.002a	2	D-1	B1041.003b	1
E-6	B1041.002a	2	D-7	B1041.003b	1
E-7	B1041.002a	1	E-3	B1041.003b	2
A-3	B1041.003b	2	E-4	B1041.003b	2
A-4	B1041.003b	2	E-5	B1041.003b	2

Table 1.
Fragility group selections for the beam/column components.

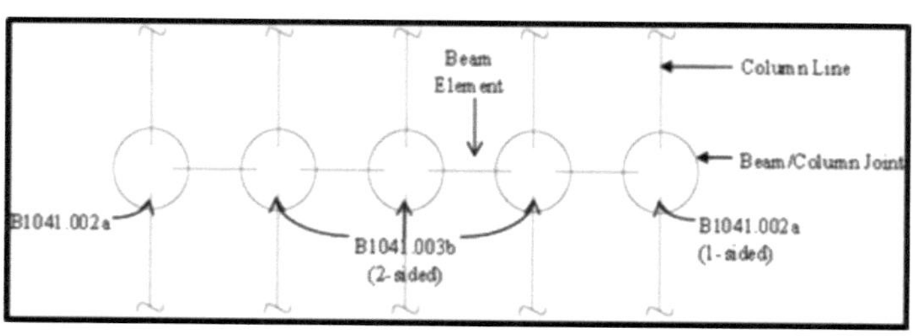

Figure 10.
Illustration of reinforced concrete elements specification selections.

Fragility Classification Number	Direction	Number Per Floor
B1041.002a	1	4
B1041.002a	2	4
B1041.003b	1	6
B1041.003b	2	6

Table 2.
Performance group quantities for reinforced concrete elements.

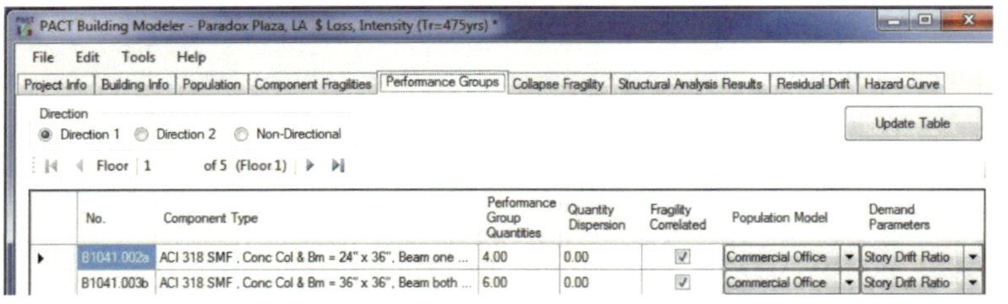

Figure 11.
PACT entries for 1st floor.

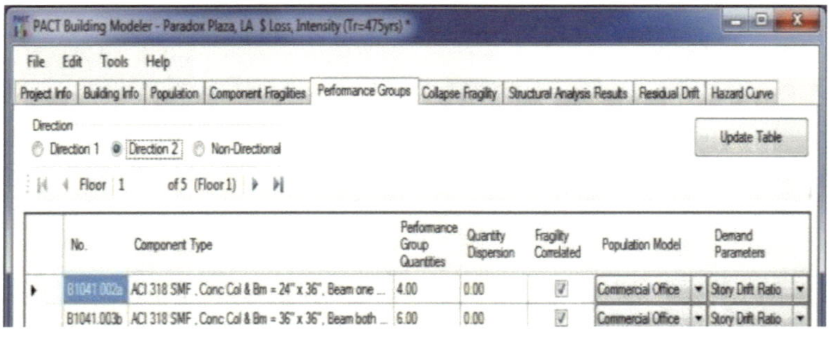

Figure 12.
PACT entries for 1st floor structural performance, direction1.Performance groups, direction 2.

5. Non-structural components

The process of identifying and selecting the type and distribution of the non-structural components can be greatly simplified using the *Normative Quantity Estimation Tool*, provided in Volume 3 of PACT. This tool can be used to generate a floor-by-floor listing of non-structural components with estimates of their performance group quantities with the simple input of building floor areas and occupancies as shown in **Figure 13**.

To use the Normative Quantity Estimation Tool, the building floors and occupancies are entered into the Building Definition Table in the Normative Quantity Estimation Tab. **Figure 14** illustrates input of this information for the example building.

Figure 15 [1] is an example of fragilities curves of reinforced concrete Special Moment Frame at different damage state details in [1] and **Figure 16 is** an example of fragility curves of different types of hospital non-structural equipment from [15].

The **Figures 17** and **18** show the overview of the in-site fragility description for Special Moment Frame (SMF) and Electrical Equipment [1].

Story	w_x (kN)	h_x (m)	wh^k x x	C_{VX}	V_x (kN)	F_x (kN)	$F_x(V)$(Kn/ml)	Computed Displacement	Corresponding drift ratio
Roof	32741.24	24.6	805434.5	0.28	19116.4	5368.30	84.10	0.1201	0.002
5th	32741.24	20.6	674469.54	0.23	19116.4	4495.41	70.42	0.1121	0.003325
4th	32741.24	16.6	543504.58	0.19	19116.4	3622.51	56.75	0.0988	0.004075
3rd	32741.24	12.6	412539.62	0.14	19116.4	2749.62	43.07	0.0825	0.00595
2nd	32741.24	8.6	281574.66	0.10	19116.4	1876.72	29.40	0.0587	0.006275
1st	32741.24	4.6	150609.7	0.05	19116.4	1003.83	15.73	0.0336	0.00730434
		<&$$$;>	2868132.6	1		19116.40	—	—	—

Table 3.
Lumped weight distribution + lateral forces and story drift ratio.

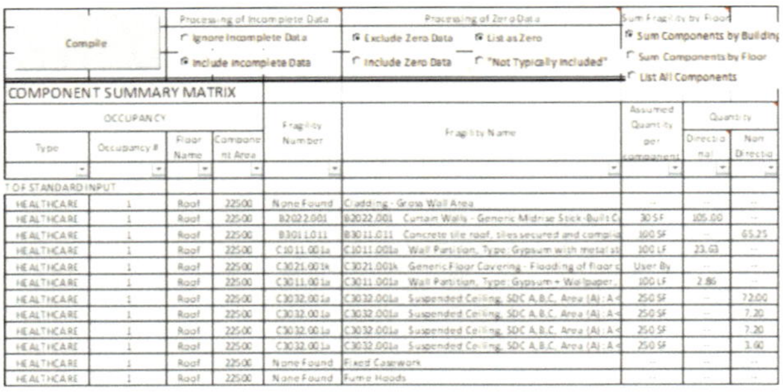

Compile	Processing of Incomplete Data: Ignore Incomplete Data / Include Incomplete Data	Processing of Zero Data: Exclude Zero Data / List as Zero / Include Zero Data / "Not Typically Included"	Sum Fragility by Floor: Sum Components by Building / Sum Components by Floor / List All Components

COMPONENT SUMMARY MATRIX

OCCUPANCY				Fragility Number	Fragility Name	Assumed Quantity per component	Quantity	
Type	Occupancy #	Floor Name	Component Area				Directional	Non Directional
T OF STANDARD INPUT								
HEALTHCARE	1	Roof	22500	None Found	Cladding - Gross Wall Area	--	--	--
HEALTHCARE	1	Roof	22500	B2022.001	B2022.001 Curtain Walls - Generic Midrise Stick-Built C	30 SF	105.00	--
HEALTHCARE	1	Roof	22500	B3011.011	B3011.011 Concrete tile roof, tiles secured and compli	100 SF	--	65.25
HEALTHCARE	1	Roof	22500	C1011.001a	C1011.001a Wall Partition, Type: Gypsum with metal st	100 LF	23.63	--
HEALTHCARE	1	Roof	22500	C3021.001k	C3021.001k Generic Floor Covering - Flooding of floor c	User By	--	--
HEALTHCARE	1	Roof	22500	C3011.001a	C3011.001a Wall Partition, Type: Gypsum + Wallpaper	100 LF	2.86	--
HEALTHCARE	1	Roof	22500	C3032.001a	C3032.001a Suspended Ceiling, SDC A,B,C, Area (A) : A<	250 SF	--	72.00
HEALTHCARE	1	Roof	22500	C3032.001a	C3032.001a Suspended Ceiling, SDC A,B,C, Area (A) : A<	250 SF	--	7.20
HEALTHCARE	1	Roof	22500	C3032.001a	C3032.001a Suspended Ceiling, SDC A,B,C, Area (A) : A<	250 SF	--	7.20
HEALTHCARE	1	Roof	22500	C3032.001a	C3032.001a Suspended Ceiling, SDC A,B,C, Area (A) : A<	250 SF	--	3.60
HEALTHCARE	1	Roof	22500	None Found	Fixed Casework	--	--	--
HEALTHCARE	1	Roof	22500	None Found	Fume Hoods	--	--	--

Figure 13.
Normative quantity estimation tool, component summary matrix showing non-structural inventories.

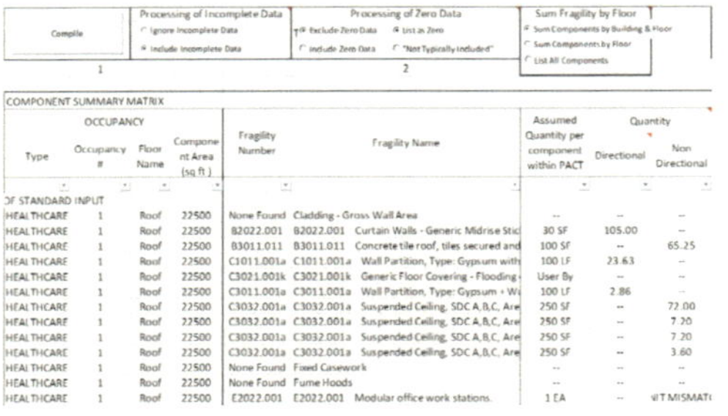

Compile	Processing of Incomplete Data: Ignore Incomplete Data / Include Incomplete Data	Processing of Zero Data: Exclude Zero Data / List as Zero / Include Zero Data / "Not Typically Included"	Sum Fragility by Floor: Sum Components by Building & Floor / Sum Components by Floor / List All Components
1		2	

COMPONENT SUMMARY MATRIX

OCCUPANCY				Fragility Number	Fragility Name	Assumed Quantity per component within PACT	Quantity	
Type	Occupancy #	Floor Name	Component Area (sq ft)				Directional	Non Directional
OF STANDARD INPUT								
HEALTHCARE	1	Roof	22500	None Found	Cladding - Gross Wall Area	--	--	--
HEALTHCARE	1	Roof	22500	B2022.001	B2022.001 Curtain Walls - Generic Midrise Stic	30 SF	105.00	--
HEALTHCARE	1	Roof	22500	B3011.011	B3011.011 Concrete tile roof, tiles secured and	100 SF	--	65.25
HEALTHCARE	1	Roof	22500	C1011.001a	C1011.001a Wall Partition, Type: Gypsum with	100 LF	23.63	--
HEALTHCARE	1	Roof	22500	C3021.001k	C3021.001k Generic Floor Covering - Flooding	User By	--	--
HEALTHCARE	1	Roof	22500	C3011.001a	C3011.001a Wall Partition, Type: Gypsum + Wa	100 LF	2.86	--
HEALTHCARE	1	Roof	22500	C3032.001a	C3032.001a Suspended Ceiling, SDC A,B,C, Are	250 SF	--	72.00
HEALTHCARE	1	Roof	22500	C3032.001a	C3032.001a Suspended Ceiling, SDC A,B,C, Are	250 SF	--	7.20
HEALTHCARE	1	Roof	22500	C3032.001a	C3032.001a Suspended Ceiling, SDC A,B,C, Are	250 SF	--	7.20
HEALTHCARE	1	Roof	22500	C3032.001a	C3032.001a Suspended Ceiling, SDC A,B,C, Are	250 SF	--	3.60
HEALTHCARE	1	Roof	22500	None Found	Fixed Casework	--	--	--
HEALTHCARE	1	Roof	22500	None Found	Fume Hoods	--	--	--
HEALTHCARE	1	Roof	22500	E2022.001	E2022.001 Modular office work stations.	1 EA	--	NIT MISMATC

Figure 14.
Normative quantity estimation tool, building definition table.

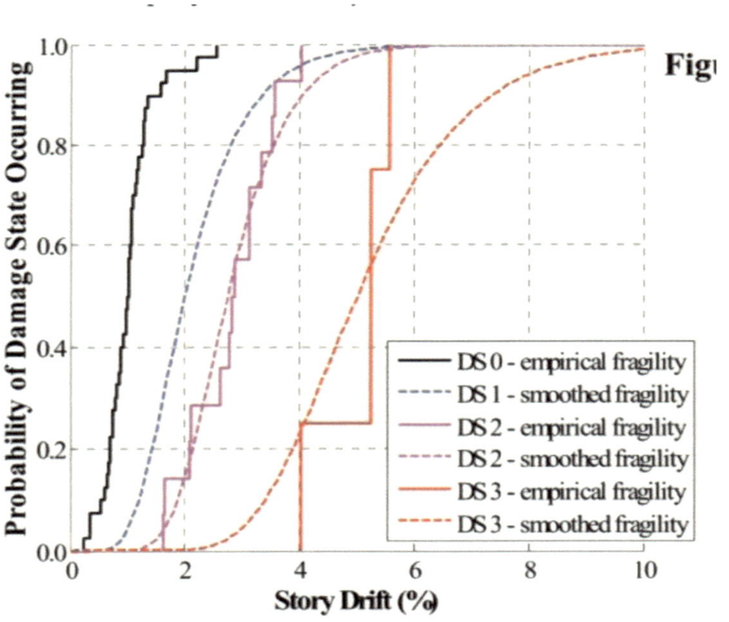

Figure 15.
Example fragility of SMF.

6. Collapse fragility and collapse mode

The collapse fragility is defined as having a median value $\hat{S}_a(T)$ and a dispersion. For this purpose, we have used the non-linear static analysis approach in SAP2000

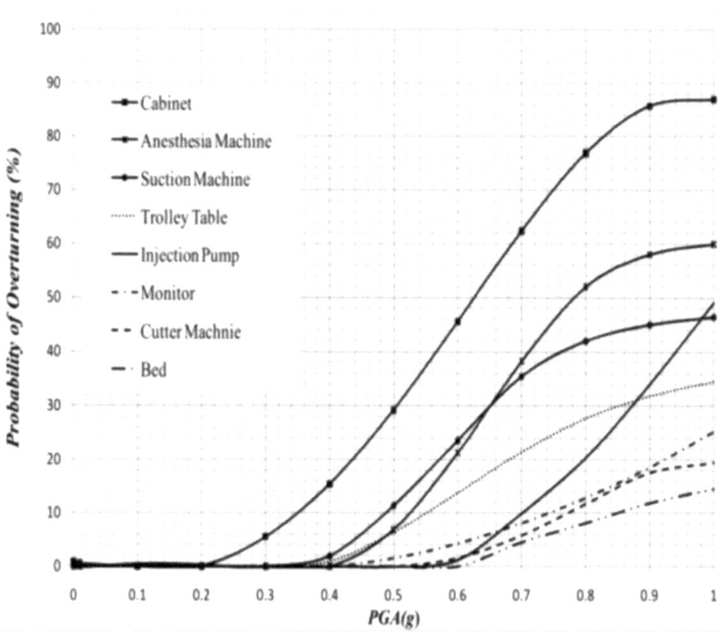

Figure 16.
Fragility curves as a function of earthquake PGA.

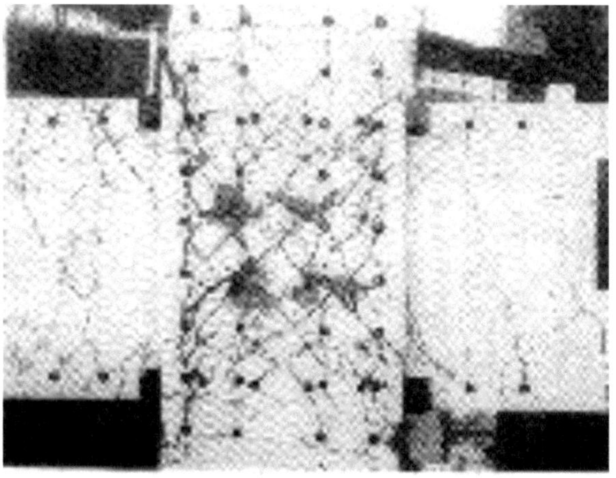

Figure 17.
B1041.002a SMF Beam-to-Column join fragility.

(**Figures 19** and **20**) and the SPO2IDA Tool in [1]. The building was modeled according to the Turkish Building Code 2018 for pushover static analysis in each building direction and the effective stiffness of reinforced concrete columns, beams and shear walls as defined in [6] Section 5 were applied.

Figure 21 illustrates the results of the pushover analysis for both building direction 1 and direction 2, which are identical.

After then, the coordinates of the pushover curves are input to the SPO2IDA Tool provided in [1] along with the building height (24.6 m or 80.71 feet), building weight (199235.85 kN or 44790.02 kips) and fundamental building period (1.96 seconds). Four control points are used to approximate the pushover curve as illustrated in **Figure 22**.

Figure 23 present the results of the SPO2IDA evaluation. The value of $\hat{S}_a(T)$ is estimated as 1.16 g.

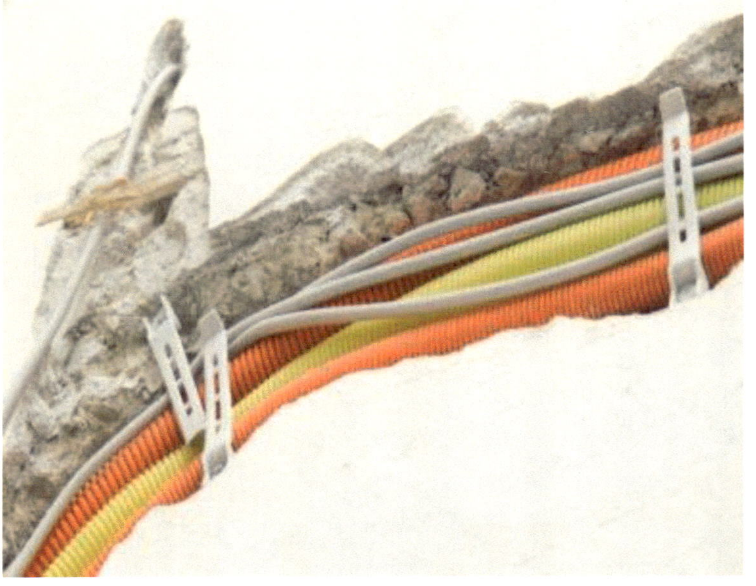

Figure 18.
Electrical distribution.

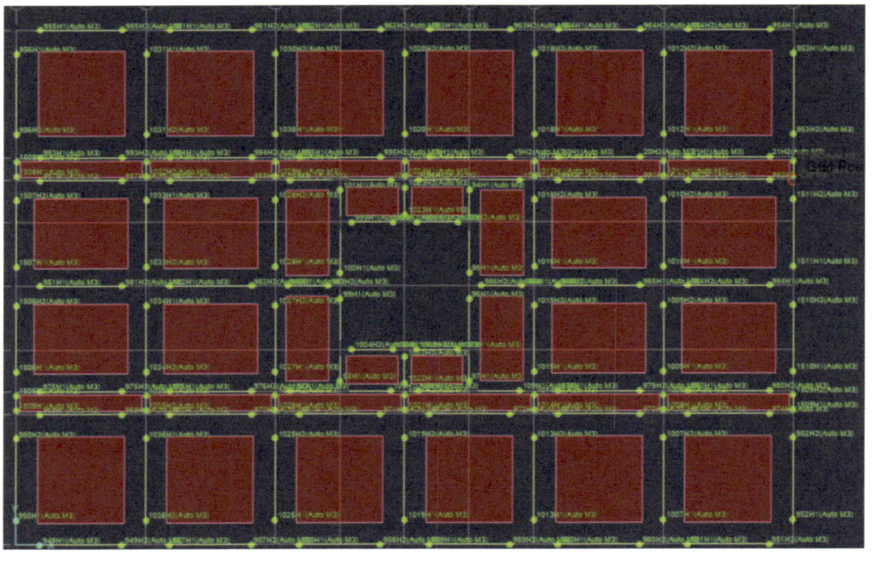

Figure 19.
SAP2000 hinges application at beam.

The collapse fragility is thus defined as having a median value of Sa(T) of 1.16 g and a dispersion of 0.6 as entered into the PACT Collapse Fragility panel (**Figure 24**).

The number of independent collapse modes which can occur and thus the probability of each is difficult to predict analytically. To figure out these data, the user must use judgment supported upon building type, structural system, experience, and analytical inferences. When using the simplified analysis approach, limited analytical information regarding potential collapse modes is out there. For this instance, just one mode of collapse is taken under consideration. More information's gained from numerous response history analyses can give additional insight into potential collapse modes.

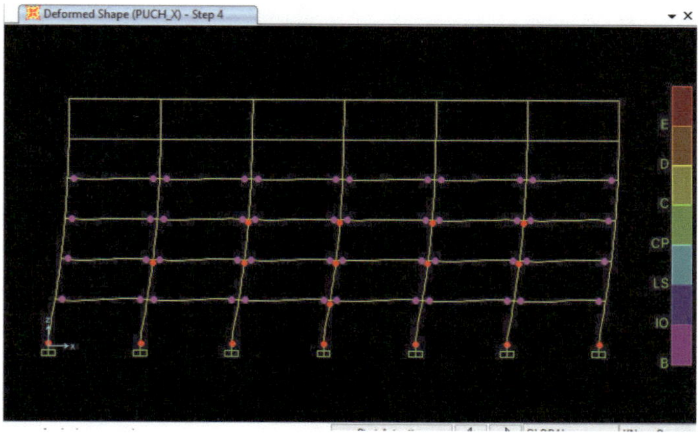

Figure 20.
Plastic hinge map in X direction earthquake loading.

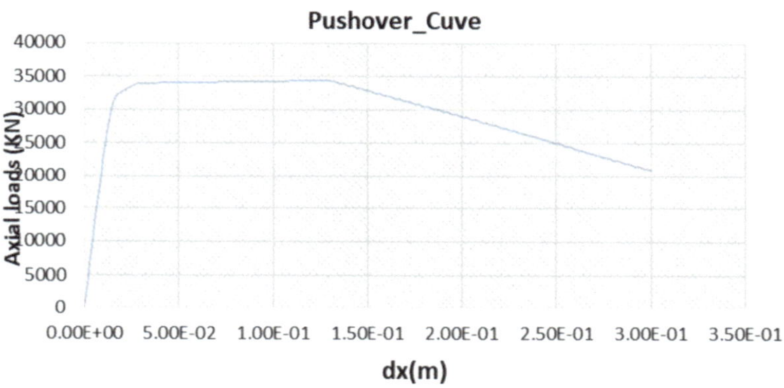

Figure 21.
Pushover curve developed by analysis.

7. Define earthquake Hazard

Ground motion prediction equations provide estimates of spectral response acceleration parameters for specified earthquake magnitude and site-to-source distance based on regression analyses of past strong motion recordings.

Most ground motion prediction equations provide geometric mean (geomean) spectral response accelerations represented by the quantity:

$$\mathbf{S_{gm}} = \sqrt{\mathbf{S_x(T) * S_y(T)}} \tag{1}$$

where S_x and Sy are orthogonal components of spectral response acceleration at period T. The x and y directions could represent the actual recorded orientations, or they could represent a rotated axis orientation.

Intensity-based assessments require a target acceleration response spectrum and suites of 11 pair of ground motions scaled for compatibility with this spectrum (see **Figure 25**). **Figure 26** represents the selected ground motion pairs with geomean spectra that are similar in shape to the target response spectrum.

To determine the building's fundamental translational periods in two orthogonal directions, modal analysis is performed. The fundamental periods in x- and y-directions are 1.94 sec. and 1.98 sec., respectively. Then, the average fundamental period of the building is considered as:

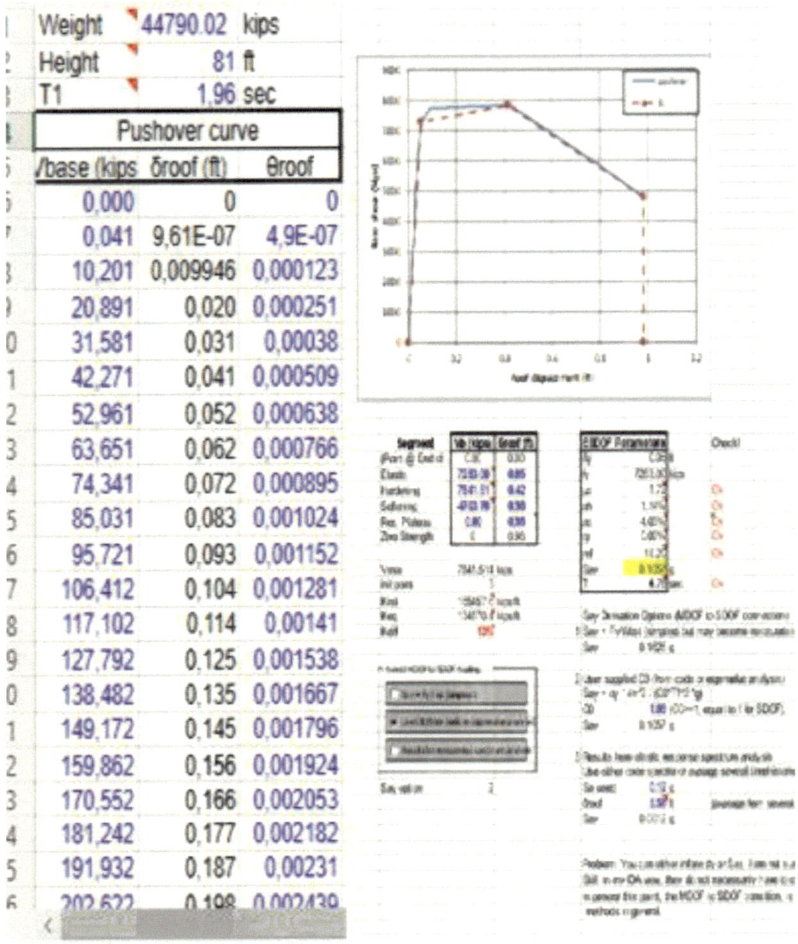

Figure 22.
SPO2IDA tool, SPO tab.

$$\overline{\mathrm{T}} = \frac{T_x + T_y}{2} = 1.96\ sec \tag{2}$$

8. Analyze building response

Simplified analysis is used to generate median estimates of peak transient drift, peak floor accelerations and residual drifts. Associated dispersions are generated using simplified analysis. Peak total floor velocities are not generated since none of the vulnerable building components use this demand parameter. A linear building model is constructed using the modeling criteria of [8], linear static procedure.

8.1 Estimate median story drift ratio

Firstly, we determined the pseudo lateral force by the formula:

$$\boldsymbol{V} = \boldsymbol{C_1 C_2 S_a (T_1) W_1} \tag{3}$$

where C1 is an adjustment factor for inelastic displacements; C2 is an adjustment factor for cyclic degradation; $Sa(T_1)$ is the 5% damped spectral acceleration at the

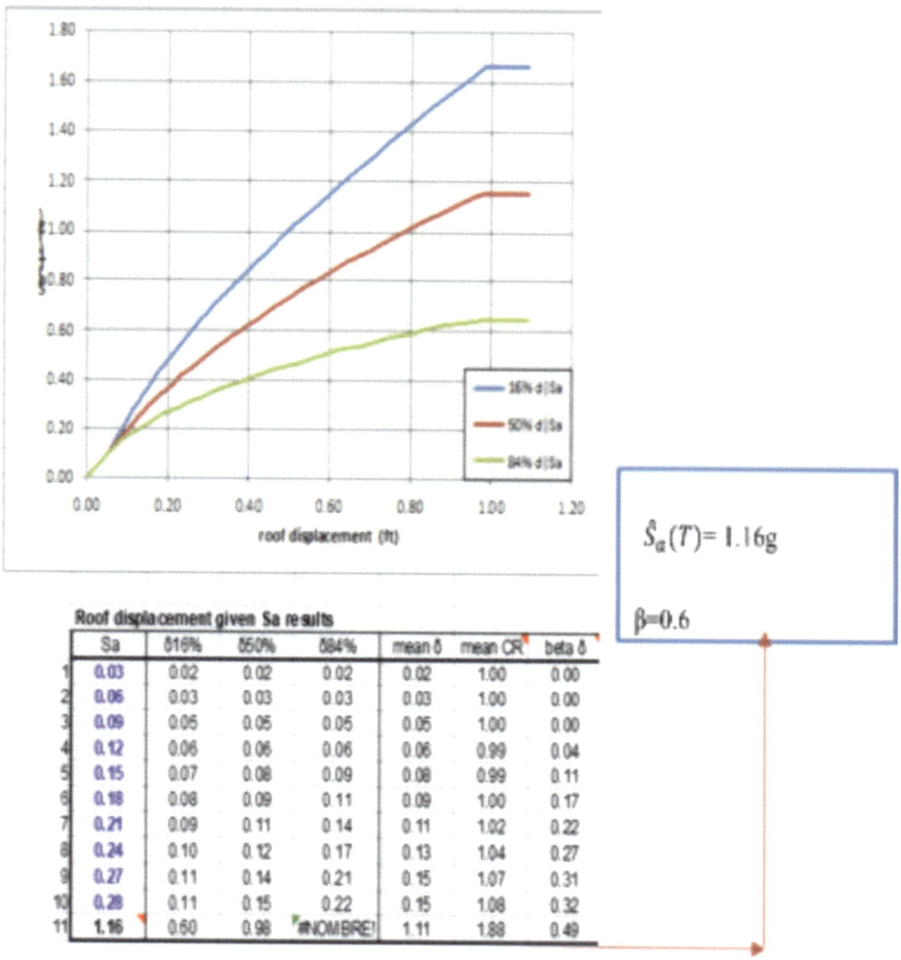

	Sa	δ16%	δ50%	δ84%	mean δ	mean CR	beta δ
1	0.03	0.02	0.02	0.02	0.02	1.00	0.00
2	0.06	0.03	0.03	0.03	0.03	1.00	0.00
3	0.09	0.05	0.05	0.05	0.05	1.00	0.00
4	0.12	0.06	0.06	0.06	0.06	0.99	0.04
5	0.15	0.07	0.08	0.09	0.08	0.99	0.11
6	0.18	0.08	0.09	0.11	0.09	1.00	0.17
7	0.21	0.09	0.11	0.14	0.11	1.02	0.22
8	0.24	0.10	0.12	0.17	0.13	1.04	0.27
9	0.27	0.11	0.14	0.21	0.15	1.07	0.31
10	0.28	0.11	0.15	0.22	0.15	1.08	0.32
11	1.16	0.60	0.98	#NOMBRE!	1.11	1.88	0.49

Figure 23.
SPO2IDA tool, IDA tab.

fundamental period of the building, in the direction under consideration, for the selected level of ground shaking; and W_1 is the first modal effective weight in the direction under consideration, taken as not less than 80% of the total weight, W.

Figures 27 and **28** and **Table 3** show the computed lateral displacement in X (2) direction and the corresponding drift ratio.

$$\Delta_i^* = H_{\Delta i}(S, T_1, h_i, H) \times \Delta_i \; i = 1 to N \quad (4)$$

where $H_{\Delta i}(S, T_1, h_i, H)$ is the drift modification factor for story i computed.

$$ln\,(H_{\Delta i}) = a_0 + a_1 T_1 + a_2 S + a_3 \frac{h_{i+1}}{H} + a_4 \left(\frac{h_{i+1}}{H}\right)^2 + a_5 \left(\frac{h_{i+1}}{H}\right)^3, S \geq 1, i = 1 \; to \; N \quad (5)$$

With

$$T_1 = 1.96s, H = 24.6m$$

Values of a_0 through a_5 for 6 stories or less in height are provided in [1] Table 5-4 by using the strength ratio given by:

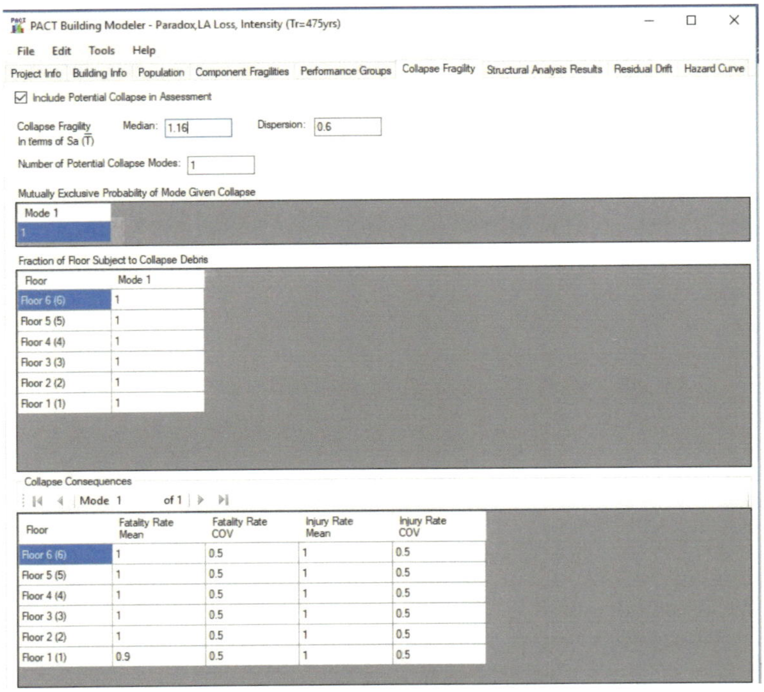

Figure 24.
PACT collapse fragility tab.

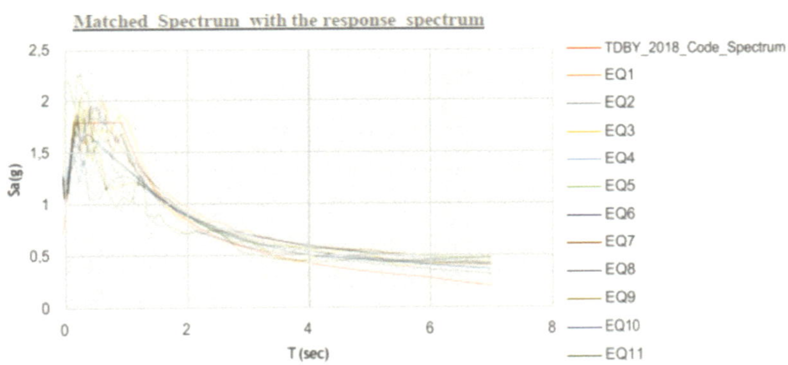

Figure 25.
Selected 11 earthquake ground motions response spectrum scaled according to the design spectra.

$$S = \frac{S_a(T)V}{V_{y1}} \tag{6}$$

The value of V_{y1} is taken from the pushover analysis used to estimate the collapse fragility (see SPO2IDA input, **Figure 22**, Elastic Segment end point) (**Table 4**).

8.2 Estimate peak floor acceleration

At the base of the building, peak floor acceleration is taken as equal to the peak ground acceleration. At other floor levels, i, the estimated median peak floor acceleration, a^*(**Table 5**) relative to a fixed point in space, is derived from the peak ground acceleration using:

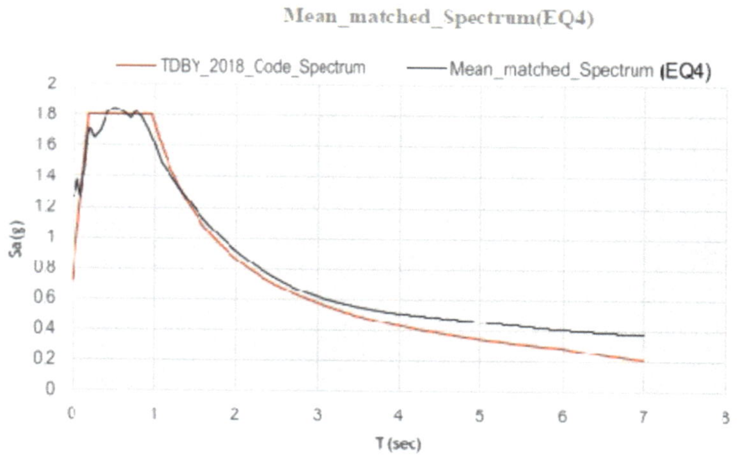

Figure 26.
Mean matched Spectrum.

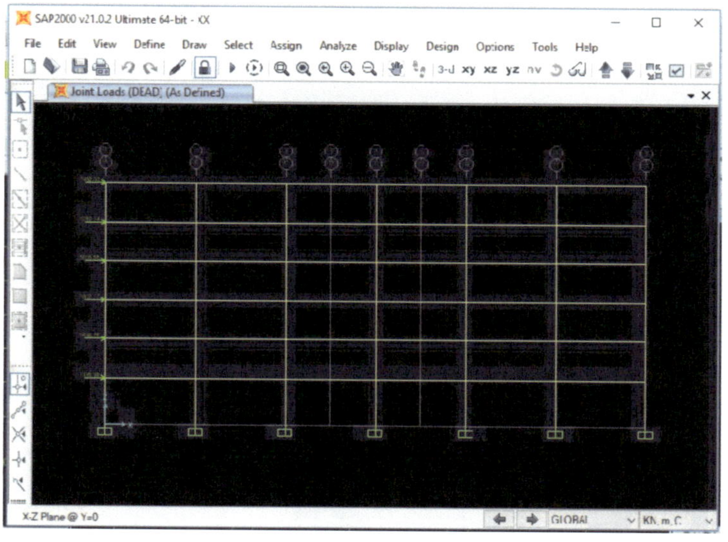

Figure 27.
SAP2000 lateral force applied.

$$a_i^* = H_{ai}(S, T_1, h_i, H) \times PGA \; i = 2 to N + 1 \tag{7}$$

$$ln\,(H_{ai}) = a_0 + a_1 T_1 + a_2 S + a_3 \frac{h_i + 1}{H} + a_4 \left(\frac{h_i + 1}{H}\right)^2 + a_5 \left(\frac{h_i + 1}{H}\right)^3, S \geq 1, i = 1 to N \tag{8}$$

The coefficients of a_0 through a_5 for 6 stories or less in height are provided in [1] **Table 5-4**.

8.3 Estimate of dispersion for median story ratio, median peak floor acceleration, and median peak floor velocity

For intensity-based, separate values of total dispersion for drift ratio, β_{SD}, floor acceleration, β_{FA}, and floor velocity, β_{FV}, are needed.

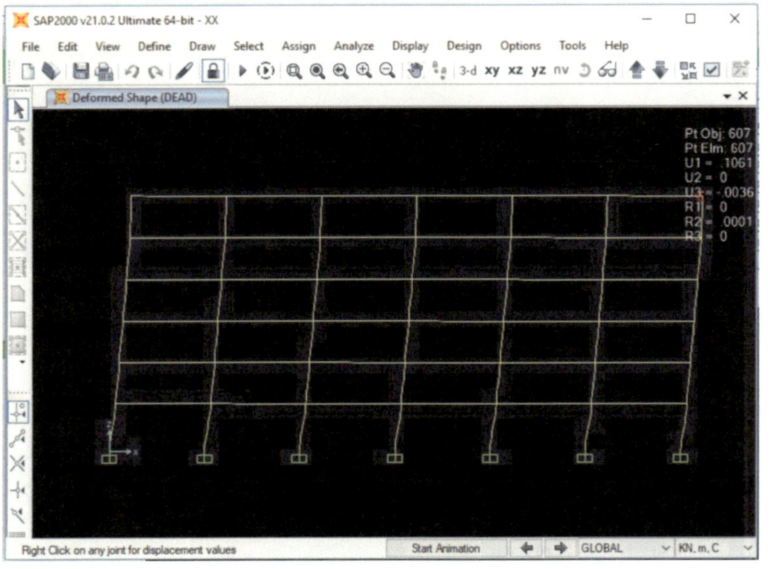

Figure 28.
X-X deformed shape.

Story	Δi	hi + l/H	lnHΔi	HΔi	Δ*i
1	0.0073043	0.1869919	0.2089634	1.2323998	0.0090019
2	0.006275	0.3495935	−0.009875	0.990174	0.0062133
3	0.00595	0.5121951	−0.107092	0.8984431	0.0053457
4	0.00595	0.6747967	−0.082688	0.9206379	0.0054778
5	0.004075	0.8373984	0.0633358	1.0653845	0.0043414
6	0.003325	1	0.3309807	1.3923329	0.0046295

Table 4.
Median story drift ratio estimates.

Story	PGA	hi + 1/H	lnHai	Hai	a*i
1	0.695	—	—	—	0.695
2	0.695	0.1869919	−0.357947	0.69911	0.4858814
3	0.695	0.3495935	−0.364289	0.6946906	0.48281
4	0.695	0.5121951	−0.37063	0.6902992	0.4797579
5	0.695	0.6747967	−0.376972	0.6859355	0.4767252
6	0.695	0.8373984	−0.383313	0.6815995	0.4737116
roof	0.695	1	−0.389655	0.6772908	0.4707171

Table 5.
Median floor acceleration estimates.

$$\beta_{SD} = \sqrt{{\beta_{a\Delta}}^2 + {\beta_m}^2} \tag{9}$$

$$\beta_{FA} = \sqrt{{\beta_{aa}}^2 + {\beta_m}^2} \tag{10}$$

$$\beta_{FV} = \sqrt{{\beta_{a\Delta}}^2 + {\beta_m}^2} \tag{11}$$

These are calculated based on [1] Table 5-6 values for analysis record- to-record dispersion for drift, $\beta_{a\Delta}$, acceleration, β_{aa}, and velocity, β_{av}, respectively, by interpolation approach.

9. Estimate median residual story drift ratio and dispersion

Since the requirements for direct simulation of residual drift are computationally complex and not practical for general implementation in design, the following equations were developed to estimate the median residual drift ratio, response of the structure:

$$\mathbf{\Delta_r = 0 \text{ for } \Delta \leq \Delta_y}$$
$$\{\Delta_r = 0.3(\Delta - \Delta_y) \text{ for } \Delta_y < \Delta < 4\Delta_y)\} \quad (12)$$
$$\Delta_r = (\Delta - 3\Delta_y) \text{ for } \Delta \geq 4\Delta_y)$$

where Δ is the median story drift ratio calculated by analysis, and Δ y is the median story drift ratio calculated at yield.

The peak transient drift ratios were estimated. The yield drift ratio is obtained from the capacity curve derived from the pushover analysis used to generate the story shear at yield (**Figure 29**).

At yield, the peak transient acceleration is determined by the equation for the building fundamental period between the range of 0.7 sec to 2 sec:

$$\boldsymbol{S_a(T) = \frac{S_a(1)}{T} = 0.755g} \quad (13)$$

From the capacity curve, the corresponding roof displacement for the peak transient acceleration at yield then is 0.0045 m. Thus, the yield drift ratio is:

$$\Delta_y = \frac{0.0045}{24.6} = 0.0002 \quad (14)$$

The maximum transient drift ratio for the building occurs at the first story (Δ= 0.0090).

$$\Delta_r = (0.0090–3 * 0.0002) = 0.0084$$

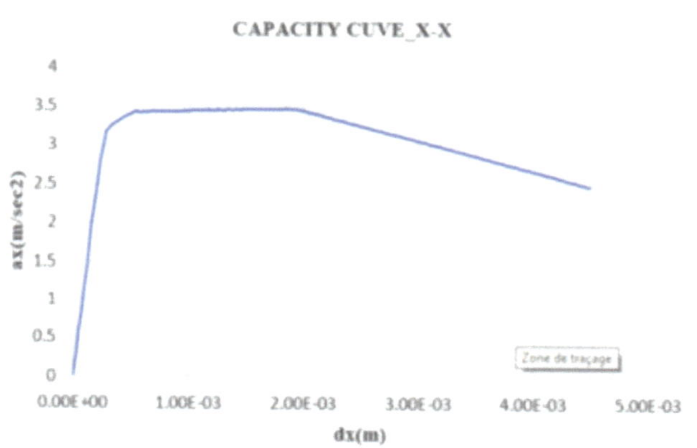

Figure 29.
Capacity curve from pushover analysis.

The median estimate of residual drift obtained from the simplified analysis method is assigned a dispersion of 0.8.

10. Input response and calculate performance

The median demand estimates for peak transient drift ratio, peak floor acceleration, and residual drift ratio are input to PACT for direction 1 and direction 2 in the Structural Analysis Results tab (see **Figures 30** and **31**). Then, PACT uses this information to make damage state assessments for all building components

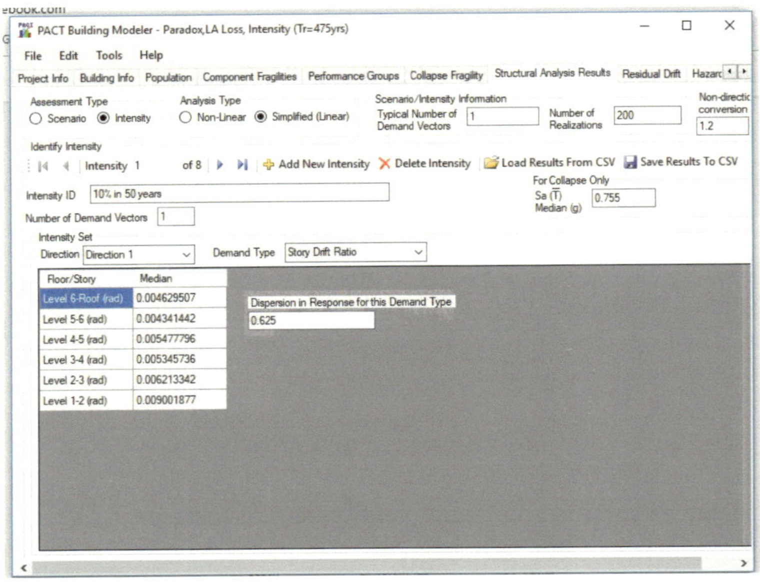

Figure 30.
PACT peak transient drift ratio input tab.

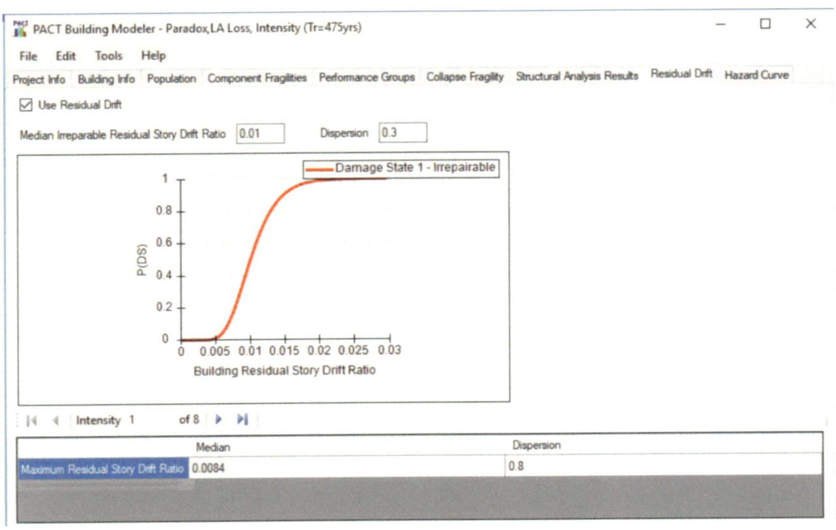

Figure 31.
PACT residual drift tab.

contained in the building model. A Monte Carlo procedure is used to assess a range of possible outcomes as repair cost and repair time.

11. Review results and comments

As showed in the **Figure 32**, the estimated median repair cost is shown as \$2.710.445, which corresponds to 8.37% of the building's total replacement cost. From the isograph on the **Figure 32**, it is seen that the yellow stick representing the performance group B2022.001 (Curtain Walls - Generic Midrise Stick-Built Curtain wall, Config: Monolithic, Lamination: Unknown, Glass Type: Unknown,

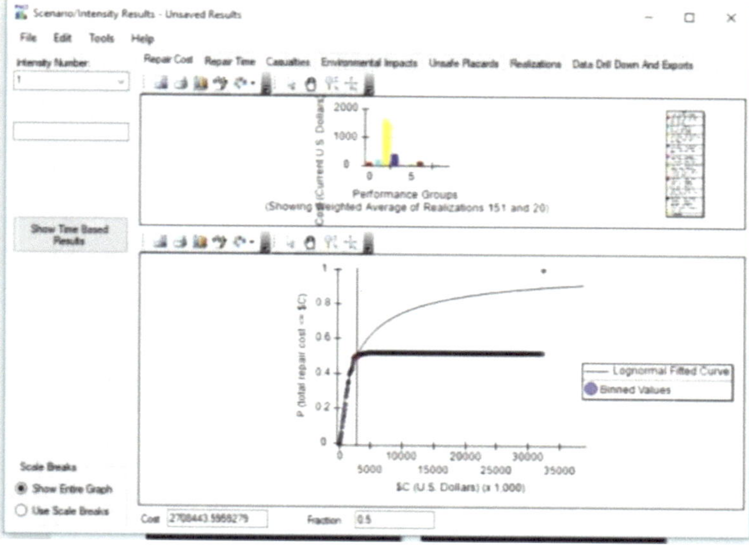

Figure 32.
PACT repair cost tab.

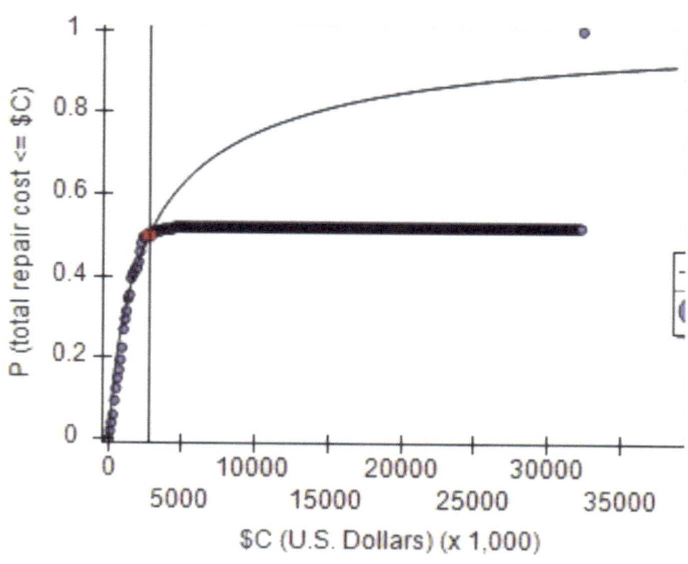

Figure 33.
PACT repair cost graph.

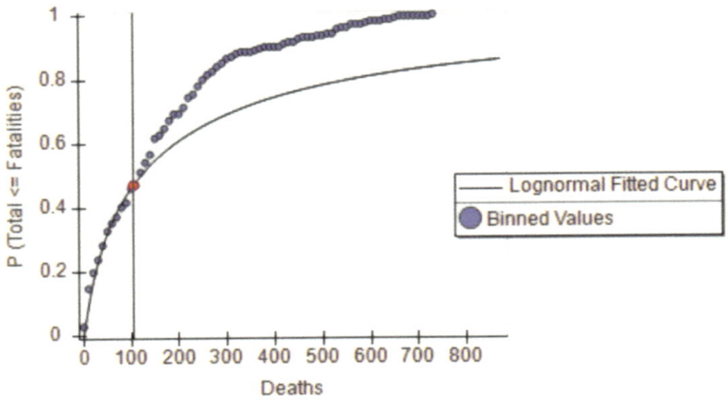

Figure 34.
PACT Casualties/ Deaths Results.

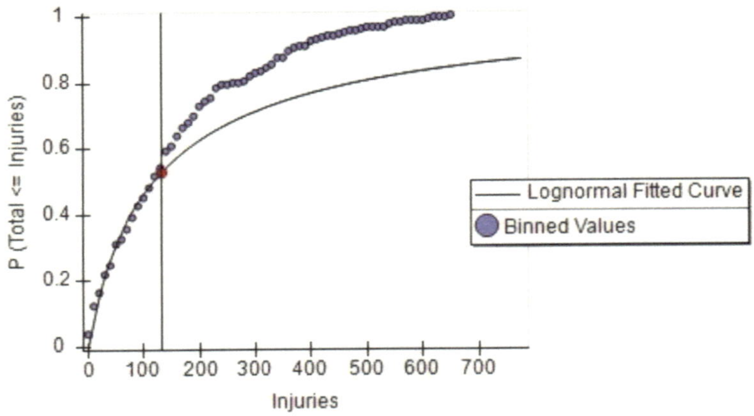

Figure 35.
PACT casualties / injuries.

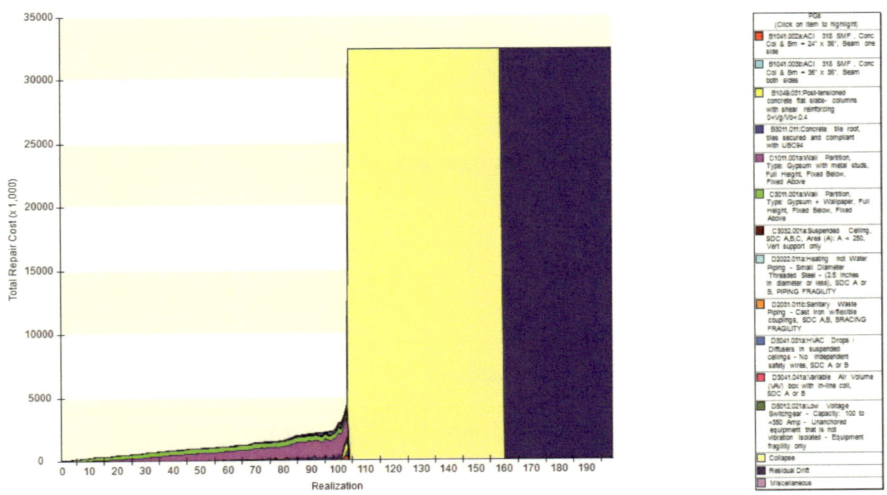

Figure 36.
PACT repair cost tab with realizations.

Details: Aspect ratio = 6:5), explain its contribution to the most of the building total repair cost with approximately $1.700.000.

Figures 33, 34, and **35** explain that 50% probability that repair cost will not exceed $2.710.445, 145 injuries, 105 deaths.

The performance results of the PACT can be examined in numerous ways. Viewing results by realization reveals that collapse plays a more significant role than residual drift (see **Figure 36**). For approximately 56 of the 200 realizations of the collapse of the component B1049.031 Post-tensioned concrete flat slabs-columns with shear reinforcing and, 40 of the 200 realization for the performance group B3011.011 Concrete tile roof tile secured and compliant for damage being judged irreparable.

12. Conclusion

Performance assessment can provide useful information for many decisions associated with real property. These include: demonstrating equivalence of alternative design approaches, selecting appropriate design criteria for new buildings, determining if an existing building constitutes an acceptable risk for a particular planned use, whether or not it should be upgraded, and if so, to what level, performing benefit–cost studies to determine a reasonable investment for improved seismic resistance in a building, determining whether or not insurance is a cost-effective risk management technique.

In this study, a case study of a typical hospital building has been analyzed. The non-linear static pushover analysis results have showed that the collapse occurs at the building base in Mode-8. The linear static analysis results have demonstrated a maximum roof displacement of 7.69 cm. Consequently, the performance assessment that used the data from these analyses have shown a low repair cost of 8.37% (less than 40%). Thus, based on the past knowledge and recommendations that suggests 40% of the total building replacement cost can be a reasonable threshold for total loss of several buildings, the decision of retrofitting can be given for this case study since the repair cost is less than 40%.

The adaptation of the current study to the typical health-care facilities in many countries as Turkey is still on-going. It is believed that the results of this study will be valuable for the building owners, managers, insurance firms and for the process of benefit–cost performance and risk management.

Author details

Wilfrid Gbenankpon Djima, Abdullah Can Zulfikar and Cüneyt Tüzün
Department of Civil Engineering, Gebze Technical University, Kocaeli, Turkey

*Address all correspondence to: aczulfikar@gtu.edu.tr

References

[1] R. O. Hamburger Simpson Gumpertz, Heger I., C. Rojahn, J. A. Heintz, M. G. Mahoney, APPLIED TECHNOLOGY COUNCIL 201 Redwood Shores Parkway, Suite 240 Redwood City, California 94065 www.ATCouncil.org, Seismic Performance Assessment of Buildings Volume 1 – Methodology, December 2018.

[2] Petal, M. Epidemiology of Deaths and Injuries In The August 17, 1999, 3:02 A.M. M=7.4, Kocaeli Earthquake Report, BU KOERI, 2013, http://www.probina.com.tr/5UDMK/PDF/AE045_FP.pdf.

[3] AHEB, Hastaneler İçin Afete (Depreme) Hazırlıklı Olma Kılavuzu B. Ü Kandilli Rasathanesi ve Deprem Araştırma Enstitüsü Afete Hazırlık Eğitim Projesi kapsamında yayınlanan kitap, 2004 www.koeri.boun.edu.tr.

[4] Diana T., Nicholas C., Riley M. Chung, H. S. Lew, 1994 Northridge Earthquake Performance of Structures, Lifelines, and Fire Protection Systems, May 1994.

[5] İPEK, C., KISTIR, M.R., KUZUCUOĞLU, A.H., “Yapısal olmayan sistemlerin deprem etkileri açısından değerlendirilmesi”, International Burdur Earthquake & Environment Symposium, 2015.

[6] R. O. Hamburger Simpson Gumpertz, Heger I., C. Rojahn, J. A. Heintz, M. G. Mahoney, APPLIED TECHNOLOGY COUNCIL 201 Redwood Shores Parkway, Suite 240 Redwood City, California 94065 www.ATCouncil.org, Seismic Performance Assessment of Buildings Volume 2 – Implementation Guide, December 2018.

[7] R. O. Hamburger Simpson Gumpertz, Heger I., C. Rojahn, J. A. Heintz, M. G. Mahoney, APPLIED TECHNOLOGY COUNCIL 201 Redwood Shores Parkway, Suite 240 Redwood City, California 94065 www.ATCouncil.org, Seismic Performance Assessment of Buildings Volume 3: Supporting Electronic Materials and Background Documentation, December 2018.

[8] Eric L., Susan P., Rob F., Ivy P., Wanda R., Design Guide for Improving Hospital Safety in Earthquakes, Floods, and High Winds, Chapter 1 and 2, June 2007.

[9] Steven L, Matthew H.,, APPLIED TECHNOLOGY COUNCIL 201 Redwood Shores Parkway, Suite 240 Redwood City, California 94065 www.ATCouncil.org, Recommendations for Improved Seismic Performance of Nonstructural Components, September 2018.

[10] Michael M., Robert D. H., Applied Technology Council. "Guidelines for Seismic Performance Assessment of Buildings, May 2007, Available from: https://www.atcouncil.org/pdfs/ATC-58-50.

[11] Djima W., A Thesis Submitted for the Degree of Master of Science/ Department of Civil Engineering/ Earthquake and Structural Engineering Program, Performance Based Design and Its implementation Tool in Healthcare Facilities, Chapter 3 and 4, June 2020.

[12] ASCE/SEI 7, American Society of Civil Engineers, Reston Virgini, Minimum Design Loads for Buildings and Other Structures, pp. 7–9–13-109-117-119-142-153-160-177-185-243, December 2005.

[13] ACI, Building Code Requirements for Masonry Structures and Specifications for Masonry Structures, 530/530.1–2008 May.

[14] ACI 318, Building Code Requirements for Structural Concrete and Commentary, pp 83–107–239-323-329, August 2011.

[15] Turkey Building Earthquake Code (TBEC-2018), Sections, 5 and 15, 2018.

Chapter 15

Evaluation of Liquefaction-Induced Settlement Using Random Forest and REP Tree Models: Taking Pohang Earthquake as a Case of Illustration

Mahmood Ahmad, Xiaowei Tang and Feezan Ahmad

Abstract

A liquefaction-induced settlement assessment is considered one of the major challenges in geotechnical earthquake engineering. This paper presents random forest (RF) and reduced error pruning tree (REP Tree) models for predicting settlement caused by liquefaction. Standard penetration test (SPT) data were obtained for five separate borehole sites near the Pohang Earthquake epicenter. The data used in this study comprise of four features, namely depth, unit weight, corrected SPT blow count and cyclic stress ratio. The available data is divided into two parts: training set (80%) and test set (20%). The output of the RF and REP Tree models is evaluated using statistical parameters including coefficient of correlation (r), mean absolute error (MAE), and root mean squared error (RMSE). The applications for the aforementioned approach for predicting the liquefaction-induced settlement are compared and discussed. The analysis of statistical metrics for the evaluating liquefaction-induced settlement dataset demonstrates that the RF achieved comparatively better and reliable results.

Keywords: liquefaction, random forest, REP tree, settlement

1. Introduction

The evaluation of liquefaction-induced settlements has become an extremely significant issue about the foundations of different buildings, nuclear power plants, and earth dams on sandy soil deposits. Saturated sand deposits when are endured during an earthquake, pore water pressures are known to develop contributing to liquefaction or loss of shear strength. The pore water pressure then begin to dissipate primarily towards the ground surface, followed by a change in the volume of soil deposits which is manifested on the ground surface as settlements. Settlements caused by liquefaction are conventionally predicted using analytical or numerical methods.

Tokimatsu and Seed [1] developed a technique for predicting ground post liquefaction settlements based on volumetric strain, SPT N-value and cyclic stress ratio (CSR) relationships in the case of completely liquefied saturated sands transformed from an experimental relationship between relative sand density, volumetric strain, and maximum shear strain. Ishihara and Yoshimine [2] used an alternative approach to estimate ground settlements based on the safety factor, by means of the maximum shear strain which is an essential factor affecting the post-liquefaction volumetric strain. The liquefaction-induced settlement during the earthquake can be identified if the safety factor and relative density are established. Furthermore, the simplified method was constructed only by a relation between relative density, the factor of safety against liquefaction (FS) and volumetric strain (ε_v) to quantify the settlement of a site where the safety factor of safety against liquefaction was obtained By combining earthquake intensity and SPT N-value with empirical equations to cause measurement error and lead to significant prediction error [3].

Analytical method used to assess liquefaction-induced settlements is based on the effective stress analysis of dynamic response which accounts for the generation and dissipation of excess pore water pressures. When used to evaluate post-liquidation settlements in saturated sand deposits, the volume compressibility coefficient of the sand is required which is very difficult to determine for the liquefied sand layer [4]. Shamoto et al., [4] suggested a simplified approach for estimating liquefaction-induced settlements of saturated sand deposits, based on the experimental evidence that there is an almost linear relationship between the function of the void ratio and the logarithm of the maximum shear strain induced during cyclic loading.

In numerical analysis, earthquake-induced liquefaction in the free-field may be interpreted as a 1D phenomenon occurring along a vertical soil column in which seismic-induced cyclic shear and compressive forces increase the pore pressure and hence cause a reduction in the transient soil strength and stiffness. Reconsolidation arises in the soil after liquefaction due to the dissipation of the excess pore pressure (Δu) by means of water flow, resulting in the vertical settlement of the ground surface [5].

Park et al. [6] established a simple and sustainable method for predicting liquefaction-induced settlement using ANN. Tang et al. [3] found that the ANN and Bayesian Belief Networks (BBN) predictive outcomes are better than the Ishihara and Yoshimine simplified approach.

Pohang earthquake (M_w = 5.4) that hit the Heunghae Basin around Pohang city had a liquefaction-induced damages—settlement and lateral displacement. In this study liquefaction-induced settlement is considered as a case of illustration. Several efforts have been made since the event to evaluate the post-earthquake damages [7–11]. Nevertheless, the liquefaction-induced settlement has received little attention. Settlement caused by liquefaction is commonly calculated by taking into account various factors and following several sophisticated analytical and numerical procedures. Nevertheless, in most cases it may not be possible to acquire such parameters in the field, as some of the required data may not be obtainable. The main purpose of this study is to evaluate liquefaction-induced settlement based on the database of field observations. To achieve this purpose, the random forest and REP tree techniques are used to develop two new models for evaluation of liquefaction-induced settlement. Although these techniques have been successfully applied in many domains, the application in geotechnical earthquake engineering is limited based on the literature surveys.

The remainder of this chapter is organized as follows: Section 2 briefly provides the description of data acquisition for liquefaction-induced settlement calculation.

Section 3 presents the methodology used to evaluate settlement caused due to earthquakes; an overview of the random forest and Rep tree techniques. Section 4 presents the development of the liquefaction-induced settlement models. Detailed results of the proposed models are discussed by performance evaluation measures are presented in Section 5, followed by conclusions in Section 6.

2. Data acquisition

In this study, Park et al. [6] collected database from the Integrated DB Centre of National Geotechnical Information, Korea [12] and the UBCSAND constitutive effective stress model [13] was used to develop predictive models. SPT data were obtained for five different borehole sites near the epicenter of the earthquake at Pohang. The input parameters for the RF and REP Tree models are depth (m), unit weight (kN/m^3), corrected SPT blow count ($N_{1(60)}$) and cyclic stress ratio (CSR) and the output is the observed settlement (mm). For details about the database, readers can refer to Park et al. [6]. The summary of the data base comprised 100 data points (20 data for each borehole) along with the corresponding settlement values is shown in **Table 1**.

Borehole	Depth (m)	Unit Weight (kN/m^3)	$N_{1(60)}$	CSR	Settlement (mm)
BH-A-1	1	20	11	0.33	0.5
	2	20	11	0.31	0.5
	3	20	14	0.29	0.8
	4	20	16	0.28	1.4
	5	20	5	0.27	3.3
	6	20	10	0.26	3.4
	7	20	5	0.27	3.4
	8	20	6	0.29	2.5
	9	20	9	0.3	1.6
	10	20	9	0.31	1
	11	18	25	0.31	0.4
	12	18	25	0.31	0.3
	13	18	25	0.32	0.2
	14	18	25	0.32	0.3
	15	18	25	0.32	0.3
	16	18	25	0.32	0.3
	17	18	25	0.32	0.3
	18	18	25	0.32	0.2
	19	18	25	0.31	0.3
	20	18	25	0.31	0.3
BH-A-2	1	20	15	0.35	0.4
	2	20	17	0.32	0.8
	3	20	17	0.3	1.6

Borehole	Depth (m)	Unit Weight (kN/m³)	$N_{1(60)}$	*CSR*	Settlement (mm)
	4	20	7	0.29	3.1
	5	20	6	0.27	2.8
	6	21	13	0.31	1.4
	7	21	18	0.34	0.8
	8	21	13	0.36	0.9
	9	21	11	0.37	1.1
	10	21	13	0.36	0.8
	11	16	2	0.36	0
	12	16	1	0.37	0
	13	16	1	0.38	0
	14	16	1	0.39	0
	15	16	1	0.39	0
	16	16	1	0.39	0
	17	16	1	0.39	0
	18	16	1	0.39	0
	19	16	1	0.38	0
	20	16	1	0.37	0
BH-A-3	1	18	6	0.24	0.6
	2	18	8	0.28	1.4
	3	18	10	0.3	2
	4	18	10	0.29	2.3
	5	18	11	0.28	2
	6	18	10	0.3	1.8
	7	18	11	0.32	1.4
	8	18	11	0.33	1.3
	9	18	12	0.34	1.2
	10	18	13	0.34	1
	11	21	25	0.34	0.7
	12	21	25	0.33	0.6
	13	21	25	0.33	0.6
	14	21	25	0.33	0.5
	15	21	25	0.32	0.5
	16	21	25	0.32	0.4
	17	21	25	0.31	0.5
	18	21	25	0.31	0.4
	19	21	25	0.3	0.4
	20	21	25	0.3	0.5
BH-A-4	1	20	5	0.23	1.1
	2	20	7	0.27	1.9
	3	20	18	0.27	1.6

Borehole	Depth (m)	Unit Weight (kN/m^3)	$N_{1(60)}$	*CSR*	Settlement (mm)
	4	20	9	0.27	2.8
	5	20	6	0.26	2.8
	6	20	11	0.31	1.6
	7	20	9	0.34	1.4
	8	21	25	0.36	0.6
	9	21	25	0.38	0.6
	10	21	25	0.38	0.6
	11	21	25	0.38	0.6
	12	21	25	0.37	0.5
	13	16	7	0.22	0
	14	16	1	0.21	0
	15	16	0	0.21	0
	16	16	2	0.22	0
	17	16	3	0.22	0
	18	16	3	0.22	0
	19	16	3	0.22	0
	20	16	3	0.22	0
BH-A-5	1	20	11	0.32	0.5
	2	20	10	0.31	1.6
	3	20	9	0.29	2.6
	4	20	11	0.3	1.9
	5	20	11	0.32	1.5
	6	20	10	0.33	1.4
	7	20	15	0.34	0.9
	8	20	15	0.35	0.8
	9	21	25	0.34	0.6
	10	21	25	0.34	0.6
	11	21	25	0.34	0.6
	12	21	25	0.33	0.6
	13	21	25	0.33	0.7
	14	21	25	0.33	0.7
	15	18	15	0.33	1.1
	16	18	11	0.33	1.4
	17	18	12	0.32	1
	18	16	14	0.32	0.1
	19	16	10	0.32	0
	20	16	7	0.32	0

Note: Borehole (BH-A-5) data comprised of 20 data points is used as testing dataset in this study.

Table 1.
Summary of liquefaction-induced settlement database.

3. Methodology

3.1 Random forest

Random Forest (RF) is an ensemble machine learning technique driven by the development of a large number of decision trees that is produced by Leo Breiman [14]. Unlike DT, which uses all the features to construct a tree-like classification graph, RF uses an "efficient bagging" learning algorithm which integrates random selection of features with bagging. If one or a few features are very good predictors for target performance, it will pick this subset of features to construct a tree-like graph. This type of sample is known as the Bootstrap Sample. Using bagging techniques, these models are fitted with the above bootstrap samples, and then combined by voting. RF improves reliability and precision, reduces uncertainty and helps avoid overfitting.

Bootstrap aggregation or bagging is used to determine an appropriate number of trees with the size and nature of the training set. The RF prediction can be expressed as: by averaging the predictions from the individual regression trees;

An optimal number of trees are calculated by bootstrap aggregation or bagging with the size and nature of the training set. By averaging the predictions from the individual regression trees; The RF prediction can be expressed as:

$$\hat{g}(x) = \frac{1}{N}\sum_{n=1}^{N} g_n(x) \tag{1}$$

where $\hat{g}(x)$represents the RF prediction from the total of N trees, and $g_n(x)$ denotes the prediction of each individual tree with the input x. In addition, an approximation of the uncertainty of the prediction can be made as the standard deviation of the predictions from all the trees, which can be expressed as:

$$\sigma = \sqrt{\frac{\sum_{n=1}^{N}\left(g_n(x) - \hat{g}(x)\right)^2}{N-1}} \tag{2}$$

Figure 1 demonstrates the method of classifying RF with the N trees. Starting from the root node (ν_n), after comparison with certain parameters or threshold values, samples are moved to the right node (ν_R) or the left node (ν_L). Repeat this partition until a terminal node is reached and get a classification tag (in this case, classes A or B). For classification task, the ensemble prediction is achieved by majority voting rule as a combination of the results of the individual trees [15].

3.2 REP tree

The reduced error pruning tree (REP Tree) is an ensemble model of decision tree (DT) and The REP Tree (Reduced Error Pruning Tree) is an ensemble model of decision tree (DT) and reduced error pruning (REP) algorithms, equally good for classification and regression problems [16]. The REP Tree algorithm generates a decision regression tree by dividing and pruning the regression tree based on the importance of the highest knowledge benefit ratio (IGR) [17]; The IGR values were determined via Eq. (3) based on the entropy (E) function.

$$IGR(x, S) = \frac{E(S) - \sum_{i=1}^{n}\frac{E(S_i)|S_i|}{|S|}}{-\sum_{i=1}^{n}\frac{|S_i|}{|S|}\log_2\frac{|S_i|}{|S|}} \tag{3}$$

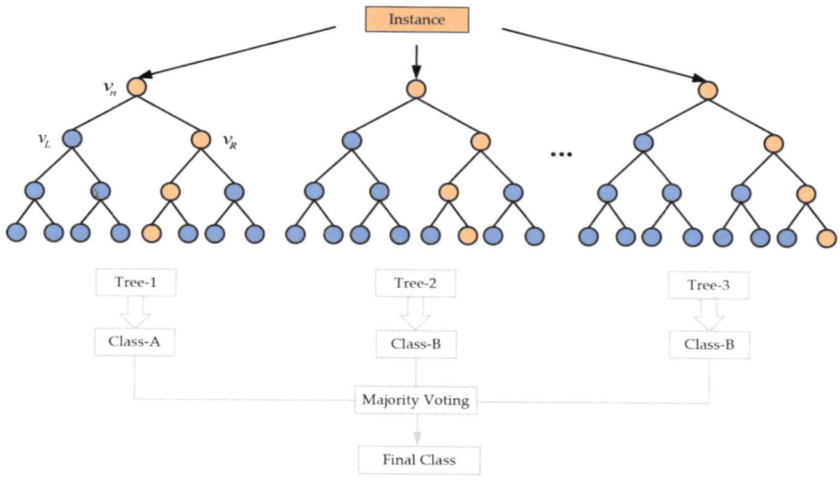

Figure 1.
Schematic representation of a RF classifier with N *trees.*

The *IGR* considers all the predictors of liquefaction-induced settlement with subset Si from the training dataset (S): $i = 1, 2,. .., n$ successive pruning steps. Since complex decision trees can result in a model being overfitted and less interpretable, REP helps to reduce complexity by removing the DT structure's leaves and branches [16, 18–20].

4. Liquefaction-induced settlement model development

4.1 Preparing training and testing datasets

The manner in which data are divided into training and test data sets in data mining procedures has a substantial effect on the results [21–23]. The statistical parameters for the input variables include the minimum, maximum, mean and standard deviation of the training and test datasets, as shown in **Table 2**. Data set splitting was done to assess the generalization efficiency and predictive ability of the developed models. The related performance of the training and testing datasets suggests that the developed models can be applied to the trained ranges. In the testing the ranges of input and output parameters often occur in the training datasets as shown in **Table 2**. The training and testing datasets' statistical consistency enhances the performance of the developed models and thus helps to properly assess them.

To ensure comparability, the RF and REP Tree models are proposed using the same training and test datasets. Using these models, liquefaction-induced settlements are predicted, and an analysis of the detailed performance of these models will find the optimum model afterwards. If the performance of this model on the training and test datasets is adequate then it can be adopted for development.

4.2 Evaluation measures

In this study, three evaluation measures, mean absolute error (MAE), root mean square error (RMSE), and correlation coefficient (r) are used to evaluate and compare the performance of the models. The MAE, RMSE and r are three useful statistical measures which provide some useful insights into the prediction model, of which the MAE is an average of the sum of the differences between the values predicted by a model and the actual values, the RMSE is a standard deviation of the

Dataset	Statistical parameter	Depth (m)	Unit Weight (kN/m^3)	$N_{1(60)}$	*CSR*	Settlement (mm)
Training	Minimum	1	16	0	0.21	0
	Maximum	20	21	25	0.39	3.4
	Mean	10.50	18.85	13.14	0.31	0.89
	Standard deviation	5.80	1.89	9.14	0.05	0.92
Testing	Minimum	1	16	7	0.29	0
	Maximum	20	21	25	0.35	2.6
	Mean	10.5	19.4	15.55	0.3255	0.93
	Standard deviation	5.92	1.76	6.66	0.01	0.65

Table 2.
Statistical parameters of the training and testing datasets.

differences, and the correlation coefficient (r) is a statistical measure representing the percentage of the variance for a model a dependent variable that's described by an independent variable, and their expressions are as follows [24]:

$$MAE = \frac{1}{n}\sum_{i=1}^{n}|y_i - x_i| \tag{4}$$

$$RMSE = \sqrt{\frac{1}{n}\sum_{i=1}^{n}(y_i - x_i)^2} \tag{5}$$

$$r = \frac{\sum_{i=1}^{n}(x_i - \overline{x})(y_i - \overline{y})}{\sqrt{\sum_{i=1}^{n}(x_i - \overline{x})^2\sum_{i=1}^{n}(y_i - \overline{y})^2}} \tag{6}$$

where y_i and x_i are the observed and predicted value of ith sample of the data respectively, $\overline{x}$ and $\overline{y}$ are the mean values of the observed and predicted values respectively, and n is the total number of samples. MAE can be given as a more natural and unambiguous index compared with RMSE to quantify errors between the estimated and actual observed values [25, 26]. RMSE was used as a standard statistical metric to assess output of a model [27]. The larger correlation coefficient (r) and lower mean absolute error (MAE) values, and the root mean squared error (RMSE) present a higher accuracy of predicted results.

5. Results and discussion

Theoretically, a specific model can be obtained when the model parameters are correctly selected and updated. The optimum values are obtained by trial and error

Algorithm	Parameters
RF	Minimum total weight of instances in a leaf: 1; minimum portion of the variance of all the data to be present in a node to be split in regression tress: 0.001; random number seed used to pick attributes: 1; K value: 0
REP Tree	Maximum tree depth: −1; minimum total instance weight in the leaf: 2; minimum likelihood of variance: 0.001; fold number: 3; seed number: 1

Table 3.
Model optimum modeling parameters.

using parameter setting. The optimum value for each machine learning parameter is illustrated in **Table 3**. In the proposed RF and REP Tree models the most significant parameters are the number of seeds and the minimum total weight of instances in a leaf during the modeling process.

The RF and REP Tree predictive results were obtained from the datasets for training and testing datasets. The MAE, RMSE and correlation coefficient (r) were subsequently determined on the basis of the Eqs. (4)–(6) shown in **Figure 2** that depicts RF and REP Tree models performance, respectively. For the RF model the

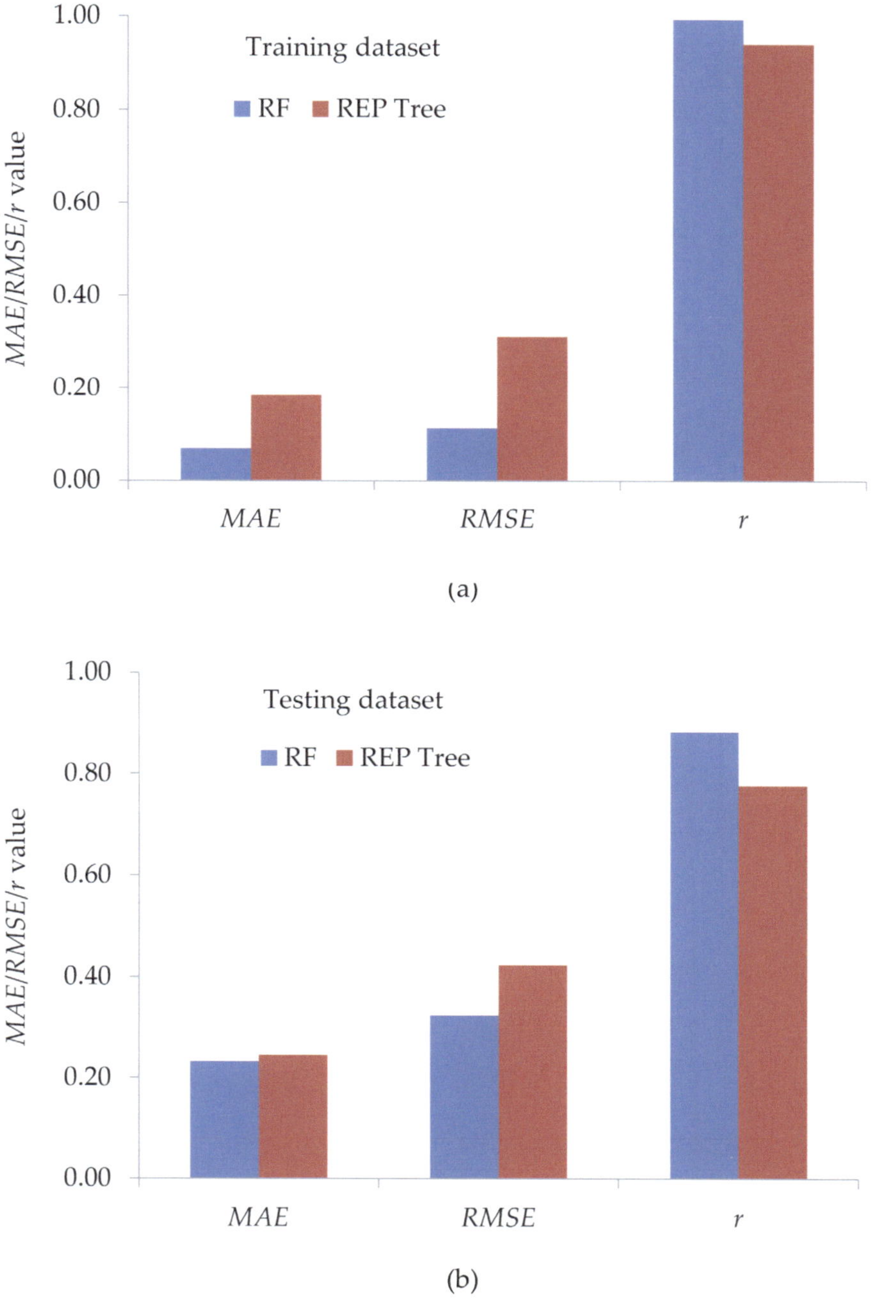

Figure 2.
Comparison of MAE, RMSE, and r *values from the RF and REP tree models.*

training data prediction is higher than the test dataset prediction. The *r* values for the training data and testing data are found 0.9935 and 0.8833, respectively. For the REP Tree model, the training data *r* value (= 0.9405) indicates marginally better results than that for the testing data (= 0.777). It is obvious to judge that the performance of RF model in training and testing datasets is higher than that of REP Tree model. **Figure 2** presents bar graphs comparing the mean absolute error (MAE), the root mean squared error (RMSE), and the correlation coefficient (*r*)

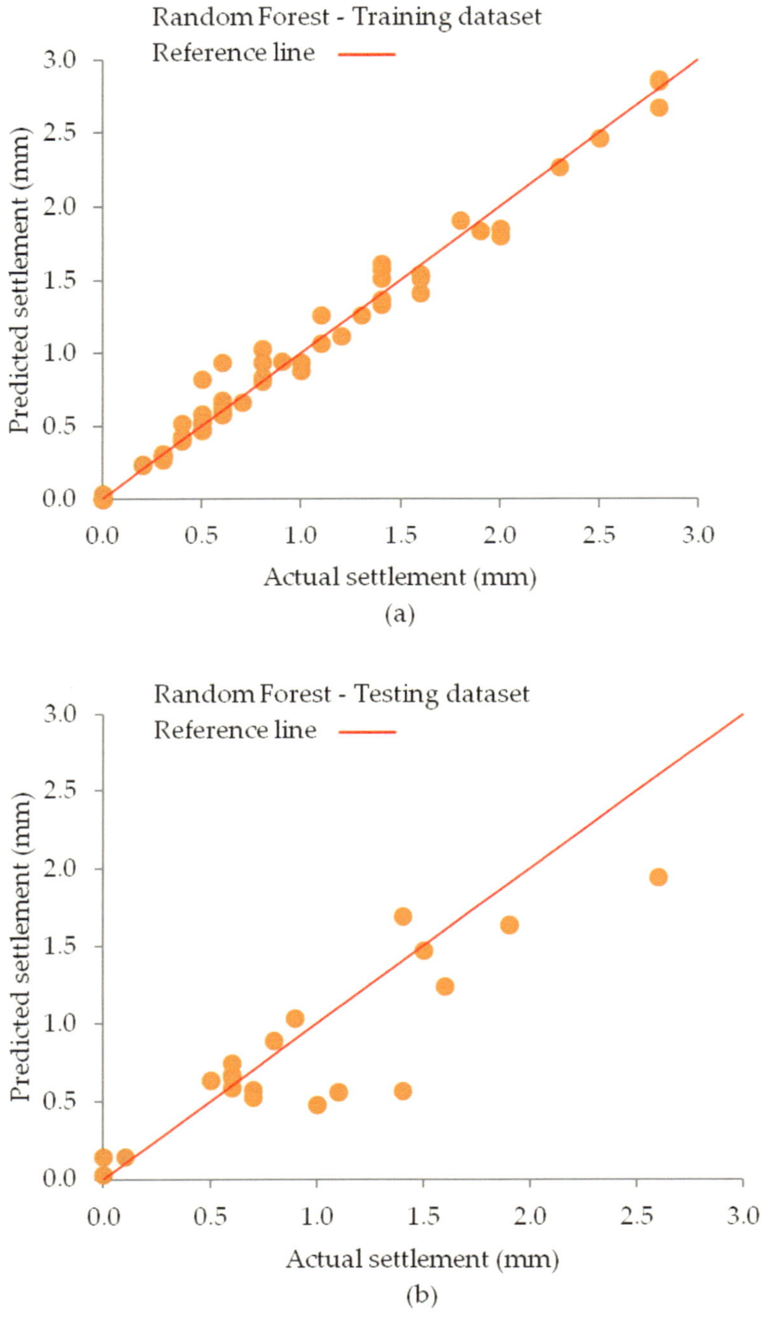

Figure 3.
Training and testing of the RF model.

for both models' training and test datasets. The MAE calculates the variance in the error term by term and reduces the significance of large errors; the *RMSE* value is more concentrated on large errors than on small ones. The RF model has lower *MAE* and *RMSE* values while higher *r* value, showing that in both training and testing datasets, the RF model provides adequate prediction of liquefaction-induced

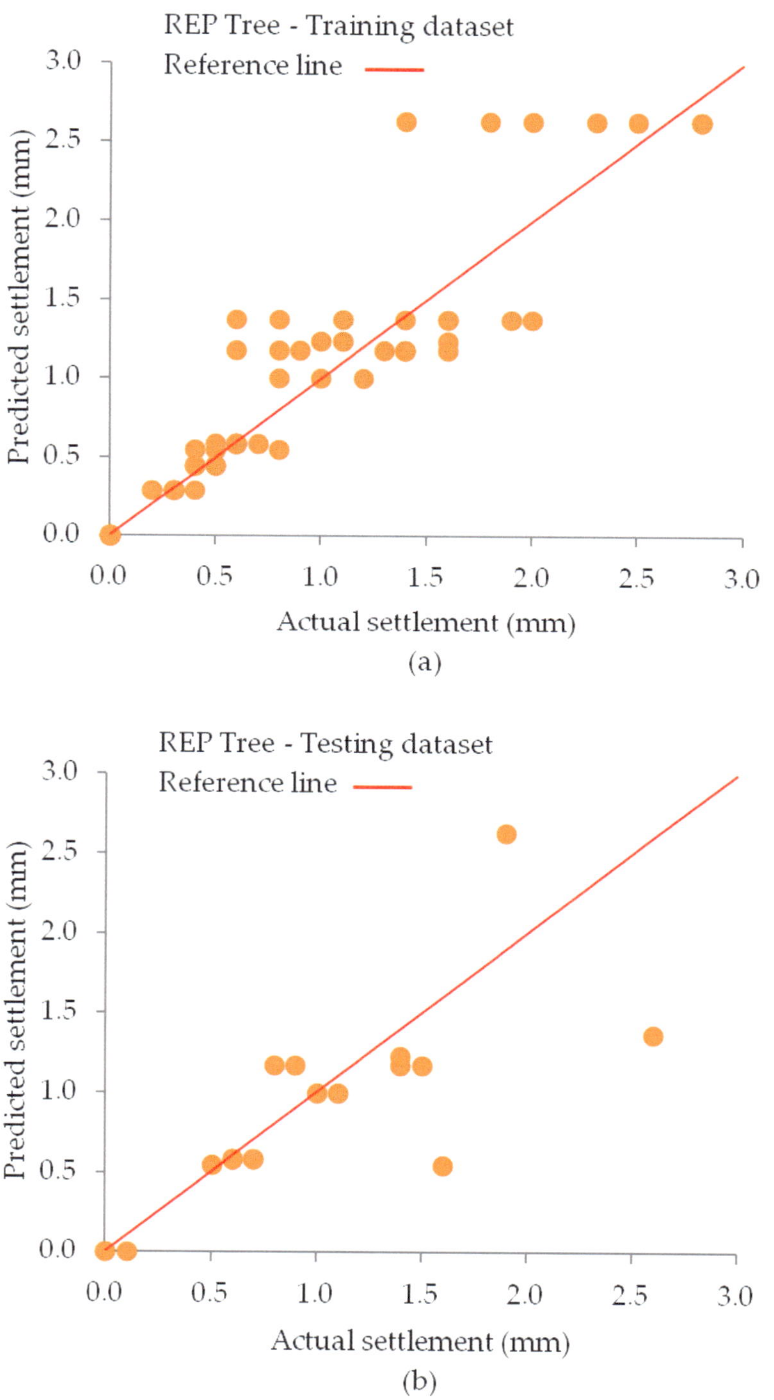

Figure 4.
Training and testing of the REP tree model.

settlement. Additionally, the results of training and testing were shown in **Figures 3** and **4**, showing the projected settlements are plotted with the actual data. One can see that settlements were predicted more accurately by the RF model than by the REP Tree model. While the REP Tree model few settlements cases are relatively under predicted as compared to the RF model.

6. Conclusions

This paper explores the potential of RF and REP Tree models for predicting liquefaction-induced settlement using field data. The models were trained and tested based on the Pohang city liquefaction-induced settlement database. Both models assess liquefaction-induced settlement with substantial contributing factors such as depth, unit weight, corrected SPT blow count and cyclic stress ratio. The performance of the models presented is measured using statistical parameters such as the correlation coefficient (r), MAE, and RMSE. The RF model indicates a better performance with respect to the training and testing datasets. From this analysis it can be inferred that the RF model works well in predicting liquefaction-induced settlement as opposed to the REP Tree model. Since, artificial intelligence-based approaches are data-dependent and their output can vary depending on the dataset, the quality and number of training datasets and the size of the experiments. Finally, it is obvious that the proposed models are open to develop and accumulation of more data will provide much better evaluation of liquefaction-induced settlements.

Acknowledgements

The work presented in this paper was part of the research sponsored by the Key Program of National Natural Science Foundation of China under Grant No. 51639002 and National Key Research and Development Plan of China under Grant No. 2018YFC1505300-5.3.

Conflict of interest

The authors declare no conflict of interest.

Author details

Mahmood Ahmad[1,2], Xiaowei Tang[1*] and Feezan Ahmad[1]

1 State Key Laboratory of Coastal and Offshore Engineering, Dalian University of Technology, Dalian, China

2 Department of Civil Engineering, University of Engineering and Technology Peshawar (Bannu Campus), Bannu, Pakistan

*Address all correspondence to: tangxw@dlut.edu.cn

References

[1] Tokimatsu, K.; Seed, H.B. Evaluation of settlements in sands due to earthquake shaking. *Journal of geotechnical engineering* **1987**, *113*, 861–878.

[2] Ishihara, K.; Yoshimine, M. Evaluation of settlements in sand deposits following liquefaction during earthquakes. *Soils and foundations* **1992**, *32*, 173–188.

[3] Tang, X.-W.; Bai, X.; Hu, J.-L.; Qiu, J.-N. Assessment of liquefaction-induced hazards using Bayesian networks based on standard penetration test data. *Natural Hazards and Earth System Sciences* **2018**, *18*, 1451–1468.

[4] Shamoto, Y.; Sato, M.; ZHANG, J. Simplified estimation of earthquake-induced settlements in saturated sand deposits. *Soils and Foundations* **1996**, *36*, 39–50.

[5] Da Fonseca, A.V.; Millen, M.; Gómez-Martinez, F.; Romão, X.; Quintero, J. DELIVERABLE D3. 1 State of the art review of numerical modelling strategies to simulate liquefaction-induced structural damage and of uncertain/random factors on the behaviour of liquefiable soils. **2017**.

[6] Park, S.-S.; Ogunjinmi, P.D.; Woo, S.-W.; Lee, D.-E. A Simple and Sustainable Prediction Method of Liquefaction-Induced Settlement at Pohang Using an Artificial Neural Network. *Sustainability* **2020**, *12*, 4001.

[7] Kang, S.; Kim, B.; Bae, S.; Lee, H.; Kim, M. Earthquake-induced ground deformations in the low-seismicity region: A case of the 2017 M5. 4 Pohang, South Korea, earthquake. *Earthquake Spectra* **2019**, *35*, 1235–1260.

[8] Choi, J.H.; Ko, K.; Gihm, Y.S.; Cho, C.S.; Lee, H.; Song, S.G.; Bang, E.S.; Lee, H.J.; Bae, H.K.; Kim, S.W. Surface Deformations and Rupture Processes Associated with the 2017 Mw 5.4 Pohang, Korea, EarthquakeSurface Deformations and Rupture Processes Associated with the 2017 Mw 5.4 Pohang, Korea, Earthquake. *Bulletin of the Seismological Society of America* **2019**, *109*, 756–769.

[9] Naik, S.P.; Kim, Y.-S.; Kim, T.; Su-Ho, J. Geological and structural control on localized ground effects within the Heunghae Basin during the Pohang Earthquake (MW 5.4, 15th November 2017), South Korea. *Geosciences* **2019**, *9*, 173.

[10] Gihm, Y.S.; Kim, S.W.; Ko, K.; Choi, J.-H.; Bae, H.; Hong, P.S.; Lee, Y.; Lee, H.; Jin, K.; Choi, S. Paleoseismological implications of liquefaction-induced structures caused by the 2017 Pohang earthquake. *Geosci. J* **2018**, *22*, 871–880.

[11] Kim, H.-S.; Sun, C.-G.; Cho, H.-I. Geospatial assessment of the post-earthquake hazard of the 2017 Pohang earthquake considering seismic site effects. *ISPRS International Journal of Geo-Information* **2018**, *7*, 375.

[12] Integrated DB Center of National Geotechnical Information, SPT Database. Availabe online: http://www.geoinfo.or.kr (accessed on August 20, 2020).

[13] Park, S.-S. Liquefaction evaluation of reclaimed sites using an effective stress analysis and an equivalent linear analysis. *Journal of the Korean Society of Civil Engineers* **2008**, *28*, 83–94.

[14] Breiman, L. *Random Forests-Random Features (# 567)*; Technical report, Dept, of Statistics, Univ. of California, Berkeley: 1999.

[15] Criminisi, A.; Shotton, J.; Konukoglu, E. Decision forests: A unified framework for classification,

regression, density estimation, manifold learning and semi-supervised learning. *Foundations and Trends® in Computer Graphics and Vision* **2012**, *7*, 81–227.

[16] Quinlan, J.R. Simplifying decision trees. *International journal of man-machine studies* **1987**, *27*, 221–234.

[17] Khosravi, K.; Pham, B.T.; Chapi, K.; Shirzadi, A.; Shahabi, H.; Revhaug, I.; Prakash, I.; Bui, D.T. A comparative assessment of decision trees algorithms for flash flood susceptibility modeling at Haraz watershed, northern Iran. *Sci. Total Environ.* **2018**, *627*, 744–755.

[18] Pham, B.T.; Prakash, I.; Singh, S.K.; Shirzadi, A.; Shahabi, H.; Bui, D.T. Landslide susceptibility modeling using Reduced Error Pruning Trees and different ensemble techniques: Hybrid machine learning approaches. *Catena* **2019**, *175*, 203–218.

[19] Mohamed, W.N.H.W.; Salleh, M.N. M.; Omar, A.H. A comparative study of reduced error pruning method in decision tree algorithms. In Proceedings of 2012 IEEE International conference on control system, computing and engineering; pp. 392–397.

[20] Galathiya, A.; Ganatra, A.; Bhensdadia, C. Improved decision tree induction algorithm with feature selection, cross validation, model complexity and reduced error pruning. *International Journal of Computer Science and Information Technologies* **2012**, *3*, 3427–3431.

[21] Ahmad, M.; Tang, X.-W.; Qiu, J.-N.; Ahmad, F. Evaluating Seismic Soil Liquefaction Potential Using Bayesian Belief Network and C4. 5 Decision Tree Approaches. *Applied Sciences* **2019**, *9*, 4226.

[22] Javadi, A.A.; Rezania, M.; Nezhad, M.M. Evaluation of liquefaction induced lateral displacements using genetic programming. *Computers and Geotechnics* **2006**, *33*, 222–233.

[23] Rezania, M.; Javadi, A.A. A new genetic programming model for predicting settlement of shallow foundations. *Canadian Geotechnical Journal* **2007**, *44*, 1462–1473.

[24] Barnston, A.G. Correspondence among the correlation, RMSE, and Heidke forecast verification measures; refinement of the Heidke score. *Weather and Forecasting* **1992**, *7*, 699–709.

[25] Chai, T.; Draxler, R.R. Root mean square error (RMSE) or mean absolute error (MAE)?–Arguments against avoiding RMSE in the literature. *Geoscientific model development* **2014**, *7*, 1247–1250.

[26] Willmott, C.J.; Matsuura, K. Advantages of the mean absolute error (MAE) over the root mean square error (RMSE) in assessing average model performance. *Clim. Res.* **2005**, *30*, 79–82.

[27] Veerasamy, R.; Rajak, H.; Jain, A.; Sivadasan, S.; Varghese, C.P.; Agrawal, R.K. Validation of QSAR models-strategies and importance. *Int. J. Drug Des. Discov* **2011**, *3*, 511–519.

Chapter 16

On Modelling Extreme Damages from Natural Disasters in Kenya

Carolyne Ogutu and Antony Rono

Abstract

We seek to develop a distribution to model the extreme damages resulting from Natural Disasters in Kenya.The distribution is based on the Compound Extreme Value Distribution, which takes into account both the distributions of the frequency of occurrence and magnitude of the events. Threshold modelling is employed, where the extreme damages are identified as the points that lie above a sufficiently high threshold. The distribution of the number of the exceedance is found to be Negative Binomial, while that of the severity is approximated by a Generalised Pareto Distribution. Maximum likelihood estimation is used to estimate the parameters, and the log-likelihood is maximised using numerical methods. Probability weighted moments estimation is used to determine the starting values for the iterations. Prediction study is then carried out to investigate the performance of the proposed distribution in predicting future events.

Keywords: natural disasters, extreme events, compound distributions

1. Introduction

A natural disaster is a sudden and adverse event caused by natural forces that significantly disrupts the workings of society [1]. Such events result to loss of life, injury or other negative health impacts, loss of livelihoods, environmental damage, social and economic disruption. These natural forces causing natural disasters are called natural hazards. A natural hazards can be formally defined as a natural process or phenomenon that may cause loss of livelihoods and services, negative health impacts such as loss of life and injury, social and economic disruption or environmental damage [2]. Therefore, a natural hazard becomes a disaster once it occurs. The Centre for Research on the Epidemiology of Disasters Database (EM-DAT) generally classifies natural disaster according to the hazards that causes them. We have geophysical (earthquakes, volcanoes landslides and tsunamis), hydrological (floods and avalanches), climatological (droughts and wildfires), meteorological (storms and cyclones) or biological (epidemics and pests/insects/plagues). Hydrological, meteorological and climatological hazards are generally termed as weather-related hazards. Natural disasters are considered extreme events and thus we consider the method of Extreme Value Theory for analysis of our data.

1.1 Extreme value modelling

Extreme value theorem (EVT) is used to develop stochastic models aimed at solving real-world problems relating to extremal and unusual events for instance stock market crashes, natural catastrophes, major insurance claims e.t.c. EVT models the tails of loss severity distributions as it only considers extreme and rare events. Extreme value modelling has been used widely in hydrology, insurance, finance and environmental application where extreme events are of interest. Due to the scope of this research, we will limit the literature to those relating to natural disasters. The relevant theories and statistical methods behind EVT are presented by [3].

[4, 5] propose the Generalised Extreme Value (GEV) distribution which is used to model the maxima of a set of independent and identically distributed (iid) events. The resulting model, referred to as the Block Maxima (BM) model, involves dividing the observations into non-overlapping blocks of equal sizes and restrict the attention to maximum observations in each period. The observations created are then assumed to follow the GEV [6]. The block sizes are vital in the EVT model as too small block sizes can lead to poor asymptotic approximation, while too big sizes will result to very little observations [7]. [8] use BM method to model hydrological floods and droughts. They find that using block sizes of one year or less to model drought leads to bias. Floods, on the other hand, is modelled using block sizes of less than one year. The same method is used by [9] to study seasonal variation of extreme precipitation across the UK. They use a one month block, having found little improvement of a longer-block. They were able to identify regions with a high risk of extreme precipitation in a given season. A major drawback of the Block Maxima is its inefficiency in terms of data usage. Dividing the dataset into blocks is wasteful of the data since not all the available data is used. To overcome this challenge, the data above some sufficiently high threshold is modelled in what is commonly referred to as excess over threshold (EOT). Several studies have been conducted to compare the performance of BM and EOT, all based on simulated data. [10] states that, in the case where the shape parameter is equal to zero, EOT estimates for a high quantile is better only if the number of exceedance is larger than 1.65 times the number of blocks. [11] find that EOT is preferable when the shape parameter is greater than zero and when the number of exceedance is greater than the number of blocks. However, when the shape parameter is less than zero, BM is more efficient. [12] carries out a vast simulation study to compare the two methods based on their accuracy as measured by mean squared errors, on the basis of time series with various characteristics and in estimating probabilities of exceedance. He finds that the EOT estimates for a sample that has an average of two or more exceedance per year are more accurate than the corresponding BM estimates. In addition, he finds that EOT should be used when the data is less than 50 years. When there is 200 years of data, both approaches have similar accuracies. In terms of the return value estimates, he finds that the EOT estimates are significantly more accurate than those obtained by BM. [6] find that the BM method at an optimal level gives lower asymptotic minimal square error than the EOT method. They conclude that BM is more efficient than EOT in practical situations.

Despite having mixed results, we can deduce that EOT is, on average, a preferred method to BM. Two features are however dominant from all these studies: first, EOT is more efficient than BM if the number of exceedance is greater than the number of blocks and secondly, the two methods have comparable performances for large sample sizes.

The biggest challenge in using EOT is selecting the appropriate threshold. The threshold selection process is always a trade–off between the bias and variance [13]. Moreover, the choice of the threshold significantly affects the tail of the distribution and hence the GPD parameters. The traditional extreme value modelling approach uses fixed threshold approach, where the threshold is chosen and fixed before the model is fitted. The threshold is usually chosen based on various diagnostics plots that is used to evaluate the model fit. The traditional EVT approach only considers the extreme data above the threshold, and discards the rest that are below. [7] states that the motivation for ignoring extreme data is that extreme and non-extreme data are caused by different processes and the latter is rarely of concern. Furthermore, there is concern that the data below the threshold may compromise the examination of the tail fit. However, there has been increasing interests in extreme value mixture models, which uses all the data for parameter estimation in the inference. This class of models has the potential to overcome some of the difficulties encountered in using the traditional EVT, in regards to threshold selection and uncertainties associated with it. The next section will highlight the development of extreme value mixture models.

1.2 Extreme value mixture models

Extreme value mixture models typically comprise of two compenents: a bulk model which describes all the non-extreme data below the threshold, and a tail model which is the traditional extreme value model to model the extreme data above the threshold. The advantage of these models is that they consider all the data without wasting information and provides an automated approach both for estimation of the threshold and quantification of the resulting uncertainties from the threshold choice [7].

[14] propose a dynamic weighted mixture model combining a light-tailed distribution for the bulk model, with GPD for the tail model. They use a Weibull distribution as the bulk distribution. Rather than explicitly defining the threshold, they adopt a Cauchy distribution function as the mixing function to make the transition between the bulk and tail distributions. They use maximum likelihood estimation for parameter estimation, and numeric iteration to calculate empirical quantiles. They then test the model using Danish fire loss dataset, and find the parameter estimates to be comparable to those reported in the literature regarding EOT inference. The quantiles are comparable as well. The model is also compared to an existing robust thresholding approach, and it is found to be better for very small percentiles in small datasets.

[15] develop an extreme value mixture model comprising of a truncated gamma to fit the bulk distributions and GPD for the tail distributions. They note that any other parametric distribution can also be used. They treat the threshold as an unknown parameter, and use Bayesian statistics for inference about the unknown parameters. Inference on the Posterior distribution is done using Markov chain Monte Carlo methods, and simulation studies is carried out to analyse the performance of the model. They find that the parameter estimates are very close to the true value for large samples. However, for small samples, they encounter problems with convergence, and consequently parameter estimation. They also apply the model to Nasdaq 100 dataset, an index in the New York Stock Exchange, and compare its performance with the traditional extreme value model. The resulting GPD estimates is close to that obtained using the traditional approach.

1.3 Compound extreme value distribution

Unlike Extreme Value Theorem which is concerned with only the tails of loss severity distributions, Compound Extreme Value Distribution (CEVD) models can simultaneously model the frequency and severity of extreme events. The idea was proposed by Tebfu and Fengshi in 1980 to model typhoon in South China [16]. They only consider the events above the threshold, as is the case with the traditional EVT, and assume that the number of exceedance (frequency) is Poisson Distributed.

The exceedance (severity) are, on the other hand, assumed to follow a Gumbel Distribution, which is a special case of the General Extreme Distribution. Hence, the model is usually referred to as Poisson-Gumbel. Tebfu and Fengshi [7] also used CEVD to model hurricane characteristics along the Atlantic coasts and the Gulf of Mexico. They assume that the number of exceedance follows a Poisson Distribution, and the exceedance is Weibull Distribution.

Initially, CEVD was mostly used in hydrology, to model wave heights and the resulting extreme events. The model has been successfully used to predict design wave height. For instance, Hurricane Katrina of 2005 corresponded to 60 years return period as predicted by the Poisson-Weibull model [17]. As a result, there has been several extensions to these class of models including the Bivariate Compound Extreme Value Distribution (BCEVD) model. [18] and Multivariate Compound Extreme Value Distribution (MCEVD) model [17]. In addition, the model has been adopted in a wider range of areas, including finance, insurance, disasters and catastrophic modelling.

[19] investigate the global historical occurrences of tsunamis. They compare the distribution of the number of annual tsunami events using a Poisson distribution and a Negative binomial distribution. The latter provides a consistent fit to tsunami events whose height is greater than one. They also also investigate the interval distribution using gamma and exponential distributions. The former is found to be a better fit, suggesting that the number of tsunami events is not a Poisson process.

[20] study tsunami events in the USA. They assume that the occurrence frequency of tsunami in each year follow a Poisson distribution. They then identify the distribution of tsunami heights by fitting six distributions: Gumbel, Log normal, Weibull, maximum entropy, and GPD. They select GPD, which had the best fit. They use MLE for parameter estimation, and investigate the fit of the Poisson Compound Extreme Value Distribution using goodness-of-fit statistics. The results is consistent with [19], that the Poisson-Generalised Pareto Distribution is appropriate for disaster modelling.

2. Classical extreme value theory

The core of the classical Extreme value theorem is the study of the stochastic behaviour of some maximum (or minimum) of a sequence of random variables. Define

$$M_n = \max\{Y_1, \dots, Y_n\} \tag{1}$$

where $\{Y_1, \dots, Y_n\}$ is a sequence of independent random variables with a common distribution function F. M_n represents the maxima (minima) of observed process over n blocks or time units. If F is known, the distribution of M_n can be derived as follows:

$$\begin{aligned} P\{M_n \le y\} &= P\{Y_1 \le y, \dots, Y_n \le y\} \\ &= P\{Y_1 \le y\} \times \dots \times P\{Y_n \le y\} \\ &= \{F(y)\}^n \end{aligned} \tag{2}$$

However, F is usually unknown in practice, and will have to be estimated from the data. This poses a problem since a small error in the estimation of F can lead to large disparities for $F^n\{y\}$. An alternative approach is to model $F^n\{y\}$ through asymptotic theory of M_n, where we study the behaviour of $F^n\{y\}$ as n tends towards infinity. Since $F(y) < 1$ for $y < y_{sup}$, where y_{sup} is the upper end-point of F, we have $F^n\{y\} \to 0$ as $n \to \infty$. We can remove the degeneracy problem by allowing some linear re-normalisation of M_n. Consider a linear re-normalisation:

$$\hat{M_n} = \frac{M_n - d_n}{c_n} \tag{3}$$

where $\{c_n\}$ and $\{d_n\}$ are sequences of constants with $c_n > 0$. Under a suitable choice of c_n and d_n, the distribution of M_n can be stabilised and which leads to extremal types theorem:

Theorem 1.1 [Extremal Types Theorem] For a non-degenerate distribution function, G, if there exists sequences of constants $\{c_n > 0\}$ and $\{d_n\}$, as $n \to \infty$, such that

$$P\left(\frac{M_n - d_n}{c_n} \le y\right) \to G(y) \tag{4}$$

then G belongs to one of the following families:

Gumbel:

$$G(y) = \exp\left\{-\exp\left[-\left(\frac{y-d}{c}\right)\right]\right\}, \quad -\infty < y < \infty \tag{5}$$

Frechet:

$$G(y) = \begin{cases} 0 & y \le d \\ \exp\left\{-\left(\frac{y-d}{c}\right)^{-\alpha}\right\} & y > d \end{cases} \tag{6}$$

Weibull:

$$G(y) = \begin{cases} \exp\left\{-\left[-\left(\frac{y-d}{c}\right)\right]^{\alpha}\right\} & y < d \\ 1 & y \ge d \end{cases} \tag{7}$$

for $c > 0$ and $d \in \mathbb{R}$.

The proof of this theory is presented in [21]. The three classes of distributions are called extreme value distributions, with type I (Gumbel), type II (Frechet) and type III (Weibull) respectively. The extremal types theorem suggests that regardless of the population distribution of M_n, if a non-degenerate limit can be obtained by linear re-normalisation, then the limit distribution will be one of the three extreme value distributions.

In modelling an unknown population distribution, we choose one of the three types of limiting distributions and then estimate the model parameters. This approach is, however, deemed to be inefficent as uncertainty associated with the choice is not considered in the subsequent inference [22]. A better approach is to combine the three models into one single family, with the distributions being special cases of the universal distribution.

2.1 The generalised extreme value distribution

Von Misses [4] and Jenkinson [5] combined the three types of extreme value distributions leading to the generalised extreme value distribution (GEV).

Theorem 1.2 If there exists sequences of constants c_n and d_n, such that

$$P\left(\frac{M_n - d_n}{c_n} \leq y\right) \quad \overset{n\to\infty}{\longrightarrow} \quad G(y) \tag{8}$$

where G is a non-degenerate distribution function, then G is a member of the GEV family:

$$G(y) = \exp\left\{-\left[1 + \zeta\left(\frac{y-\nu}{\sigma}\right)\right]^{-1/\zeta}\right\} \tag{9}$$

defined on y such that $1 + \zeta\left(\frac{y-\nu}{\sigma}\right) > 0$ and with parameters: scale $\sigma > 0$, location $\nu \in \mathbb{R}$ and scale $\zeta \in \mathbb{R}$.

The shape parameter determines the tail behaviour under the same values of location and scale, and thus indicates the type of extreme value distribution:

- If $\zeta = 0$, the GEV becomes a Gumbel distribution and the tail decays exponentially.
- If $\zeta < 0$, the GEV becomes a negative Weibull distribution with the upper endpoint being finite.
- If $\zeta > 0$, the GEV simplifies to a Frechet distribution with a heavy tail and which decays polynomially.

3. Threshold models and the generalised Pareto distribution

Modelling using block maxima is inefficient as it is wasteful of the data, considering that the complete dataset is available. An alternative approach is to model all the data above some high threshold, in what is commonly referred to as threshold modelling or Excess over Threshold (EOT) modelling.

Given a set of independent and identically distributed random variables $Z_1, \ldots, Z_n$, having a common distribution function,F, we are interested in estimating the conditional excess distribution function, F_v, of random variable X above a high threshold v:

$$F_u(y) = P(Z - v \leq y | Z > v), \quad 0 \leq y \leq z_+ - v \tag{10}$$

where $y = z - v$ are the exceedances and z_+ is the right endpoint of F. We can express $F_v(y)$ in terms of z as:

$$F_v(y) = \frac{F(v+y) - F(v)}{1 - F(v)} = \frac{F(v+z-v) - F(v)}{1 - F(v)} = \frac{F(z) - F(v)}{1 - F(v)} \tag{11}$$

Piklands [23] posed that if the underlying distribution F(z) is in the maximum domain of attraction of extreme value distribution, then the conditional excess distribution function $F_v(z)$ for a large v, can be approximated by:

$$F_v(z) \approx F_{(\zeta,\nu,\sigma)}(z), v \to \infty \tag{12}$$

where $F_{(\zeta,\nu,\sigma)}(z)$ is the Generalised Pareto Distribution (GPD).

3.1 Generalised Pareto distribution (GPD)

Definition 1The random variable Z has a Generalised Pareto Distribution if the cumulative distribution function of Z is given by:

$$F_{(\zeta,\nu,\sigma)}(z) = \begin{cases} 1 - \left(1 + \frac{\zeta(z-\nu)}{\sigma}\right)^{-1/\zeta} & \text{for} \, \zeta \neq 0 \\ 1 - \exp\left(-\frac{z-\nu}{\sigma}\right) & \text{for} \, \zeta = 0 \end{cases} \tag{13}$$

for $z > \nu$ and $\left(1 + \frac{\zeta(z-\nu)}{\sigma}\right) > 0$ and parameters: location $\nu > 0$, scale $\sigma > 0$, shape $\zeta \in \mathbb{R}$.

Remark 1 (Special Cases) Under specific conditions, the GPD simplifies to other continuous distributions:

- When $\zeta = 0$, the GDP simplifies to the exponential distribution
- When $\zeta > 0$, the GDP becomes an ordinary Pareto distribution, and
- when $\zeta < 0$, the GDP simplifies to a Pareto type II distribution.

The mean of the GPD is:

$$E(Z) = \nu + \frac{\sigma}{1-\zeta}, \text{for } \zeta < 1 \tag{14}$$

The Variance of the GPD is:

$$Var(Z) = \nu + \frac{\sigma^2}{(1-\zeta)^2(1-2\zeta)}, \text{for } \zeta < \frac{1}{2} \tag{15}$$

In general, the $r-th$ moment of the GPD only exists if $\zeta < \frac{1}{r}$.

The shape parameter, ζ, determines the tail distribution of the GPD as indicated in remark 1. When $\zeta = 0$, there exists a decreasing exponential tail, when $\zeta > 0$, there is a heavy tail and when $\zeta < 0$, the tail is short, with finite upper end point $\nu - \frac{\sigma_\nu}{\zeta}$.

3.2 Relationship between GEV and GPD

Theorem 1.3 Let $Z_1, \dots, Z_n$ be a sequence of independent and identically distributed random variables with a common cumulative distribution function F, and let $M_n = \max\{Z_1, \dots, Z_n\}$ satisfying the conditions to be approximated by GEV, i.e., for large n:

$$P\{M_N \leq z\} = G(z), \text{where} \, \mathrm{G(z)} = \exp\left\{-\left[1 + \zeta\left(\frac{\mathrm{z}-\nu}{\sigma}\right)\right]^{-1/\zeta}\right\} \tag{16}$$

Then, for a sufficiently high threshold v, the conditional distribution function of $(Z-v)$, conditioned to $Z > v$ is approximately given by

$$H(y) = 1 - \left(1 + \frac{\zeta y}{\delta}\right)^{-1/\zeta} \tag{17}$$

where $y = z - v > 0$, $\left(1 + \frac{\zeta y}{\delta}\right) > 0$ and $\delta = \sigma + \zeta(v - \nu)$.

Proof. Denote the distribution function of the random variable Z by F. By theorem 1.2, for large n,

$$\begin{aligned} G(z) = F^n(z) &\approx \exp\left\{-\left[1 + \zeta\left(\frac{z - \nu}{\sigma}\right)\right]^{-1/\zeta}\right\} \\ n \log F(z) &\approx \left\{-\left[1 + \zeta\left(\frac{z - \nu}{\sigma}\right)\right]^{-1/\zeta}\right\} \end{aligned} \tag{18}$$

For large values of z, Taylor series expansion implies:

$$\log F(z) \approx -\{1 - F(z)\} \tag{19}$$

Hence,

$$\begin{aligned} n\{1 - F(z)\} &\approx \left[1 + \zeta\left(\frac{z - \nu}{\sigma}\right)\right]^{-1/\zeta} \\ \{1 - F(z)\} &\approx \frac{1}{n}\left[1 + \zeta\left(\frac{z - \nu}{\sigma}\right)\right]^{-1/\zeta} \end{aligned} \tag{20}$$

So, for $z = v + y$

$$\{1 - F(v + y)\} \approx \frac{1}{n}\left[1 + \zeta\left(\frac{v + y - \nu}{\sigma}\right)\right]^{-1/\zeta} \tag{21}$$

Thus,

$$\begin{aligned} P\{Z > v + y | Z > v\} &= \frac{1 - P(Z < v + y)}{1 - P(Z < v)} \\ &\approx \frac{[1 + \zeta(v + y - \nu)/\sigma]^{-1/\zeta}}{[1 + \zeta(v - \nu)/\sigma]^{-1/\zeta}} \\ &= \left[\frac{1 + \zeta(v + y - \nu)/\sigma}{1 + \zeta(v - \nu)/\sigma}\right]^{-1/\zeta} \\ &= \left[\frac{1 + \zeta(v - \nu)/\sigma + \zeta y/\sigma}{1 + \zeta(v - \nu)/\sigma}\right]^{-1/\zeta} \\ &= \left[\frac{1 + \zeta(v - \nu)}{1 + \zeta(v - \nu)} + \frac{\zeta y/\sigma}{1 + \zeta(v - \nu)/\sigma}\right]^{-1/\zeta} \\ &= \left[1 + \frac{\zeta y/\sigma}{\frac{\sigma + \zeta(v - \nu)}{\sigma}}\right]^{-1/\zeta} \\ &= \left[1 + \frac{\zeta y}{\sigma + \zeta(v - \nu)}\right]^{-1/\zeta} \\ &= \left[1 + \frac{\zeta y}{\delta}\right]^{-1/\zeta} \end{aligned} \tag{22}$$

Therefore,

$$P\{Z \leq v + y | Z > v\} = 1 - \left[1 + \frac{\zeta y}{\delta}\right]^{-1/\zeta} \tag{23}$$

where, $y > 0$, $\left(1 + \frac{\zeta y}{\delta}\right) > 0$ and $\delta = \sigma + \zeta(v - \nu)$.

Eq. (23) is the generalised Pareto family of distributions.

Theorem 1.3 implies that we can use GPD as an approximation to the distribution of maxima using EOT as alternative to GEV in block maxima. We can observe how the parameters of the GPD are uniquely determined by those of the GEV. In particular, the shape parameter is equal in both cases. When the block sizes change in the GEV, the parameters ν and σ change, while the shape parameter remains constant.

Eq. (23) clearly indicates the dependence of the scale parameters σ on the threshold. The threshold choice is an important part of threshold modelling. As with the case with the block sizes in GEV, the choice of threshold is a trade-off between the bias and variance [13]. If the threshold is too low, we violate the asymptotic arguments underlying the GPD, whereas, if the threshold is too high, we will generate few exceedances to estimate the parameters, resulting to a large variance. We now describe the process of selecting the appropriate threshold, to be used in modelling extreme events using GPD.

3.3 Threshold selection

In modelling using EOT, the threshold is usually chosen before the model is fitted. We will present three tools that will help in identifying the threshold:

3.3.1 Mean residual life plot

This method is based on the mean of the GPD: $E(Z) = \frac{\sigma}{1-\zeta}$, when $\zeta < 1$. If the excesses $z - v$ are approximated by a GPD for a high threshold v, the mean excess, for $\zeta < 1$ is:

$$e(v) = E(Z - v | Z > v) = \frac{\sigma_v}{1 - \zeta} \tag{24}$$

For any higher threshold $r > v$:

$$e(r) = E(Z - r | Z > r) = \frac{\sigma_r}{1 - \zeta} = \frac{\sigma_v + \zeta(r - v)}{1 - \zeta} \tag{25}$$

Therefore, the mean excess function $e(r)$ is a linear function of v, once a suitable high threshold has been reached. The sample mean residual life plot is drawn using:

$$\left(v, \frac{1}{n_v}\sum_{i=1}^{n_v}\left(z_{(i)} - v\right)\right) \tag{26}$$

where $z_{(i)}$ is the observation, i, above the threshold, v, and n_v is the total number of observations above v. For a high v, all the exceedances, $r > v$, in the mean residual life plot changes linearly with v.

Using this result, the procedure for estimating the threshold is as follows:

- Draw a mean residual life plot using (26)
- Choose a threshold above which the plot is approximately linear. We use confidence interval to determine whether the plot looks linear.

3.3.2 Parameter stability plot

Assuming that the exceedances, $(z - v)$, over a threshold v follow a GPD (ζ, σ_v), the exceedances will still follow a GPD for any higher threshold $r > v$, with the same ζ, but with scale parameter of:

$$\sigma_r = \sigma_v + \zeta(r - v) \tag{27}$$

(This follows from theory 1.3).
Let us re-parametrise the scale parameter σ_r:

$$\sigma^* = \sigma_r - \zeta_r \tag{28}$$

such that:

$$\begin{aligned} \sigma_r &= \sigma_v + \zeta r - \zeta v \\ \sigma_r - \zeta_r &= \sigma_v - \zeta v \\ \sigma^* &= \sigma_v - \zeta v \end{aligned} \tag{29}$$

The parameter σ^* now only depends on a sufficiently high threshold v. The parameter stability plot involves plotting the GPD parameter estimates for a range of values v. The threshold is chosen to be the point where the shape and the modified scale parameters become stable (that is, the parameter estimates is constant above the threshold at which the GPD becomes valid). We use confidence interval to select this point.

3.3.3 Gertensgarbe and Werner plot

The test was proposed by Gertensgarbe and Werner (1989) and is used to select a threshold by detecting the starting point of the extreme region. The idea behind the test is that the behaviour of a series of differences that correspond to the extreme observations is different from the one corresponding to the non-extreme observations. So, given a series of differences $\Delta_r = z_{(r)} - z_{(r+1)}, i = 2, 3, \cdots, n$, of the order statistics, $z_{(1)} \leq z_{(2)} \leq \cdots \leq z_{(n)}$, the starting point of the extreme region, and hence the threshold, will be the point at which the series of differences exhibit a significant statistical change.

To detect this point, we apply a sequential version of the Mann-Kendall test to check the null hypothesis that there is no change in the series of differences. Define a series:

$$U_r = \frac{U_r^* - \frac{r(r-1)}{4}}{\sqrt{\frac{i(r-1)(i+5)}{72}}} \tag{30}$$

where $U_r^* = \sum_{j=1}^{r} n_j$ and n_j is the number of values in the the series of differences $\Delta_1, \cdots, \Delta_j$ less than Δ_j. We also define another series, U_p, by applying the same procedure to the series of differences from the end to start,$\Delta_n, \cdots, \Delta_1$, rather than from the start to the end. The starting point of change in the series of differences, and hence the threshold, is the the intersection point of the series U_r and U_p.

4. Data analysis

We use data for all the natural disasters that has been recorded in Kenya in the period 1964 – 2018. The data is obtained from CRED database, which is the currently the most comprehensive database for natural events. The data was also cross-referenced with that from other sources including UN-agencies and NDMU. The impact of natural disasters is quantified in terms of the total number of people affected on an annual basis, which we deemed to be more reliable than the the total damage in monetary terms. The total number of people affected includes those who were injured, died, left homeless or affected in any other way by natural disasters. A total of 112 events have been recorded over that period of time with a resulting damage of approximately 62 million people affected.

Descriptive statistics for both the annual occurrence and the impacts are provided in **Table 1**. The minimum number of disasters and the resulting impact is zero, which corresponds to those years where no natural disasters occurred. A total of 22 years recorded no natural disaster events. The average annual number of natural disaster occurrences in Kenya is two, and about 1.3 million people are affected every year. The maximum number of natural disaster occurrences observed in any year is 9 and the worst disaster recorded in any year affected approximately 23, 331, 469 people. The mean is greater than the median for both variables, indicating that the data is right-skewed. We can also observe that the spread of the impacts is large as suggested by both the standard deviation and the inter-quantile range (about 3.5 million and 252718 respectively), as opposed to that of the occurrence.

We are interested in the distribution of the number of occurences and the corresponding impact of natural disasters in Kenya in the period of study. **Figure 1** shows the distribution of the number of natural disaster occurrences in the last 55 years. The number of natural disasters between 1964 and the late 1990s was fairly low, with no year experiencing more than two events. We can then observe a sharp

Statistic	Annual Occurrence	Annual Impact
N	55	55
Total	112	62,160,910
Mean	2	1,130,198
Std. Dev	2.26	3,494,302
Minimum	0	0
st Quantile	0	0
Median	1	15000
rd Quantile	4	252718
Maximum	9	23,331,469
Inter Quantile Range	4	252,718

Table 1.
Descriptive Statistics.

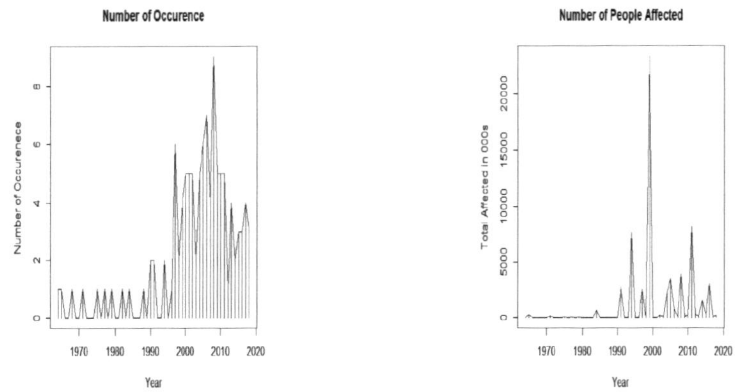

(a) Number of Occurrence (b) Time plot of Number of People Affected

Figure 1.
Distribution of Natural Disasters in Kenya in the period: 1964–2018.

increase in the annual occurrences in the last 20 years. During this period, all the years have experienced at least two events.

A similar trend is observed for the impacts of the natural disasters, as shown in **Figure 1b**. Very few people were being affected in the years between 1964 and 1990. Then from 1990, the severity of the natural disasters increased and a lot more people were being affected.

4.1 Check for Independence assumption

We assumed that the annual occurrences are independent of each other as with is the case with annual impact. We also assumed that the occurrences and impact are independent of each other. We can observe that the scatter plots 2,3 and 4 indicate that there is no serious violation of these assumptions (**Figures** 2-4).

4.2 Detection of heavy-tailed behaviour

The statistical methods used in this project rely on the heavy-tailedness of the underlying distribution. We observed that the data is right skewed as the mean is

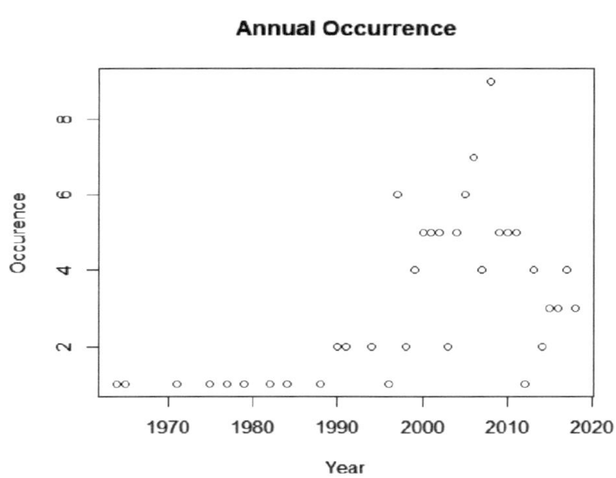

Figure 2.
Scatter plot of Annual Occurrence.

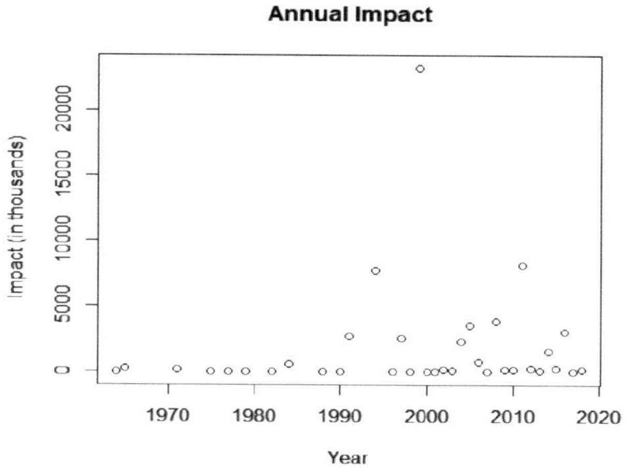

Figure 3.
Scatter plot of Annual Impact.

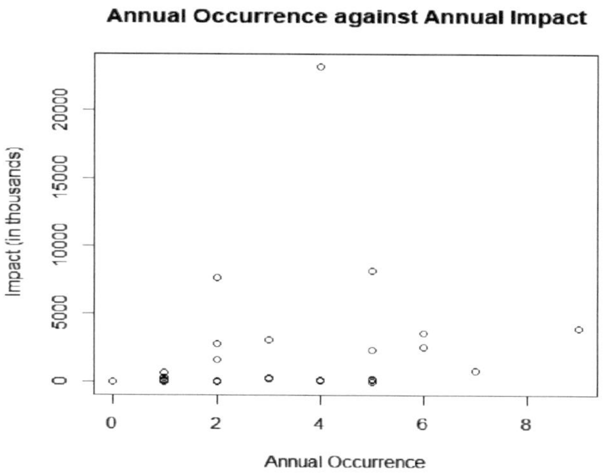

Figure 4.
Scatter plot of Annual Occurrence against Annual Impact.

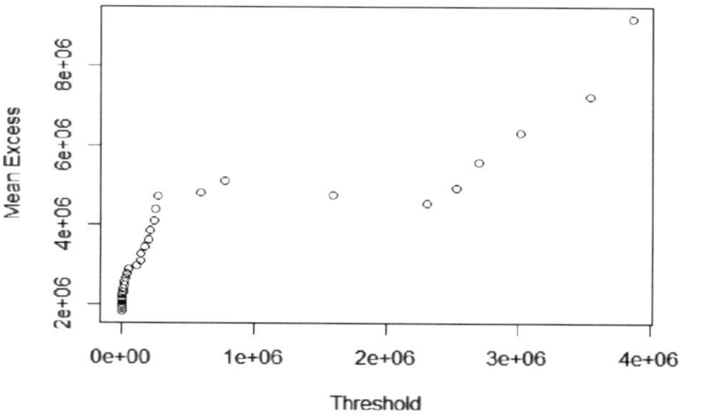

Figure 5.
Mean Excess Plot.

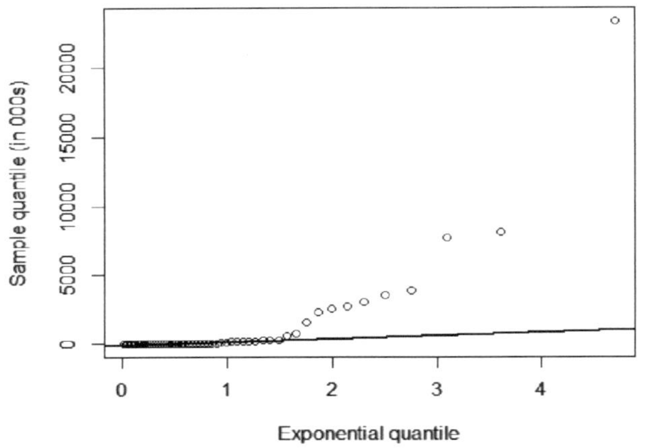

Figure 6.
Exponential Q-Q plot.

greater than the median (**Table 1**), indicating that the underlying distribution has a long tail. We will further investigate this by plotting the empirical mean excess function against different threshold values. We test the assumption using an empirical mean excess plot.

An upward trend in the mean excess plot indicates heavy-tailed behaviour, as explained in Section 4.2. From **Figure 5**, we can observe a general upwards trend in the graph, except for the area between one million and two million.

To get more conclusive results, we also plot an exponential Q-Q and observe the pattern of the points in relation to the straight line. Heavy-tailed behaviour will be indicated by a convex departure from the straight line as explained in Section 4.2. Shorter tailed-distribution will have a concave departure and if the data are a sample from an exponential distribution, the points should be approximately linear.

We can observe the convex behaviour of the exponential Q-Q plot in **Figure 6**. Thus, we can conclude that the underlying distribution is heavy-tailed.

5. Threshold selection

Having proved that the underlying distribution is heavy-tailed, we now select a threshold, above which the distribution of the impact is approximated by the GPD. We will use the techniques provided in Section 3.3 to estimate the appropriate threshold. First, we plot the mean excesses for each value of 200 different thresholds across the whole dataset, against their corresponding thresholds, with a significance level of 0.05. The goal is to find the lowest threshold such that the graph is linear with increasing thresholds, within uncertainty.

The mean residual life plot in **Figure 7** looks like a straight line right from the beginning except at the very right end. The plot suggests that the threshold should be between zero and around 250, 000.

The threshold selection can be carried out more carefully through a parameter stability plot. Recall that if the threshold is too low, the assumption underlying the GPD will be violated. On the other hand, if the threshold is too high, we will have too few observations to effectively fit the distribution. Given that natural disasters can be regarded, as extreme and rare events, we will limit the range of the threshold to within the upper and lower quantiles of the impact dataset: $0 \leq v \leq 250,000$, which is surprisingly in alignment to the values indicated by the mean excess plot. The resulting range of the number of exceedance is then 13 and 42. **Figure 8** shows

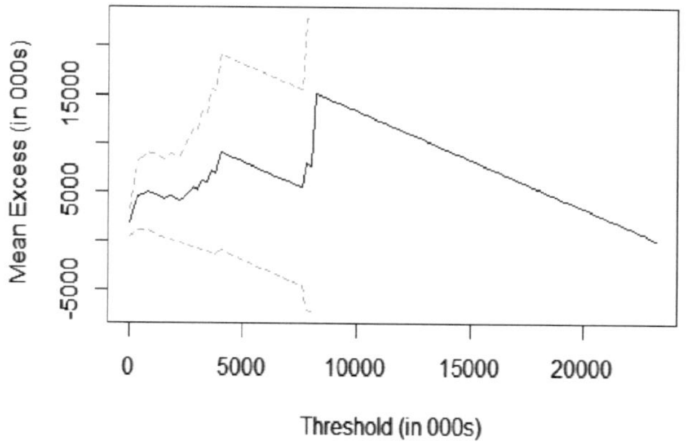

Figure 7.
Mean residual Life Plot.

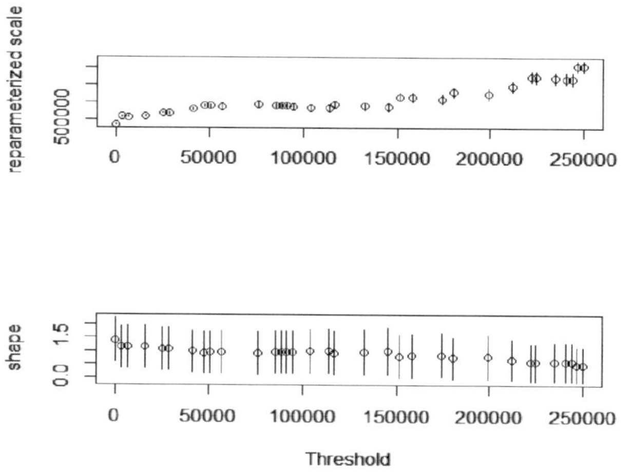

Figure 8.
Parameter Stability Plot.

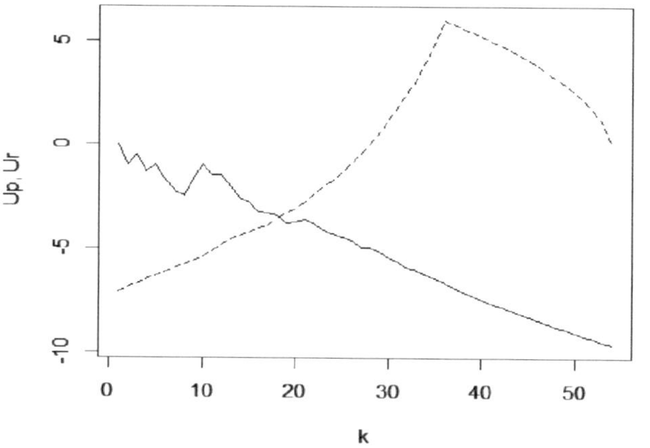

Figure 9.
Gertensgarbe Plot.

Statistic	Value
k_0	19
p-value	0.00014
threshold	150000
tail index	0.48507

Table 2.
Mann-Kendall Test Results.

the plot the MLE estimates for the parameters of the GPD against their 80 corresponding thresholds, within the range $0 \leq v \leq 250,000$, together with a 95% confidence intervals.

We can tell that the estimates of the shape parameter appear to be constant for the whole dataset (when the threshold,v, is zero). The re-parametrized scale parameter estimates on the other hand, seem to be stable beyond 50,000 but not so beyond 225,000.

The information provided by these plots can however, be rather approximative. The Gertensgarbe plot provides a more powerful procedure for threshold estimation [24].

The cross point of the Gertensgarbe graph (**Figure 9**) is at the observation numbered $k = 19$, which corresponds to a threshold of 150,000. The null hypothesis

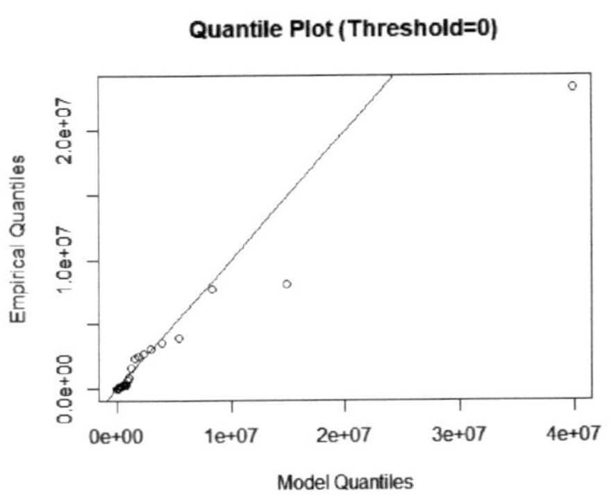

(a) GPD Q-Q Plot (threshold=0)

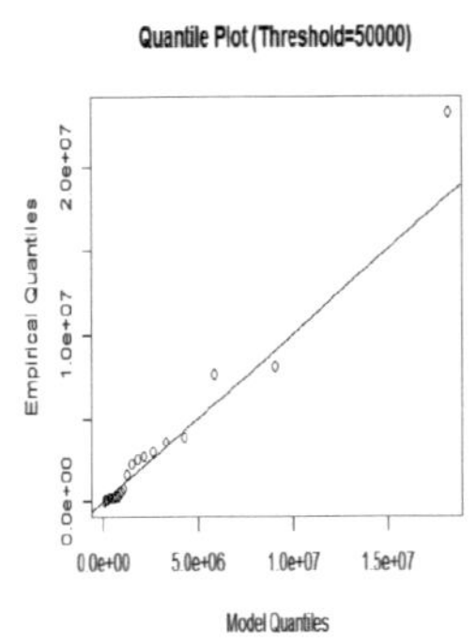

(b) GPD Q-Q Plot (threshold=50, 000)

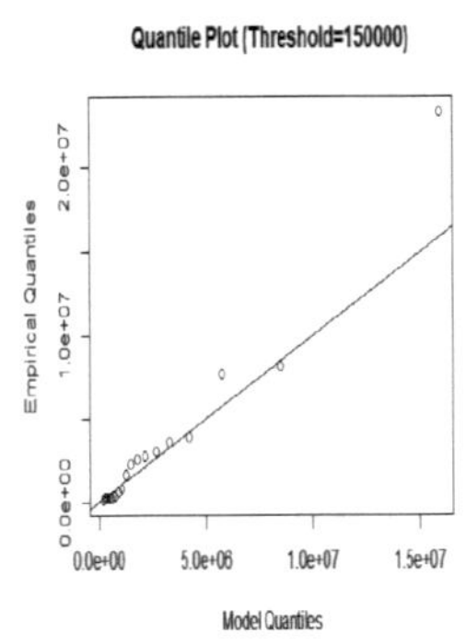

(c) GPD Q-Q Plot (threshold=150, 000)

Figure 10.
Q-Q Plot of GPD under different thresholds.

that there is no change in the series of differences is rejected with a p-value less than 0.001 (see **Table 2**). Therefore, all the three techniques detect the threshold to be between $0 \leq v \leq 250,000$. The parameter stability plot estimates the threshold to be 50,000 and the Gertensgarbe plot 150,000. We will investigate the goodness-of-fit of the GPD to each of the three cases.

Figure 10 shows that GPD fits the data best when we set the threshold at 50,000 and 150,000. The differences between the fit in the two cases also appears to be very small. Since we want to have as many exceedances as possible, we choose the threshold to be 50,000.

6. Distribution of the exceedances

Given a threshold of 50,000, we are interested in the distribution of the excesses. The number of exceedances is 22, and **Figure 11** shows the plot of the kernel density estimates of the number and value of the exceedances.

The number of exceedances is assumed to be Negative-binomial-distributed. We carry out a chi-squared test using the MLE estimates to test its goodness of fit:

Table 3 shows the output of the chi-squared test carried out using "fitdistrplus" package in R. The p-value is greater than 0.01 hence, we fail to reject the null hypothesis that the data follows a Negative binomial distribution.

In section 3, we saw that the distribution of the exceedances can be approximated by a GPD. We will test whether this theorem is justified in our dataset. We use the "bootstrap goodness-of-fit test for the GPD" [25] provided in R package "gPdtest. This test investigates the goodness-of-fit of the GPD, for cases where the distribution is heavy-tailed (shape parameter $\zeta \geq 0$) and non-heavy tailed ($\zeta < 0$).

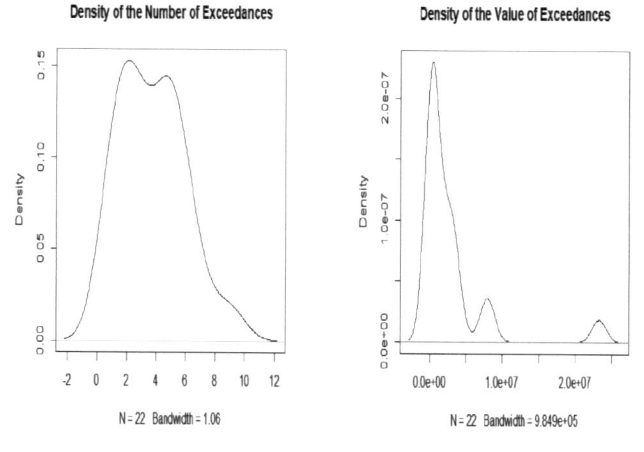

Figure 11.
Density of the Exceedances.

Name	Value
Ch–squared statistic	3.01
Degree of Freedom	2.00
Chi-squared p-value	0.22

Table 3.
Goodness of fit of Negative Binomial to the distribution of the number of Exceedances.

Item	P-value	R-statistic
H_0^-: X has a GPD with negative shape parameter	0.0961	0.9581
H_0^+:X has a GPD with positive shape parameter	0.3874	0.9720

Table 4.
Goodness of fit test for GPD.

Table 4 shows that the p-value in both cases is greater than 0.01. Hence, we conclude that the exceedances do indeed follow a GPD.

7. Fitting the negative binomial-generalised Pareto compound extreme value distribution

7.1 Parameter estimation

We now fit the proposed distribution to the Kenyan data on natural disasters using MLE. We derived the log-likelihood function of the NB-GP CEVD in Section 7.1, and found that the function is not in closed form. Therefore, we will use iterative methods to estimate the parameters.

First, we have to select an appropriate initial value for the parameter estimates. These values represents the initial guess for the estimates. The parameters of the NB-GP CEVD are supported at $\alpha > 0, \sigma > 0, 0 < \theta < 1, 0 < a < 1$ and $\zeta \in \mathbb{R}$. We will study how the choice of the starting value affects the parameter estimates by choosing different starting points, within the support of the parameters.

From **Table 5**, we can observe how the starting values of the parameter estimates significantly affects the estimates. To get more reliable estimates, we use PWM estimates, as discussed in subsection X, to be the starting values. The fist five sample PWMs are found to be:

$$\hat{p_0} = 2817753, \hat{p_1} = 2440020, \hat{p_2} = 2185841, \hat{p_3} = 1997587, \hat{p_4} = 1851745 \quad (31)$$

The resulting system of equations is then:

$$\begin{aligned} \frac{\sigma}{\zeta}\left\{\alpha\left(\frac{a\theta}{1-\theta}\right)^{\zeta} B((1-\zeta),(\alpha+\zeta)) - 1.0\right\} &= 2817753 \\ \frac{\sigma}{\zeta}\left\{\alpha\left(\frac{a\theta}{1-\theta}\right)^{\zeta} B((1-\zeta),(2\alpha+\zeta)) - 0.5\right\} &= 2440020 \end{aligned} \quad (32)$$

Parameter	α	θ	σ	ζ	a
Starting Values	**4.5**	**0.1**	**1**	**0.5**	**0.4**
Estimates	24.2281	0.9228	0.9757	0.5039	0.3900
Starting Values	**0.5**	**0.1**	**1**	**0.5**	**0.4**
Estimates	3.3001	0.9898	1.8597	0.7754	0.9016
Starting Values	**0.1**	**0.1**	**0.1**	**0.1**	**0.1**
Estimates	0.5632	0.9781	0.2440	0.2226	0.3991

Table 5.
MLE estimates of the parameters using different starting values.

$$\begin{aligned}
&\frac{\sigma}{\zeta}\left\{\alpha\left(\frac{a\theta}{1-\theta}\right)^{\zeta}B((1-\zeta),(3\alpha+\zeta))-0.33\right\}=2185841\\
&\frac{\sigma}{\zeta}\left\{\alpha\left(\frac{a\theta}{1-\theta}\right)^{\zeta}B((1-\zeta),(4\alpha+\zeta))-0.25\right\}=1997587\\
&\frac{\sigma}{\zeta}\left\{\alpha\left(\frac{a\theta}{1-\theta}\right)^{\zeta}B((1-\zeta),(5\alpha+\zeta))-0.2\right\}=1851745
\end{aligned} \quad (33)$$

We use "nleqslv" package in R to solve the resulting system of non-linear equations:

Using the values in **Table 6** as the initial values, the MLE estimates of the NB-GP CEVD distribution is [26]:

The likelihood function is also found to be −192.9873.

7.2 Goodness-of-fit tests

Next, we test the goodness-of-fit of the NB-GP CEVD to the data. We create a sample of NB-GP CEVD with the parameters given in **Table 7**. Since this sample is the basis of comparison, we create a large sample of size 10,000 to closely approximate the distribution. Two-sample KS test and two-sampled AD test is then carried out to test the null hypothesis that the two samples have the same distribution.

As observed in **Table 8**, the p-values in both tests are greater than 0.01. Thus, we fail to reject the null hypothesis that the two samples come from the same population at 1% level of significance. Alternatively, for the KS-test, we have:

$$d(\alpha)\sqrt{\frac{n+m}{nm}}=\sqrt{-0.5\ln(0.1)}\sqrt{\frac{10022}{220000}}=0.3239 \quad (34)$$

Parameter	Estimate
$\hat{\alpha}$	9,000,000
$\hat{\theta}$	0.8509
$\hat{\sigma}$	0.2297
$\hat{\zeta}$	0.8378
$\hat{a}$	0.3836

Table 6.
Probability-weighted Moments Estimates.

Parameter	Estimate	Standard Error
$\hat{\alpha}$	9,000,000	2.9659
$\hat{\theta}$	0.0000178	—
$\hat{\sigma}$	1268.1920	0.0100
$\hat{\zeta}$	1.2336	0.0111
$\hat{a}$	0.4000	0.0341

Table 7.
Maximum Likelihood Estimates.

Test	statistic	p-value
Kolmogorov–Smirnov	0.3010	0.0375
Anderson-Darling	2.8140	0.0346

Table 8.
Goodness-of-fit Test results.

The test-statistic

$$D_{m,n} = 0.3010 < 0.3239 \tag{35}$$

so we fail to reject the null hypothesis. We conclude that the data of natural disasters in Kenya follow a NB-GP CEVD.

7.3 Prediction study

From the estimation results in **Table 7**, we can tell that the probability of damages exceeding 50,000 in any year is approximately 0.4. We now want to predict the expected maximum and minimum among those exceeding the threshold in any year with different certainties as explained in section X.

Using Eq. (X), the minimum damage that is expected to be exceeded among those exceeding the threshold with probability p, is predicted using:

$$\hat{x}_p = 50000 + \frac{\hat{\sigma}}{\hat{\zeta}}\left\{\left[\frac{\hat{a}\hat{\theta}}{(1-\hat{\theta})\left((1-p)^{-1/\hat{\alpha}}-1\right)}\right]^{\zeta} - 1\right\} \tag{36}$$

The corresponding minimum equation, x_p,among those exceeding 50000 that is expected not to be exceeded at any given year under different probabilities can be predicted using:

$$\hat{x}_p = 50000 + \left\{\frac{\hat{\sigma}}{\hat{\zeta}}\left[\frac{\hat{a}\hat{\theta}}{(1-\hat{\theta})(p^{-1/\hat{\alpha}}-1)}\right]^{\zeta} - 1\right\} \tag{37}$$

Using the parameter estimates in **Table 7**, the prediction results for $p = 0.9, 0.95, 0.99, 0.995, 0.999$ are given in **Table 9**.

The predicted maximum damage increases as the probabilities increase while the predicted minimum damage decreases as the probabilities increase. In other words, the range becomes larger at higher certainty levels. We can determine the accuracy of the prediction by comparing the prediction results to the statistics of the exceedances (**Table 10**). This can be done using two measures:

1. Absolute Error = Predicted Value - Actual Value

2. Relative Error = $\frac{\text{AbsoluteError}}{\text{ActualValue}}$

From **Table 11**, all the error measurements for the minimum values are positive, suggesting that the predicted values are higher than the actual value. This implies that the NB-GP CEVD tends to overestimate the minimum damage among the exceedances in any given year. We can also observe how the relative minimum error for the the minimum damage decreases as the probability increases.

Probability	0.9	0.95	0.99	0.995	0.999
Predicted Maximum Value	2,788,420.00	6,689,195.00	49,35,2186.00	116,056,448.00	842,007,196.00
Predicted Minimum Value	110,633.00	93,581.66	75,257.41	71,093.15	64,934.09

Table 9.
Prediction Results for Extreme Damages in Kenya.

Statistic	Value
Minimum	60,340
Median	692,142
Mean	2,817,753
Maximum	23,331,469

Table 10.
Statistics of the Exceedances.

Probability	0.9	0.95	0.99	0.995	0.999
Absolute Error for the Minimum	50,293.000	33,241.663	14,917.407	10,753.149	4,594.092
Relative Error for the Minimum	0.833	0.551	0.247	0.178	0.076
Absolute Error for the Maximum	−20,543,049	−16,642,274	26,020,717	92,724,979	818,675,727
Relative Error for the Maximum	−0.880	−0.713	1.115	3.974	35.089

Table 11.
Prediction Accuracy for Extreme Damages.

The Relative error is highest at 0,9 level of certainty and lowest at 0.999 certainty level. So, accuracy of the predictions increases as the level of certainty increases. Generally, the distribution performs well at predicting the minimum values of those exceeding the threshold.

For the maximum values, the predicted values are lower than the actual values at 0.9 and 0.95 probabilities, and higher at $p \geq 0.99$. So, the proposed distribution underestimates the maximum values at probability levels less than 0.95, and overestimates them at probability levels greater than 0.95. In addition, the relative maximum errors are smaller when the predicted values are lower than the actual ones, as compared to when they are greater. So, the distribution overestimates the maximum values by a greater margin than it tends to underestimate. IN general, the NB-GP CEVD performs poorly at predicting the maximum values of the damages exceeding the threshold.

In overall, the absolute errors are smaller for the minimum values than the maximum values. This implies that the NB-GPD CEVD performs better at predicting the minimum values among those exceeding the threshold, than it does for the maximum values.

8. Conclusions

The main goal for this research is to model the extreme damages in Kenya resulting from natural disasters by considering both occurrence and the magnitude of the events. The magnitude of the natural disasters was quantified in terms of the total number of people affected. We use Excess over Threshold modelling, where the extreme damages are identified as the points beyond a sufficiently high threshold. The choice of the threshold significantly affects the parameter estimates, so we use three methods to identify the threshold: mean excess plot, parameter stability plot and Gertensgarbe and Werner plot. The threshold is found to be at 50,000.

In order to capture both the frequency and magnitude of the natural disasters, we use Compound Extreme Value Distribution that was proposed by Liu et.al in 1980. We identify the distribution of the number of exceedances to be Negative Binomial Distribution, while that for the magnitude of the exceedances is approximated by a GPD. We estimate the parameters of the CEVD using maximum likelihood estimation. The log-likelihood function is not in closed form, so we use PWM to determine the starting value for the iterations. We then carry out goodness-of-fit tests using a two-sample Kolmogorov-Smrinov test and a two-sample Anderson-Darling Test. We find that the NB-GP CEVD is a good fit for the data.

Finally, we carry out a prediction study for the NB-GP CEVD. The probability of exceeding the threshold in any year is found to be 0.4. We also predict, with different levels of certainty, the minimum and maximum values expected to be exceeded and not to be exceeded, respectively in the event of the occurrence of an extreme damage. We study the accuracy of these predictions by comparing the predicted values to the actual maximum and minimum. We find that the proposed distribution tends to overestimate the minimum damage at all the certainty levels. On the other hand, the distribution tends to underestimate the maximum damage at $p < 0.95$ certainty levels, and overestimate them at $p \geq 0.99$ levels. Generally, we find that the proposed distribution performs better at predicting the minimum damages among those exceeding the threshold, than it does for the corresponding maximum damages.

We conclude that the NB-GP CEVD is a good fit for the distribution of the extreme damages resulting from natural disaster in Kenya. It performs better at predicting the minimum value that is expected to be exceeded by extreme damages as compared to maximum damages expected not to be exceeded.

Acknowledgements

The authors would like to thank the School of Mathematics, University of Nairobi for providing a conducive environment to conduct this research.

Conflict of interest

The authors declare no conflict of interest.

Author details

Carolyne Ogutu*,† and Antony Rono†
University of Nairobi, Nairobi, Kenya

*Address all correspondence to: cogutu@uonbi.ac.ke

† These authors contributed equally.

References

[1] E. Montz, A. Tobin and R. Hagelman, (2017). Natural hazards: explanation and integration

[2] Zlateva and Velev (2016). *A Method for Risk Assessment from Natural Disasters Using an Actuarial Model*, Journal of Economic, Business and Management, Vol. 4.

[3] P. Embrechts, C. Klüppelberg and T. Mikosch (2013), *Modelling extremal events: for insurance and finance*, vol 33, Springer.

[4] R. von Mises, (1954). La distribution de la plus grande de n valeurs. *American Mathematical Society: Providence RI II*, 271–294.

[5] A. F. Jenkinson (1955). The frequency distribution of the annual maximum (or minimum) value of meteorological events. *Quarterly Journal of the Royal Meteorological Society* 81 (348), 158–172.

[6] A. Ferreira and H. De (2015), *On the block maxima method in extreme value theory: PWM estimators*, The Annals of Statistics, volume 43, number 1, pages 276–298

[7] Y. Hu (2013), *Extreme Value Mixture Modelling with Simulation Study and Applications in Finance and Insurance*, University of Canterbury.

[8] K. Engeland, H. Hisdal, and A. Frigessi (2004). Practical extreme value modelling of hydrological floods and droughts: a case study. *Extremes*, 7(1), 5–30.

[9] H. Rust, D. Maraun and T. Osborn (2009), *Modelling seasonality in extreme precipitation*, The European Physical Journal Special Topics, Volume 174, number 1, page 99–111.

[10] C. Cunnane, (1973). *A particular comparison of annual maxima and partial duration series methods of flood frequency prediction*, Journal of Hydrology, Volume 18, number 3–4, pages 257–271.

[11] H. Madsen, C. Pearson and D. Rosbjerg (1997), *Comparison of annual maximum series and partial duration series methods for modeling extreme hydrologic events: 2. Regional modeling*, Water Resources Research, volume 33, number 4, pages 759–769.

[12] S. Caires (2009), *A comparative simulation study of the annual maxima and the peaks-over-threshold methods*, Deltares report 1200264–002 for Rijkswaterstaat, Waterdienst.

[13] S. Coles, J. Bawa, L. Trenner, and P. Dorazio, (2001). *An introduction to statistical modeling of extreme values* (Vol. 208). Springer.

[14] A. Frigessi, O. Haug and H. Rue (2002), *A dynamic mixture model for unsupervised tail estimation without threshold selection*, Extremes, volume 5, number 3, pages 219–235.

[15] C. Behrens, H Lopes and D. Gamerman (2004), *Bayesian analysis of extreme events with threshold estimation*, Statistical Modelling, volume 4, number 3, pages 227–244.

[16] T. F. Liu, and F. S. Ma(1980). Prediction of extreme wave heights and wind velocities. *Journal of the Waterway, Port, Coastal and Ocean Division*, 106 (4), 469–479.

[17] D. Liu, L. Wang and L. Pang (2006), *Theory of multivariate compound extreme value distribution and its application to extreme sea state prediction*, Chinese Science Bulletin, volume 51, number 23, pages 2926–2930.

[18] D. Liu, S. Wen and L. Wang (2002), *Poisson-Gumbel mixed compound distribution and its application*, Chinese Science Bullettin, volume 47, number 22, pages 1901–1906.

[19] E.L. Geist and T. Parsons (2011), Assessing historical rate changes in global tsunami occurrence, *Geophysical Journal International*, 187(1):497–509.

[20] S. Dong, J. and Zhai, S. Tao. (2017), Long-term statistics of extreme tsunami height at Crescent City, *Journal of Ocean University of China*, 16(3):437–46.

[21] M.R. Leadbetter, G. Lindgren, and H. Rootzén (1983), *Extremes and Related Properties of Random Sequences and Processes*. Springer Series in Statistics [Internet]. Springer New York.

[22] E. Omey, F. Mallor, and E. Nualart (2009). *An introduction to statistical modelling of extreme values*. Application to calculate extreme wind speeds.

[23] J. Pickands III et al (1975), *Statistical inference using extreme order statistics*, The Annals of Statistics, volume 3, number 1, pages 119–131.

[24] Cebrián, A. C., Denuit, M., & Lambert, P. (2003). Generalized pareto fit to the society of actuaries' large claims database. North American Actuarial Journal, 7(3), 18–36.

[25] Villaseñor-Alva, J. A., & González-Estrada, E. (2009). A bootstrap goodness of fit test for the generalized pareto distribution. Computational Statistics & Data Analysis, 53(11), 3835–3841.

[26] J. Liu, D. Shi, and X. Wu (2008). Estimation of poisson-generalized pareto compound extreme value distribution by probability-weighted moments and empirical analysis. *Transactions of Tianjin University*, 14(1), 50–54.

Chapter 17

Effects of Earthquakes on Buildings in the Ibero-Maghrebian Region

José Antonio Aparicio Florido

Abstract

The types of damage caused by earthquakes in buildings are closely related to the design and building techniques with which they have been built and the quality of the construction materials used. Most of countries with moderate to high seismic risk areas have implemented earthquake-resistant standards to prevent the collapse of buildings and minimize the severity of the damage. However, every new strong shake that occurs around the world reveals bad construction practices that could have been avoided, and the inadequacy or non-existence of earthquake-resistant standards aimed at reducing vulnerability to non-catastrophic levels. Based on the EMS-98 scale, in this chapter we will analyze three case studies of the Ibero-Maghrebian region that have been using similar construction patterns with similar catastrophic results for buildings despite the different dates in which they occurred and the different earthquake-resistant standards: SW Cape St. Vincent earthquake of February 28, 1969; Al Hoceima earthquake of February 24, 2004; and Lorca earthquake of May 11, 2011.

Keywords: earthquakes, damage, buildings, standards, shear cracks, X-shaped cracks, plastic hinges, soft-story, vulnerability

1. Introduction

The Ibero-Maghrebian region comprises the southern part of the Iberian Peninsula, (including Andalusia, Murcia and Alicante) and North Africa from the Atlas range to the Mediterranean sea, bordering Tunisia to the east and the Atlantic coast of Portugal to the west, embracing Morocco and Algeria [1]. The seismicity of this region is characterized by the N-S to NW-SE convergence between Eurasian and Nubian plates [2]——with a possible rotational axis to the north of Canary Islands—— and by the occurrence of shallow earthquakes (<30 km).

Along this area, between January 1, 1950 and December 31, 2019, 198 earthquakes of magnitude M≥5 took place and, of these, 42 shocks (21%) reached intensities ≥VI (**Table 1**). The most destructive event was that of El Asnam (Algeria) on September 9, 1954 with intensity X-XI [3]. However, the largest earthquake in magnitude was that of February 28, 1969 (M_w = 7.8), located in the Horseshoe Abyssal Plain, southwest of Cape St. Vincent [4]. This one caused a widespread destruction in Portugal, Spain and Morocco, and a moderate tsunami recorded in the tide gauges of Cascais (93.2 cm), Lagos (84.3 cm), Cádiz (28.4 cm) [5], La Coruña (12 cm),

Date	Time	Lat.	Long.	Depth (km)	Int.	Mag.	Epicenter
1950/04/20	17:19:14	339.000	21.000	—	VI	5.1 M_0	El-Gheicha.AL
1951/03/10	10:38:26	375.950	−39.750	15	VII	5.2 Mw	Castillo de Locubán.SP
1951/05/19	15:54:26	375.670	−39.170	19	VII	5.3 Mw	Castillo de Locubán.SP
1953/08/29	14:08:50	358.000	50.000	—	IX	5.2 M_0	Hodna.AL
1954/09/09	09:28:42	360.000	15.000	—	X	6.0 M_0	El Asnam.AL
1954/09/09	01:04:37	362.833	14.667	—	X-XI	6.7 M_0	El Asnam.AL
1954/09/10	05:44:05	366.000	13.000	—	VIII	6.0 M_0	Tenes.AL
1954/10/10	06:01:48	363.000	18.000	—	VI	5.5 M_0	Kerba.AL
1954/10/12	19:23:29	362.500	17.000	—	VII	6.0 M_0	Fodda.AL
1955/06/04	03:41:35	371.333	−36.467	5	VI-VII	5.1 M_0	Armilla.SP
1956/04/19	18:38:54	371.917	−36.833	5	VII-VIII	5.0 M_0	Purchil.SP
1956/08/16	02:09:40	369.100	−86.067	5	VI	5.0 M_0	Gulf of Cádiz
1959/05/24	13:19:38	363.383	45.317	5	VIII	5.1 M_0	Bordj Bou Arreridj.AL
1959/08/23	22:21:30	355.133	−32.267	20	VI	5.4 M_0	South Alboran
1959/11/07	02:32:08	364.000	25.000	—	IX	5.1 M_0	Bou Medfa.AL
1960/02/21	08:13:33	356.517	42.500	5	VIII	5.5 M_0	M'Sila.AL
1960/02/29	23:40:14	304.500	−96.167	—	X	6.0 M_0	Agadir.MR
1961/02/10	18:52:01	417.250	−61.967	—	VI	5.2 M_0	Zamora.SP
1964/03/15	22:30:26	361.317	−77.500	30	VII	6.2 MbLg	Gulf of Cádiz
1965/01/01	21:38:26	357.000	45.000	—	VIII	5.2 MbLg	M'Sila.AL
1967/07/13	02:10:21	355.300	−0.1267	5	VII	5.0 MbLg	Mascara.AL
1967/08/13	22:07:47	432.950	−0.6767	5	VIII	5.3 MbLg	Navarrenx.FR
1969/02/28	02:40:32	359.850	−108.133	20	VIII	7.8 Mw	SW Cape St. Vincent
1973/11/24	15:22:09	361.000	44.000	—	VII	5.1 MbLg	Mansourh.AL
1975/05/26	09:11:49	359.000	−176.000	—	VI	6.7 MbLg	Athlantic Ocean
1980/10/10	12:25:23	361.533	14.467	5	IX	6.5 MbLg	Chlef.AL
1988/10/31	10:12:59	364.433	26.083	13	VII	5.4 MbLg	Blida.AL
1989/10/29	19:09:14	367.483	24.333	5	VIII	5.7 MbLg	Mediterranean-Algeria
1989/12/20	04:15:05	372.250	−73.917	23	VI	5.0 MbLg	Ayamonte.SP
1992/10/23	09:11:08	312.200	−43.567	7	VI-VII	5.3 MbLg	Morocco
1992/10/30	10:44:01	314.117	−43.833	21	VII	5.1 MbLg	Morocco
1993/12/23	14:22:35	367.800	−29.367	8	VI-VII	5.0 MbLg	Berja.AL
1994/08/18	01:13:07	354.783	−0.1417	5	VII	5.7 MbLg	Mascara.AL
1997/05/21	23:50:45	427.833	−72.583	13	VI	5.1 MbLg	Triacastela.SP
2003/05/21	18:44:19	368.187	37.203	—	IX-X	6.6 Mw	Boumerdès.AL
2004/02/24	02:27:46	351.563	−39.841	—	VIII	6.2 Mw	Al Hoceima-Tamassint.MR

Date	Time	Lat.	Long.	Depth (km)	Int.	Mag.	Epicenter
2004/12/04	10:29:59	349.796	−29.722	6	VI	5.0 Mw	Hasssi Berkane. MR
2006/03/20	19:44:23	366.736	55.604	—	VII	5.1 Mw	Bejaia.AL
2006/04/02	06:44:31	349.373	37.942	—	VI	5.4 mb	Bou Saada.AL
2008/02/01	07:33:41	368.223	35.288	12	VI	5.0 mb	Boumerdès.AL
2011/05/11	16:47:26	377.175	−17.114	4	VII	5.1 Mw	Lorca.SP
2016/01/25	04:22:01	356.004	−38.056	12	VI	6.3 Mw	South Alboran

Table 1.
Earthquakes felt with intensity ≥ VI in the Ibero-Maghrebian region between January 1, 1950 and December 31, 2019. Source: Instituto Geográfico Nacional (IGN).

Chipiona (30 cm), Santa Cruz de Tenerife (0.17 cm) and Casablanca (120 cm) [6]. On the other hand, the shallowest earthquake (depth = 1.4 km) and the second most damaging in this list, with ≈15,000 dead and 70% of all new buildings ruined [7], was that of February 29, 1960 in Agadir (Morocco), considered as a moderate earthquake compared to its magnitude M_0 = 6.0.

These initial considerations show us that the effects of earthquakes are not merely a proportional relationship between magnitude and intensity; moreover, there are some other parameters and more deterministic in terms of vulnerability, such as the distance from the epicenter to populated areas, the depth of the hypocenter, the geological characteristics of the soils and, mainly, the quality of the buildings regarding to the construction techniques, pattern, materials and antiseismic design, which is the subject of this chapter.

2. Method of analysis

To explain the effects of earthquakes on buildings using the same analysis criteria, we will use as reference the 1998 European Macroseismic Scale (EMS-98) [8], based on the MSK scale, consisting of twelve incremental levels of intensity in Roman numerals (I-XII) and six vulnerability classes identified by the first letters of the alphabet (A, B, C, D, E and F). The scale also distinguishes and describes four types of structures (masonry, reinforced concrete, steel and timber) of which we will focus on those of masonry and reinforced concrete (hereinafter, RC) without earthquake-resistant design (ERD), the most common in the Ibero-Maghrebian region. Both types of construction, masonry (non-engineered) and reinforced concrete (engineered), present a vulnerability class A, B or C; classes D, E or F, require the implementation of moderate or high ERD measures or the use of steel or timber as the main construction material, unusual within the study area. Here, as we will see later, the buildings share similar construction patterns and offer up the same level of resistance: vulnerability class A or B for masonry buildings, with the exception of those made of massive stone or ashlars (class C), and vulnerability class C for RC buildings. We will prove, for example, how a building of class A located in the Algarve (Portugal) is very similar in terms of construction pattern to another of equal vulnerability in Ait Kamra (Morocco); or how a building of class C in Lorca (Spain) can be damaged in the same way as a similar one in Al Hoceima.

With respect to the classification of damage, which is carried out by direct observation in the field, the EMS-98 scale considers five levels: grade 1, negligible to slight damage; grade 2, moderate damage; grade 3, substantial to heavy damage;

grade 4, very heavy damage or partial collapse; and grade 5, destruction or total collapse. Buildings with a grade of damage 1 to 3 do not result in structural or significant structural damage, being easy to repair and recover, while those of grade 4 or 5 have severe or complete structural damage and are difficult or impossible to recover. In this scientific contribution we will explain these effects in detail based on three of the most destructive and best documented earthquakes recorded in the study area: southwest Cape St. Vincent, 1969; Al Hoceima, 2004; and Lorca, 2011.

3. Documentary sources

The analysis and description of damage from these selected seismic events is based on several documentary sources: available unofficial field reports, official reports from public authorities, photographs showing damage to buildings, and information provided by news agencies and newspapers.

In the case of the 1969 earthquake, our main source for the area of Portugal is the technical report carried out by the seismologist Mário de Vasconcelos Trêpa [9] and, for the case of Huelva (Spain), the official reports of Isla Cristina Town Council. Regarding the 2004 Al Hoceima earthquake, the analysis will be based on the field report made by the technical architect Patrick Murphy Corella [10]. Finally, with respect to the 2011 Lorca earthquake we will take as documentary source our own field report and photographs taken by emergency managers of the Instituto Nacional para la Reducción de los Desastres (IERD) [11].

4. The February 28, 1969 earthquake

The 1969 SW Cape St. Vincent earthquake took place on Saturday, 28th February, at 02.40.32 UTC (t_0), and was located in the Horseshoe Abyssal Plain, coordinates 35.98° N 10.81° W, 200 km away to the SW of Cape Saint Vincent, with hypocenter at 20 km depth. Its high magnitude Mw = 7.8 (IGN) caused seismic waves to heavily shake Portugal, Spain, and Morocco. Moreover, this seismic event is associated with a reverse faulting between Azores and the Gulf of Cadiz in the western part of Africa-Eurasia plate boundary, which generated a moderate tsunami recorded in several tide gauges, strongly damaging the Norwegian tanker "Ida Knutsen" while sailing in deep waters [12].

4.1 Effects in Portugal

4.1.1 Algarve: an example of bad behavior of buildings of vulnerability A or B

The greatest damage to buildings in Portugal was registered in the Algarve region, with total or partial collapses spread over different locations, amplified by the so-called *site effect* [13, 14], that is, the amplification of the seismic waves due to the geological characteristics of soils in superficial layers and the soil-structure interaction of buildings. Settlements built on younger tectonic formations, such as poorly consolidated alluvial soils, felt the shakes more strongly [15], increasing the vulnerability of traditional constructions made with extremely poor materials and techniques. Most of them had one or two floors supported on load-bearing walls made of fieldstone or adobe masonry bound with mud mortar, clay or without mortar. These external walls, quite thick and most often strengthened with double-leaf shape (two walls facing each other, barely joined by a poor mortar), supported a gable roof made of wooden logs or struts under a bed of reeds covered with tiles or zinc plates. Therefore, they must be classified as structures of vulnerability class A, and hence the effects in Castro Marim, Barão de São Miguel, Vila do Bispo, Bensafrim, Fonte de Louzeiros or Silves.

The damage reached greater severity and geographical extension within a radius of 50 km around Cape St. Vincent. In Silves, several houses collapsed, and lateral loads produced many shear cracks, overturning or toppling of load-bearing walls, corner failures and large cracks in the walls of the castle. **Figure 1** shows shear cracks, with detachment of little pieces of plaster and partial toppling of external wall from the upper floor, after loss of connection and possible mutual pounding of adjacent buildings with roofs at different levels, in a three-story building of class B (grade 3). Bensafrim, 26 km from Silves and closer to Cape St. Vincent, was one of the towns hardest hit by the earthquake. More than twenty houses were destroyed and around forty suffered the typical damage of buildings of class A: loss of connection between load-bearing walls, corner failure, overturning or toppling of external walls, detachment and fall of outer leaf, X-shaped or diagonal cracks, fall of plaster, and roof collapse. The primary school, recently built with RC (class C), had cracks in several parts (grade 3). An example of corner failure with partial roof collapse is described in **Figure 2**, which should be assessed as grade 4 in building of class A. The same description can be made for Odiáxere, a *freguesia* (municipality) of Lagos council, at the distance of 7 km from Bensafrim.

In Vila do Bispo, 8 km from Sagres, many houses of class A were partially or completely collapsed and turned into rubble (**Figure 3**), with a level of destruction similar to Bensafrim. **Figure 4** shows an example of façade overturning pulling down the roof and causing a complete collapse of the structure (grade 5). In other photographs we can clearly observe damage of grade 3 in corner failures (**Figure 5**) and grade 4 in the toppling of external wall, seriously affecting the roof (**Figure 6**). Also, the upper body of the church tower of Nossa Senhora da Conceição, built with massive stone, ashlars and burnt clay/sand bricks showed a progressive X-shaped

Figure 1.
Silves, building of class B with damage of grade 3: (a) shear cracks with detachment of little pieces of plaster; (b) partial toppling of external wall. Source: IPMA.

Figure 2.
Bensafrim, building of class A with damage of grade 4: (a) advanced corner failure and toppling of external wall with partial roof collapse; (b) loss of connection and drift between load-bearing walls; (c) generalized detachment of plaster. Source: IPMA.

Figure 3.
Vila do Bispo, general overview: (a) loss of the outer leaf of load-bearing walls and detachment of plaster; (b) partial toppling of façade wall, pulling down part of the roof; (c) good behavior of a recent construction (class B); (d) simple overturning of a freestanding masonry wall; (e) adobe house without apparent damage. Source: IPMA.

Figure 4.
Vila do Bispo, total collapse of building of class A made of mixed fieldstone and adobe masonry, mud or clay mortar and roof supported by wooden logs and bed of reeds (grade 5). See the doble-leaf shape on an interior load-bearing wall remaining standing. Source: IPMA.

Figure 5.
Vila do Bispo, building of class A with damage of grade 3 due to loss of connection between load-bearing walls and corner failure (left). This two-story house was repaired and is currently in use (right). Sources: IPMA/ Google Earth.

crack and fall of pinnacles [15], which means damage of grade 3 in buildings of class B. In summary, the traditional construction techniques based on adobe masonry and clay mortar showed a high vulnerability and weakness (**Figure** 7).

Several houses were totally collapsed in Sagres and many others suffered damage of grade 2 or 3 in buildings of vulnerability class A. In **Figure 8**, one single-story house with gable roof shows shear cracks, detachment of cornices and tiles, partial roof collapse, loss of connection between load-bearing walls and drift from the façade with risk of overturning (grade 3). Also, in Raposeira, 5 km east from Vila do

Figure 6.
Vila do Bispo, building of class A with damage of grade 4: (a) loss of connection between load-bearing walls; (b) toppling of external wall; (c) collapse of roof made of wooden logs and bed of reeds. Source: IPMA.

Figure 7.
Vila do Bispo, typical house of vulnerability class A built with raw materials: load-bearing walls made with adobe bricks, doors and windows with lintels, and wooden logs supporting a tiled roof over a bed of reeds. Source: Google Earth.

Bispo, a crack arose on the façade of the hermitage of Nossa Senhora de Guadalupe, running from the gable to the apex of the ogival arch of the entrance, from top to bottom, crossing the rose window [15]. This chapel dates from 14th century and was built with load-bearing walls made of massive stone, with pillars of ashlars embedded and supporting a ribbed vault built with the same material. The entire structure was reinforced with solid buttresses around the side walls and toothed ashlars in the corners; therefore, it should be assessed as damage of grade 2 in a building of vulnerability class C.

Figure 8.
Sagres, building of class A with damage of grade 3: (a) loss of connection of load-bearing walls with drift of façade, and (b) detachment of cornices and tiles with partial roof collapse. Source: IPMA.

A similar description could be made for Portimão, located 40 km to the east. According to Trêpa's report, this town was "undoubtedly, the most affected city in the whole territory". However, there is no specific information of damage that would allow us to confirm this statement. **Figure 9** shows the partial detachment and toppling of the outer leaf of the external wall from the second floor of a two-story building of class B, with the subsequent fall of a cornice section and detachment of tiles (grade 3) due to a hammering effect between the wooden joists supporting the roof and the top of the load-bearing walls.

4.1.2 Fonte de Louzeiros: the maximum intensity in the whole macroseismic area

Fonte de Louzeiros is a small rural hamlet located 7 km at east of Silves. In this settlement of sixteen houses at the time, the most of buildings were total or partially collapsed. The earthquake caused all type of damage mentioned above: X-shaped cracks, corner failures, overturning of load-bearing walls and roof collapses; almost no house could be rebuilt. As an exceptional example, in **Figure 10** we can see one of the few buildings left standing and showing the patched damage of a corner failure. About 95% of these single-story buildings of vulnerability class A suffered damage of grade 5.

Some seismologists have assigned to this place intensity VIII, the maximum level of destruction in the whole macroseismic area [16], although other researchers extend this isosist, in general, to the Algarve region [4] or enclose it to the Cape St. Vincent and surroundings areas [9, 15, 17, 18]. But this particular case of Fonte de Louzeiros, with a total destruction of buildings of vulnerability class A, does not correspond to an area of intensity VIII but rather to intensity IX, or at least VIII-IX [19]. Therefore, it is consistent to raise the damage recorded in Fonte de Louzeiros to intensity VIII-IX, where "a bombing would not have been worse", literally "wiped off the map", as defined by the *Diario de Lisboa* on March 2, 1969.

4.1.3 Alentejo coast, Setúbal and Lisbon: the transition to vulnerability class C

Outside the Algarve region, the intensity of the damage decreases slightly, except along the Atlantic coast between Aljezur and Setúbal, where the effects reached

Figure 9.
Portimão, two-story building of class B with detachment and toppling of the outer leaf of the external wall on the second floor, due to a hammering effect between the wooden joists supporting the roof and the top of the load-bearing walls (grade 3). The lack of shear cracks indicates a ground displacement from back to front. Source: IPMA.

Figure 10.
Fonte de Louzeiros, building of class A rebuilt after the 1969 earthquake, showing the repair mark of a corner failure (dotted line). Inside the white circle is visible the red color of the clay used to join the adobe masonry. Source: Google Earth.

intensity VII, or VI-VII in the Lisbon district. In São Teotónio, Longueira, Vila Nova de Milfontes, Cercal and Sines, midway between Cape St. Vincente and Setúbal, many houses of class A collapsed. The rural village of Longueira, completely built with the same construction techniques of the Algarve, was razed; and the same situation occurs in Vila Nova de Milfontes and Cercal, where only the RC buildings (class C) performed well. In Odemira, partial roof collapse can be observed in buildings of class A (grade 4), due to the hammering effect of the wooden logs supporting the roof against the top of the load-bearing walls, to which they were not well anchored (**Figure 11**).

Compared to the previous locations, the city of Sines felt abnormally strong vibrations of soils, probably influenced by geological conditions and site effect. Some houses of class A or B total or partially collapsed (grade 4 or 5) and other buildings of class C as the City Hall, the church of São Salvador and the hospital suffered significant damage with fall of apparently non-structural elements. Further north along the coast, in Grândola, 38 km northeast, buildings of class B (simple stone) had a good behavior with some shear cracks and detachment of non-structural elements (grade 1 to 3), but twenty houses of class A (adobe masonry) had to be rebuilt of demolished because of corner failures, overturning of load-bearing walls, roof collapse and fall of plaster and cornices (**Figure 12**).

Finally, in Setúbal and Lisbon the structures of vulnerability A and B had been widely replaced a long ago by those of class C (monumental and RC buildings), contributing to a lower level of destruction. Even so, in Setúbal some old houses collapsed, and shear cracks arose everywhere. Ceramic shops were specially shaken, with high material losses of facilities and merchandise. The factory of Pinhal Novo, built of RC supporting structure with hollow clay floor slab blocks over RC beams (class C), became a mountain of debris (grade 5), due to the scarce rigidity of column-beam joints and the use of smooth steel rods (not corrugated) within the core of RC frames. The information available for Lisbon allows us to assess damage of grade 2 in buildings of class C and grade 3 for class B, with a lot of cars buried in rubbles by the fall of plasters, bricks, cornices, eaves, chimneys, balconies, windows and other non-structural elements form buildings of class C, with similar consequences as we will see later when analyzing the 2011 Lorca earthquake.

Figure 11.
Odemira, damage of grade 4 in adobe masonry house of vulnerability class A: (a) detachment of plaster; (b) partial roof collapse due to the strong pounding of wooden logs against the top of the load-bearing wall. Source: IPMA.

Figure 12.
Grândola, damage of grade 4 in a house of class A (adobe masonry): (a) advanced corner failure not compromising the roof; (b) general detachment of plaster. Source: IPMA.

4.2 Effects in Spain

4.2.1 Severe damage limited to Guadiana/Guadalquivir strip

In Spain, as happened during the earthquake of November 1, 1755, the shocks were felt most strongly in western Andalusia, near the border with Portugal, along the strip between the rivers Guadiana and Guadalquivir. It is undoubtedly due to the increasing human settlement on young soils formed by sedimentary alluvial deposits from both rivers, tributaries streams and other intermediate rivers such as Tinto and Odiel. In the city of Seville, the seismic event caused a widespread partial fall of non-structural elements in buildings of class B or C (as in Setúbal, class A is not very representative, practically nonexistent) and moderate damage not higher than grade 3 to monumental buildings (class C) as La Giralda, Torre del Oro, Alcázar, Museum, Navy Command Headquarters, City Hall, Telefónica building, Post Office building and Cathedral.

However, in the western half of the province of Huelva, the level of damage was remarkably similar to that of the Algarve region, due to the analogy between the construction patterns, especially with regard to buildings of class B or monumental buildings of class C (adobe masonry structures were less common in this geographical area). About 274 buildings in the city of Huelva were damaged, 1 completely ruined and 18 partially collapsed. Several photographs published in newspapers allow us to diagnose grade 3 in residential buildings of class B and grade 2 and 3 in monumental buildings of class C (Cathedral, Iglesia de la Milagrosa, the Town Hall, etc.). In the surrounding villages little damage of grade 4 or 5 was recorded, with the exception of Ayamonte and Isla Cristina.

4.2.2 Widespread damage to buildings of class A and B in Isla Cristina

Isla Cristina and Ayamonte are two Spanish towns located in the province of Huelva, at the mouth of the Guadiana River, very near the border with Portugal just 10 km from Castro Marim. Obviously, there was a very close relationship between the damage caused by the earthquake and the fact that both towns were settled over sedimentary soils. In Ayamonte, a few houses that already showed a previous state of abandonment collapsed, and only were recorded a few cases of fall of beams and roofs, or cracks in some load-bearing walls and infill partitions. But in Isla Cristina, 7 km away, damage of grade 4 and 5 was widespread.

The historical and archival reports provided to us by the Town Council of Isla Cristina ——unfortunately, there are no photographic documents—— are not very descriptive in relation to the damaging effects in this place. Three expert teams that carried out the disaster evaluation limited their analysis to estimating that 65% of the total number of buildings in this municipality were damaged (≈ 350 houses), resulting 21 of them in partial or total collapse and completely uninhabitable (grade 4 or 5). With this scarce official information available and based on the photos published by the press, we can compare this level of destruction with this one observed in Bensafrim or Vila do Bispo, in the Algarve region, but in an area 130 km further away from the epicenter with respect to the Portuguese towns here mentioned. Therefore, the site effect is also evident in the Guadiana estuary.

4.2.3 Clocks

The earthquake took place in origin (t_0) at 02.40.32 UTC (IGN), but P waves reached the seismographic stations of Spain and Portugal at different times depending of the epicentral distance (UTC): Lisbon, 02.41.20; Coimbra, 02.41.41; Porto, 02.41.52; San Fernando, 02.41.30.5; Toledo, 02.42.07; Canarias, 02.42.08; Granada, 02.42.09; and Barcelona, 02.43.10. In San Fernando we take as reference the record of the official seismic bulletin of the Real Observatorio de la Armada (ROA).

Just at 03.45 suddenly stopped the clocks of the City Halls of Cádiz, El Puerto de Santa María and the Compañía de Seguros building located in Plaza de Neptuno, Madrid. In Lisbon, the clock on the main façade of the Estação do Sul (railway station) stopped at 03.44 local time (+1 UTC), although the time gap is probably due to a lack of synchronization between local and UTC time. In Huelva, the clock of the Catedral de la Merced stopped at 03.41 and in Palencia, 860 km from the epicenter, the City Hall clock also stopped at 03:45, because the machinery had come out of its gear due to the strong motion. Surely this was the same cause that made all the other clocks to stop.

4.3 Effects in Morocco

The earthquake caused strong shakes in Marrakech, Safi, Casablanca, Tétouan, Rabat, Salé and Tangier, among other sites in Morocco. In Salé, two houses that already had a previous state of ruin collapsed, and in Rabat some houses in the surrounding area (class A) were ruined. However, in Agadir the earthquake had not been felt.

The available macroseismic data from Morocco is very scarce, mainly due to the coincidence of an episode of severe flooding caused by a powerful storm that left torrential rain, flash floods and rivers bursting their banks along the Atlantic coasts and in the Moroccan Atlas. In the region of Casablanca-Settat (formerly named Dukala-Abda) large areas of crop fields were flooded and affected by landslides, with three villages submerged under waters. In the city of Salé, a short distance north of Rabat, a house collapsed due to the riverine flooding of the river Bou Regreg, forcing the evacuation of its residents. In Kenitra, 35 km north of Rabat, the roads were cut by flooding and 100,000 hectares of crop fields were devastated by the rainfall. Also, 39 houses were washed away by the floodwaters 120 km north of Rabat, leaving some villages isolated. Further inland, in the middle of the Atlas, 95 houses were left with water over rooftops. These widespread floods in much of the

Moroccan territory made communications difficult, and the authorities and rescue teams focused on the effects of the floods rather than the effects of the earthquake. As a consequence, under the impact of two natural phenomena at the same time and along de same disaster area, it is difficult to distinguish which damages are due to the floods, to the earthquake or to the combined effects of both episodes.

5. The Al Hoceima earthquake of February 24, 2004

The Al Hoceima earthquake of February 24, 2004 took place at 02:27:46 UTC, at coordinates 35.1563 N -3.9841 W, south of the Alboran Plain, in the Mediterranean Sea near the Strait of Gibraltar, with epicenter north of Tamassint (Morocco). The magnitude of this seismic event was Mw = 6.2 (IGN) and there are no data about its depth, so it is estimated that it must have been very superficial <5 km. To have an approximate idea of the superficiality and energy released, the main shock was followed by more than 760 aftershocks over Mw>3 during the following two months.

5.1 The failure of traditional construction patterns

This earthquake hit the province of Al Hoceima, especially the city so named, the municipality of Imzourem and the villages of Ait Kamra and Izemmouren. The rural houses in this geographical area of Morocco, most of them of vulnerability class A, were built using the traditional construction pattern, very similar to those we have just highlighted in the most rustic areas of the Algarve, with more emphasis in Fonte de Louzeiros: single-story load-bearing walls made with unworked raw materials and unskilled masonry techniques, supporting a brittle roof of wooden logs in parallel under a bed of reeds tied together with rope or wire, all covered by a layer of water-repellent mortar and, occasionally, with a plastic film interposed acting as waterproof element (**Figure 13a**). The load-bearing walls were made up

Figure 13.
Ait Kamra, traditional house pattern: (a) partial collapse showing the doble-leaf load-bearing walls, fieldstone masonry bonded with mud mortar and roof of wooden logs and canes covered with a plastic film (blue) and layer of water-repellent mortar (grade 4, class A); (b) handmade joist of hollow bricks assembled with a single steel rod. Source: P Murphy.

of two faces of field stone masonry (double-leaf) not anchored between them, and poorly bonded with high porosity and low adhesion mud mortar. The bending of the walls in both directions due to reverse horizontal forces caused the inner detachment of the double-leaf masonry and the subsequent bowing, peeling, and overturning or collapse of the external skin. Shear cracks also caused widespread wall collapses and massive corner failures and fall of roofs, pulling down huge amounts of debris (**Figure 14**). The use of hollow bricks assembled with a single steel rod to form a handmade joist as reinforcing element for the roof did not provide greater rigidity or better connection between roofs and load-bearing walls (**Figure 13b**).

5.2 X-shaped damage, a seismic evidence

Regardless of the function of the walls (infill, load-bearing, retaining, etc.), type of building materials (adobe, stone, brick, gray concrete block...) and vulnerability class (A, B, C...), not all cracks should be exclusively interpreted as a result of seismic shaking. In general, wall cracks are caused by the effect of five recognized forces: tension, compression, bending, torsion, or shearing. But only when cracks have a diagonal shape it can be inferred that they have been generated by shear stress or a combined effect tension-shearing or bending-shearing. Shear or diagonal cracks may be due to earthquakes, but also——and most frequently——to a slope in the ground, poor foundations, landslide, soil composition (i.e., expansive clays), geological conditions, etc., which can lead to subsidence or settlement processes.

In April 2019, during a brief field visit to the Roman ruins of Baelo Claudia (Tarifa, Spain), an ancient city where some geologists have believed to find evidence of great destruction caused by a sequence of earthquakes between the 1st and 3rd centuries BC, we certainly observed shear cracks in the *summa cavea* of the theatre, bowing of a load-bearing wall in a *vomitorium* arch, and expulsion of voussoirs in another arch near the *parascenium* that looked like piano keys shifted (**Figure 15**). We can firmly state that these structural damages were not induced by earthquakes because they occurred after the reconstruction works of this archeological site carried out in the 1980s and without seismic events to justify them. Rather, these

Figure 14.
Ait Kamara, corner failure pulling down the roof slab (grade 4, class A). Source: P Murphy.

Figure 15.
Baelo Claudia, roman ruins (theatre): bowing of a vomitorium *wall (left) and expulsion of voussoirs near the* parascenium *(right), which show a tensile stress in the direction of the slope. Source: J A Aparicio.*

effects can be explained by the instability and slope of the ground, as well as a bad combination between the weight of the structure, the insufficient foundations, and the progressive and slow landslide.

A single shear crack only proves the existence of a one-way displacement force. However, X-shaped cracks are unmistakable signs of lateral and reverse forces that result in a very characteristic diagonal crack pattern. The best candidates are earthquakes due to the effect of S-waves and surface waves, particularly Love waves, which transmit loads in opposite directions. In Al Hoceima, this type of damage is found in any vulnerability class buildings (A, B or C). **Figure 16** shows a very striking case, where the X-shaped damage completely surrounds the load-bearing walls on the ground floor of a four-story residential building of vulnerability class B in Imzourem (grade 3). There are no shear cracks on the upper floors, which is a clear indication

Figure 16.
Imzourem, X-shaped damage completely surrounding the load-bearing walls on the ground floor of a four-story residential building of vulnerability class B (grade 3). Source: P Murphy.

of a soft-story damage, which will be discussed later. X-shaped cracks are sometimes blurred by the loss of outer leaf or toppling of the external walls (**Figure 17**).

The effects of this diagonal tension cracking are also common in infill walls of RC structures where, except in case of collapse, are clearly visible. They spread from lower to upper floors, with a more severe impact on the ground floor after receiving the loads of seismic shaking. In this case, the positive aspect is that these infill walls are not structural elements, unlike in buildings of class A or B, and must be assessed as damage of grade 3 in buildings of vulnerability class C, unless the RC frame structure has been seriously damaged. When X-shaped cracks cross external infill walls located between discontinuities or openings arranged for windows and doors, and the location and depth of these cracks coincide with a RC frame column that ends up being damaged, it can result in a very characteristic type of damage called "short column" or "captive column" (**Figure 18**). This effect is caused by the modification of the expected proportional distribution along the column body of its deformation ability under the influence of lateral loads [20], due to a partial confining of RC frames and a lesser stiffness of a free portion of the column less supported by partitioning brickworks. The consequence is a shorter column that concentrates most of the shear stress, i.e., a major part of the column ductility is lost.

Figure 17.
Examples of X-shaped damage clearly visible in a single-story class A house (above) and blurred by loss of outer leaf in a four-story residential block (below). Source: P Murphy.

Figure 18.
Imzourem, short column effect in a three-story residential block of vulnerability class C (grade 4). On the right, a detail of the damage is shown. Source: P Murphy.

Figure 19.
Alhucemas, soft-story failures: (a) tilting of RC frames on the ground floor; (b) soft-story damage with clear X-shaped crack on the infill wall; (c) collapse of ground floor due to a previous soft-story damage. Source: P Murphy.

5.3 Bad practices in RC frames and soft-story damage

The housing blocks in Al Hoceima were conceived on the basis of a wrong construction pattern very widespread throughout the Ibero-Maghrebian region, with some particularly distinctive features. The partial and total collapse of grade 4 and 5 observed in

RC structures are due to several key defects for the results. The use of non-corrugated steel rods in the vertical and horizontal elements of RC frames is one of the most important errors, because it does not prevent the rebar from sliding inside the concrete core; the longitudinal steel rebars were thick enough to support the weight and height of the structure but not the lateral deformations and bending from additional loads; and the stirrups were not hooked in the correct way to prevent the openning and separation from the steel mesh. Moreover, in the column-beam joints, the stirrups are placed at the same equidistance as in the rest of the column body, not providing the necessary rigidity to prevent it from plastifying when the columns are forced to tilting.

But the most damaging and characteristic construction pattern of the 2004 Al Hoceima earthquake is undoubtedly the soft-story failure. The distribution of the residential buildings was as follows: four or five floors high, densely partitioned in upper flats by interior infill walls embedded between the RC frames and a ground floor less partitioned or completely diaphanous for use as garage in most cases or as small stores. These walls help to stiff the RC frames, reducing the bending of the columns under horizontal displacements of the structure. In areas of moderate to high seismic risk, the upper floors, which are highly rigidized, perform like an unique and solid block that sends all the elastic stress to the ground floor frames. As a consequence, the conections between the column heads on the first floor and the slab on the second floor come under load. If the column-slab joint is not reinforced with the stirrups needed, the column would be excesively bended, producing a type of failure called "plastic hinge" [21]. As a consequence, irreversible tilting of the structure or collapse can occur (**Figure 19a**), transferring the damage in many cases to the upper floors, due to the strong impact during the fall (**Figure 19c**).

6. The Lorca earthquake of May 11, 2011

The epicenter of the Lorca earthquake (Spain) of May 11, 2011, Mw = 5.1 (IGN), was located about 4.5 km from downtown and was preceded by another less violent event of Mw = 4.5, widely felt one hour and 45 minutes before the main shock. The very short distance, shallow depth of the hypocenter (5 km) and geological conditions such as soft soils and, probably, progressive subsidence due to the massive exploitation of aquifers in the Guadalentín Valley, caused or increased a widespread and rough damage not precisely to the most vulnerable structures of class A or B, but to monumental and, mainly, residential buildings of class C which, as in the case of Setúbal, had been replacing traditional construction patterns. Two buildings collapsed and in the following weeks and months almost 1,164 houses and 45 industrial facilities and warehouses had to be demolished, with unrecoverable structural damage. The two collapsed buildings were the church of Santo Domingo (class C, but of low resistance) and a four-story residential block of RC frames (three housing floors and basement for garage), that had been evacuated after the first shaking.

This seismic event in a relatively modern city with a majority presence of RC structures is a powerful evidence that moderate earthquakes of magnitude M<6 can cause a great destruction and that the failure of non-structural elements can also lead to catastrophic consequences. In this case study of seismic behavior of buildings, we will only focus on the effects on class C residential buildings.

6.1 The absence or inadequacy of earthquake-resistant standards

Since the adoption of RC structure as the most widespread construction practice, four earthquake-resistant national standards have been approved in Spain: 1968, 1974, 1995 and 2002. The continuous updating of these legal provisions in a relatively very short period of time shows that the heavy damage caused during the succesive seismic events occurred in this country with intensities ≤VII (**Table 2**), have highlighted the inadequacy of each previous earthquake-resistant standard. All of them have been imposing more rigorous technical requirements and guidelines for the construction of new buildings and major rehabilitation works but have not prevented structures built under the previous rules from being exposed to the risk of collapse or seismic damaging. In addition, the rule currently in effect also do not guarantee the absence of damage to new buildings raised after the effective date of the law and only aims to avoid the immediate collapse of the structures.

Date	Time	Lat.	Long.	Depth (km)	Int.	Mag.	Epicenter
1950/04/04	03:06:22	433.000	−60.000	—	VI	4.6 M_D	Teverga
1950/05/02	07:37:46	381.500	−13.333	—	VI	4.0 M_D	Archena
1950/07/01	12:19:44	371.000	−25.333	—	VI	3.8 M_D	Gergal
1951/03/10	10:38:26	375.950	−39.750	15	VII	5.2 M_D	Castillo de Locubín
1951/05/19	15:54:26	375.670	−39.170	19	VII	5.3 M_D	Castillo de Locubín
1953/09/28	21:41:10	411.333	−15.833	—	VII	4.7 M_D	Used
1954/01/08	16:33:50	369.333	−38.833	—	VII	4.2 M_D	Arenas del Rey
1955/11/27	20:30:08	373.017	−24.583	5	VI	4.1 M_D	Bayarque
1956/04/19	18:38:54	371.917	−36.833	5	VII-VIII	5.0 M_D	Purchil
1956/04/22	15:56:14	372.800	−36.100	5	VI	3.7 M_D	Calicasas
1956/04/29	14:54:29	371.833	−36.833	—	VI	—	Albolote
1956/05/03	01:03:43	373.867	−35.967	5	VI	4.3 M_D	Iznalloz
1956/05/14	09:57:32	371.833	−36.833	—	VI	—	Albolote
1956/06/05	11:41:24	371.717	−70.983	5	VI	4.2 M_D	Punta Umbría
1956/08/16	02:09:40	369.100	−86.067	5	VI	5.0 M_D	Golfo de Cádiz
1958/01/16	15:13:38	381.000	−0.6000	—	VI	—	Guardamar del Segura
1958/02/05	10:18:25	384.583	−0.7350	5	VI	4.7 M_D	Petrer
1958/06/18	14:24:17	389.000	−15.250	10	VI	4.3 M_D	Hoya-Gonzalo
1958/12/22	02:48:16	381.833	−11.167	—	V-VI	4.0 M_D	Fortuna
1959/08/23	22:21:30	355.133	−32.267	20	VI	5.4 M_D	Alborán Sur
1960/06/01	06:18:54	380.967	−0.9117	5	VI	4.4 M_D	Redován
1960/11/14	20:10:26	370.283	−53.283	5	VI	4.5 M_D	Pruna
1960/12/05	21:21:47	356.900	−66.217	5	VII	4.9 M_D	Golfo de Cádiz
1961/02/10	18:52:01	417.250	−61.967	—	VI	5.2 M_D	Zamora
1961/09/03	23:33:13	419.333	−20.833	—	VI-VII	4.6 M_D	Aguilar Río Alhama
1962/05/03	23:27:22	438.850	−70.150	5	VI	4.3 MbLg	Cantábrico

Date	Time	Lat.	Long.	Depth (km)	Int.	Mag.	Epicenter
1963/01/19	20:50:29	382.167	−10.500	—	VI	3.3 MbLg	Abanilla
1964/01/29	01:47:53	370.583	−36.233	5	VI	3.7 MbLg	Dílar
1964/03/15	22:30:26	361.317	−77.500	30	VII	6.2 MbLg	Golfo de Cádiz
1964/06/09	02:33:35	377.367	−25.667	5	VII	4.8 MbLg	Galera
1964/09/09	09:39:45	370.850	−36.200	5	VII	4.3 MbLg	Otura
1967/08/03	00:34:13	383.567	−12.883	5	VI	3.9 MbLg	Jumilla
1969/02/28	02:40:32	359.850	−108.133	20	VIII	7.8 Mw	SW Cape St. Vincent
1970/03/14	15:48:09	424.800	16.800	—	VI	4.3 MbLg	Ílles de Cerdanya
1972/03/16	21:31:32	374.200	−22.450	5	VII	4.8 MbLg	Partaloa
1976/09/26	04:29:20	388.867	−0.5933	5	VI	4.0 MbLg	Aielo de Malferit
1977/06/06	10:49:12	376.450	−17.283	9	VI	4.2 MbLg	Lorca
1979/01/16	00:55:16	428.883	−71.517	80	VI	3.6 MbLg	Becerreá
1979/03/20	21:53:56	371.633	−38.017	5	VI	4.1 MbLg	Chimeneas
1979/06/19	03:55:53	371.483	−35.967	5	VI	3.2 MbLg	Cájar
1979/06/20	00:09:06	372.483	−34.917	60	VI	4.5 MbLg	Beas de Granada
1979/07/30	00:55:25	371.133	−36.733	5	VI	3.7 MbLg	Alhendín
1979/07/31	21:43:20	371.167	−36.033	5	VI	3.9 MbLg	Gójar
1979/11/25	01:56:27	368.650	−37.733	5	VI	3.4 MbLg	Lentegí
1979/12/18	05:47:34	428.883	−71.633	20	VI	4.2 MbLg	Becerreá
1980/11/11	10:59:46	378.333	−52.150	5	VI	4.1 MbLg	Hornachuelos
1988/08/20	13:03:03	372.067	−37.667	2	V-VI	3.9 MbLg	Chauchina
1989/12/20	04:15:05	372.250	−73.917	23	VI	5.0 MbLg	Ayamonte
1991/08/14	10:32:08	387.550	−0.9600	2	VI	4.1 MbLg	Caudete
1993/12/23	14:22:35	367.800	−29.367	8	VI-VII	5.0 MbLg	Berja
1995/11/26	05:39:40	380.383	−12.700	2	V-VI	4.1 MbLg	Alguazas
1995/11/29	23:56:28	428.167	−73.033	9	VI	4.6 MbLg	Triacastela
1995/12/24	14:29:21	428.600	−73.150	15	VI	4.6 MbLg	Baralla
1997/05/21	23:50:45	427.833	−72.583	13	VI	5.1 MbLg	Triacastela
1999/02/02	13:45:17	380.963	−15.014	1	VI	4.7 MbLg	Mula
2005/01/29	07:41:32	378.535	−17.555	11	VII	4.8 Mw	Aledo
2011/05/11	15:05:13	377.196	−17.076	2	VI	4.5 Mw	Lorca
2011/05/11	16:47:26	377.175	−17.114	4	VII	5.1 Mw	Lorca
2016/01/25	04:22:01	356.004	−38.056	12	VI	6.3 Mw	Alborán Sur

Table 2.
Earthquakes felt with intensities VI to VII in Spain between January 1, 1950 and December 31, 2019. Note that earthquakes of magnitude M < 5 can also cause effects of intensity VII in direct relation to the shallowness. Source: Instituto Geográfico Nacional (IGN).

In Lorca, most of the damaged RC housing blocks had been built in three different construction periods: before the first anti-seismic standard of 1968, during the period from 1968 to 2002, and after 2002 standard, in use on the date of this earthquake. But none of these technical rules devoted enough extension to deal with the coupling conditions of non-structural elements, with the consequences that will be discussed below. The first RC housing blocks in Lorca began to be built from the early 1950s onwards, in a period of time marked by the international blockade and economic situation of autarchy (self-production) of the Spanish government after the Second World War. Faced with the impossibility of importing raw materials, steel rebars of low quality and scarce quantity were used for the two essential elements of RC frames: beams and columns. In fact, several patents were developed in Spain to create RC beams containing up to only two steel rebars [22].

However, although it seems difficult to explain, older RC buildings did not suffer a greater damage than those built after the 2002 standard. In **Figure 20** we can appreciate slight shear cracks in the façade of pre-1968 RC buildings, not showing apparent structural damage or deformations of RC frames on the ground floor, despite the lower stiffness due to the lack of infill walls (grade 3, class C). Paradoxically, in **Figure 21** we have the complete collapse of the four-story residential block mentioned above, built during the transition period to the 2002 earthquake-resistant standard. This means that improving rules does not always implies improving resistance.

6.2 The hazard of non-structural elements

The typical design of residential buildings in Lorca is formed by a RC supporting structure (columns and beams) and double-leaf external walls separated by an inner cavity, used as thermal insulation, filled with expanded polystyrene sheets. The internal wall is made of hollow brick, and the external wall of unplastered solid red brick. This external panel was not confined into the RC frame, but externally attached with mortar to the structure, acting rather as cladding or load-bearing wall and not properly as infill wall. In addition, the two brick walls were not tied together with any kind of coupling element, and the lack of plaster, which serves as cohesive mesh and grip for the bricks, provided less resistance to the external face.

The 2011 Lorca earthquake caused nine fatalities due to the fall of these non-structural elements from façades (**Figure 22**) and the evident cause-effect relationship led to a conclusion that became an axiom: non-structural elements also

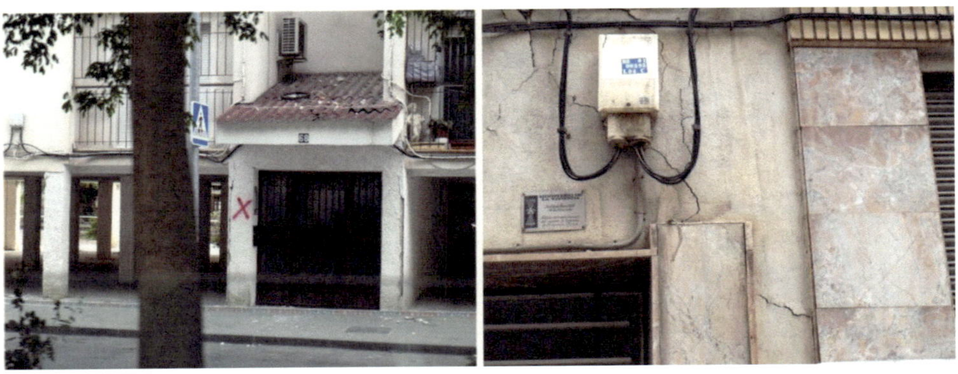

Figure 20.
Lorca, shear cracks in pre-1968 RC buildings of class C (grade 3). There is apparently no evidence of structural damage, and on the ground floor (left) the columns forming an evident soft story have no sign of tilting. Source: IERD.

Figure 21.
Lorca, complete collapse of RC residential building (grade 5) designed under the earthquake-resistant standard prior to the 2002 standard: (a) the three slabs are completely collapsed without any gap to allow survival; (b) plastic hinges and overturning in the direction of the slope of the street; and (c) plastic hinge in the adjacent building caused during the collapse, wedging against the base of the column. Source: IERD.

kill [23]. This damage took place in several steps: (1) formation of shear cracks in the infill walls caused by cyclic reverse movements of the ground due to seismic shaking; (2) detachment and drift of external walls from RC frames; (3) bending and fall of cladding and external walls. **Figure 23** shows the different types of damage here described: shear cracks cause the detachment and loss of stone or marble cladding and the toppling of the outer leaf of the masonry wall, made in this case with hollow bricks (grade 3, class C). After the collapse of the wall, the expanded polystyrene sheets are exposed outside the inner cavity of insulation. The drift between brick infills, with no connecting elements between them, is also clear. However, there are no shear cracks in the cantilever plaster at the top of the building entrance, which proves the high vulnerability of the infill wall design. The same situation occurs in **Figure 24**, at a moment prior to the external wall detachment, appearing X-shaped cracks by reversal of shear forces.

The lateral loadings were stronger in ground floors, due to the same widespread and wrong conception of structural design exposed in the case of Al Hoceima: the soft story effect. In this construction pattern, the upper floors are more partitioned for residential use than the ground floor, commonly used for open-plan commercial premises, garage, or other purposes. Given this so unbalanced stiffness distribution, the building performs like a rigid block that swings over RC frames of the ground floor, resulting in shear cracks, corner failure, X-shaped or diagonal cracks, partial or complete overturning of infill walls or, in the worst case, plastic hinges that do involve structural damage.

Figure 22.
Lorca, debris fallen on vehicles and sidewalk. The center of the street is free of danger. In Lorca, people died while walking or looking for protection near the façades. Source: IERD.

Figure 23.
Lorca, shear cracks caused the detachment and loss of stone cladding and the toppling of the outer leaf of the brickwork wall (grade 3, class C). Source: IERD.

Figure 24.
Lorca, loss of stone cladding, fall of window lintel and X-shaped crack on external wall about to fall down. Source: IERD.

6.3 Miscalculations lead to catastrophic results

The only collapsed building in Lorca was designed under the earthquake-resistant standard of 1995. Following the technical requirements of this regulation, the RC frames, columns and beams, were reinforced by four longitudinal steel bars enclosed by vertical stirrups placed at regular intervals, including the column-beam or column-slab joints. In addition, the stirrup ends were bent at 90°, without forming a hook around the longitudinal rebars. In **Figure 25a–c**, each RC column is made up of four longitudinal steel rebars and stirrups with 90° hooks (**Figure 25d**). The 1994 earthquake-resistant standard required four rebars per column in areas with expected gravity acceleration values <0.16 g and eight rebars per column in areas with acceleration ≥0.16 g (m/s^2). The maximum estimated value for the municipality of Lorca was 0.12 g; therefore, the collapsed building studied here, built in 2001, fulfilled the construction requirements at the time. Afterwards, the 2002 standard reduced to 0.12 g the acceleration value needed to force the implementation of eight rebars per column, but it was too late. Obviously, this building ——and many others of similar design—— had used inadequate construction parameters that were to be approved the following year, stablishing stricter technical requirements more consequent with the behavior of RC frames under acceleration values ≥0.12 g.

As said above, the effective acceleration value was 0.37 g, that is, three times higher than the value for which the structure had been engineered. With this level of vulnerability, these RC structures, with four rebars per column arranged for maximum accelerations of 0.12 g, would hardly have been able to withstand the effective accelerations of 0.37 g without suffering severe damage. The design miscalculation of RC frames was 2/3 lower than the real value. If initial calculations had been overestimated, the damage would have been considerably reduced; but reinforcing structures with steel material means a significant increase in construction costs.

Several photographs taken in the affected area show the difficulty of the RC frames in other residential buildings to resist the violence of the horizontal

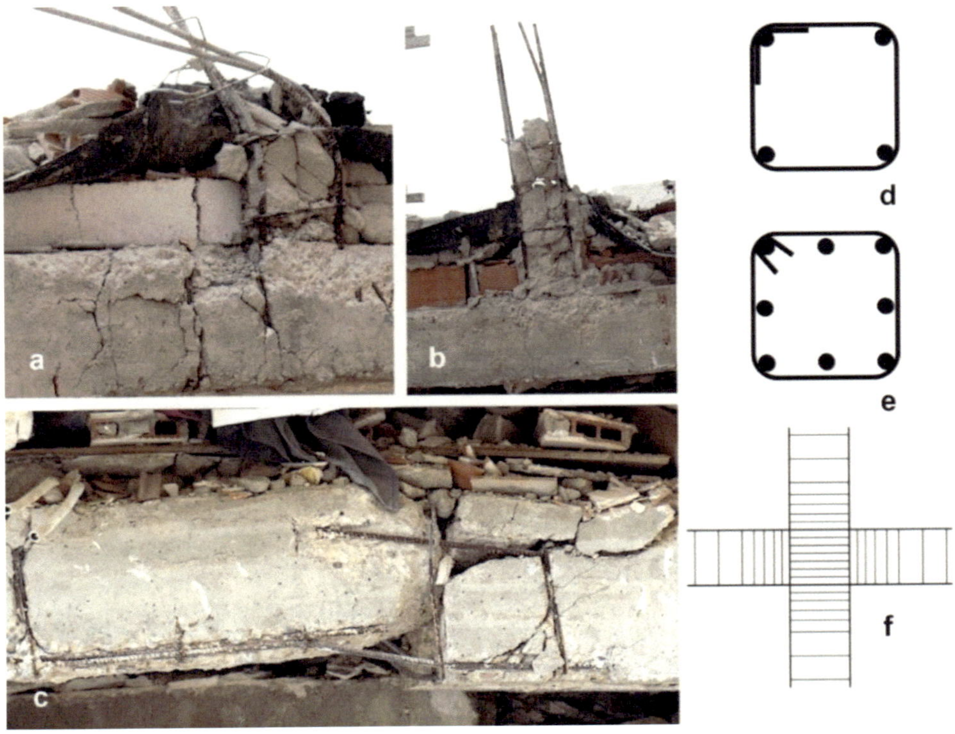

Figure 25.
Lorca, details of the RC residential block collapsed in Lorca, showing the four-rebar cores of beams and columns: (a) and (b) show the regular equidistance between the stirrups near the column-beam joints; (c) beam with similar arrangement of rebars; (d) drawing of the four-rebar longitudinal distribution used in this building with stirrup ends bent at 90°; (e) more adequate design for seismic resistance, with eight rebars and stirrup bent at 135°; (f) correct way to reinforce the RC column-beam joints. Source: IERD.

forces. As an example of better structural behavior, in **Figure 26** we can observe a complete damage in a double-leaf infill wall in a housing block with overturning of internal and external faces, but without apparent cracks in the RC frame and with no signs of plastic hinges or deformation in the column-beam joint (grade 3, class C). On the contrary, in **Figure 27** the loss of cladding allows us to clearly appreciate the occurrence of shear cracks at the base of the column-beam connection of the second floor, implying in all cases a moderate structural damage. It is due to the much more stiffness of the slab, which overloads the column resistance by sending all the energy of horizontal forces to the column base, not reinforced with an adequate seismic-resistant design. This damage, although less frequently, is not exclusive to the column-beam or column-slabs joints and can also occur in the middle body of the column (**Figure 28**), especially in case of short-column effect.

The damage to the column-beam joints of RC frames is consequence of the lack of steel reinforcement at the coupling node. As shown in **Figure 25**, the stirrups have a regular equidistance along the column as in the beam-column joints; as a result, the RC structure plastifies. A greater number of stirrups in the coupling node, progressively reducing the distance between them towards the intersection core, would increase the structure resistance and distribute the displacement energy along the full length of the columns, not transferring all the cyclic reversal of loads to the column-beam joints (**Figure 25f**). In addition, in areas with a moderate to high level of seismicity the stirrups should be anchored to the longitudinal rebars, overlapping 135° hooks [24] to avoid the opening of the closed-loop and outward bending of the rebars (**Figure 25e**). The Turkish Earthquake Code 2007 and Indian Standard IS13920-1993 are two examples of the implementation of these earthquake-resistant requirements.

Figure 26.
Lorca, complete overturning of double-leaf wall with overturning of both brickwork faces, but without apparent cracks in the RC frame. Source: IERD.

Figure 27.
Lorca, cracks in the column-beam base suggest the incipient formation of a plastic hinge. The loss of cladding is due to a hammering effect with the adjacent building. Source: IERD.

Figure 28.
Lorca, X-shaped cracks in infill walls and loss of concrete in the middle body of the RC column suggest a soft story effect. Source: IERD.

7. Conclusions

After the occurrence of earthquakes it is necessary to carry out a detailed field report, taking all the detailed photographs possible, to document and analyze the behavior and vulnerability of buildings according to the magnitude, distance and depth of the seismic event, in relation to the construction techniques used in the past and the adequacy of the current construction standards. The reconstruction of a territory severely affected by earthquakes cannot be planned using the same wrong techniques or with the standards in force that have demonstrated its inadequacy. Otherwise, a future earthquake with identical or higher parameters will again cause the same level of destruction.

For a resilient and sustainable reconstruction of devastated areas, a thorough review of the earthquake-resistant building standards and construction requirements is needed, aiming to ensure the resistance of structures to the higher stresses expected in the macroseismic area. A more precise seismic microzonation for smaller sectors is also required to get accurate structural strength calculations; the case of Lorca studied in this chapter shows that within the same city considered as a whole within the same level of seismic acceleration, specific points can undergo much greater accelerations, probably due to geological anomalies or discontinuities of the ground. Lessons learned like the ones we have discussed in this chapter should help to avoid repeating the same errors in other places. For this reason, it will be necessary to rethink whether the current construction patterns are a good example of good practices, whether they require a modification or reinforcement of the structural design, or whether they should be changed by other patterns more consistent with seismic activity in the area. It is also very important to note that the increase in costs of implementing earthquake-resistant systems in buildings will normally be much less than the volume of economic losses caused by earthquake damage.

It is evident that the buildings designed under previous or obsolete constructive standards will continue to be vulnerable even when reinforcement works are carried out, so it would be a good preventive task to carry out or review the seismic vulnerability maps, identifying the construction patterns used in each point of the map, their level of vulnerability and types of possible damage here described. It is also very important that the population be aware of the behavior of each type of structure in each city or neighborhood where they live, and customize the recommendations to the public on how they should respond during and after an earthquake at each time and place. Finally, it should be said that no country is exempt from being affected by a seismic disaster and therefore we must continue making efforts to increase the buildings resistance. Because inadequate construction patterns can be repeated and many buildings are exposed to the risk of collapse in very large geographic areas that can encompass several countries, it would be a great opportunity to improve earthquake-resistant designs by applying the goals of the 2030 Agenda, especially the goal 11 "cities", to significantly reduce the number of deaths and people affected, and substantially decrease the direct economic losses caused by disasters.

Acknowledgements

We want to thank here all those people and institutions that have contributed with photographs to describe the damage of earthquakes on buildings, without which it would be very difficult to see and understand the characteristic effects of these natural phenomena on the different building structures. In the case of the earthquake in San Vicente del Cabo SW, our thanks to the Instituto Português do Mar e da Atmosfera and especially to Antonio Caneira, from the Projects and Contracts Division (DivPC), for his cordiality; with respect to the photographs of the earthquake in Al Hoceima, to the technical architect Patrick Murphy Corella for authorizing their use in this book; and our most sincere gratitude and debt to Antonio Peláez Pérez and Antonio Enrique Peláez Fuertes for the great photographic work they did in the city of Lorca the same afternoon the earthquake occurred and for giving them up for use in this book.

Nomenclature

ERD	Earthquake-resistant design
IERD	Instituto Español para la Reducción de los Desastres
IGN	Instituto Geográfico Nacional
IPMA	Instituto Português do Mar e da Atmosfera

Author details

José Antonio Aparicio Florido
Instituto Español para la Reducción de los Desastres (IERD), San Fernando, Cádiz, Spain

*Address all correspondence to: apariciofflorido@gmail.com

References

[1] Buforn E, Sanz de Galdeano C, Udías A. Seismotectonics of the Ibero-Maghrebian region. Tectonophysics. 1995;248:247-261. SSDI: 0040-1951(94)00276-2

[2] Bezzeghoud M, Ayadi A, Caldeira B, Fontiela J, Borges J F. The largest earthquakes in Algeria in the modern period: the El Asnam and Zemmouri-Boumerdès faults. Física de la Tierra. 2017;29:183-202. DOI: https://doi.org/10.5209/FITE.57618

[3] Aparicio JA. 1755: El maremoto que viene. Cádiz: Q-book; 2017. 251 p 978-84-15744-50-4

[4] Grandin R, Borges JF, Bezzeghoud M, Caldeira B, Carrilho F. Simulations of strong ground motion in SW Iberia for the 1969 February 28 (Ms = 8.0) and the 1755 November 1 (M ~ 8.5) earthquakes - II. Strong ground motion simulations. Geophysical Journal International. 2007;**171**:807-822. DOI: 10.1111/j.1365-246X.2007.03571.x

[5] Baptista MA, Miranda PMA, Victor LM. Maximum entropy analysis of Portuguese tsunami data: the tsunamis of 28.02.1969 and 26.05.1975. Science of Tsunami Hazards. 1992;**10**: 9-20 ISSN: 0736-5306

[6] Martínez Solares JM. Tsunamis en el contexto de la Península Ibérica y del Mediterráneo. Enseñanzas de las Ciencias de la Tierra. 2005;**13**(1):52-59 ISSN: 1132-9157

[7] Paradise TR. Perception of seismic risk in a Muslim city. The Journal of North African Studies. 2006;**11**(3):243-262. DOI: 10.1080=13629380600802961

[8] Grünthal G, editor. European Macroseismic Scale 1998. Luxembourg: Centre Europèen de Géodynamique et de Séismologie; 1998. 99 p 2-87977-008-4

[9] Trêpa MV. Relatório do inquérito levado a cabo nas provincias de Estremadura, Alto Alentejo, Baixo Alentejo e Algarve sobre o sismo de 28 de fevereiro de 1969. Serviço Meteorológico Nacional: Portugal; 1969 41 p

[10] Murphy P. Al-Hoceima earthquake on February 24, 2004. 2004. Available from: https://www.emsc-csem.org/Files/news/Earthquakes_reports/Alhucemas_040224_eng.pdf [Accessed: 2020/09/05]

[11] Aparicio J A, Peláez A, Peláez A E. El terremoto de Lorca del 11 de mayo de 2011. 2011. Available from: http://ierd.es/wp-content/uploads/2014/06/informeIERD-terremoto-lorca-11052011.pdf [Accessed: 2020/09/05]

[12] Aparicio JA. Effects of the 28 February 1969 Cape Saint Vincent Earthquake on Ships. Pure and Applied Geophysics. 2020;**177**:1801-1808. DOI: 10.1007/s00024-019-02368-0

[13] Borges JF, Bezzeghoud M, Caldeira B, Carvalho J. Ground-Motion Simulation in the Lower Tagus Valley Basin. Pure and Applied Geophysics. 2015;**172**:2411-2420. DOI: 10.1007/s00024-015-1060-7

[14] Carvalho J, Dias R, Pinto C, Leote J, Mendes-Victor L. A soil classification for seismic hazard assessment and mitigation of the Algarve. The 14th World Conference on Earthquake Engineering (WCEE '08). 12-17 October 2008; Beijing, China.

[15] Teixeira C. Alguns aspectos geológicos dos efeitos do sismo de 28 de fevereiro de 1969. Boletim da Sociedade Geológica de Portugal. 1975;**XIX**(III):239-250

[16] Paula A, Sousa C. Evaluation of 1947-1993 macroseismic information in

Portugal using the EMS-92 scale. Annali di Geofisica. 1996;**XXXIX**(5):989-1003. DOI: 10.4401/ag-4029

[17] Mendes AS. Notícia acerca do sismo de 28 de Fevereiro de 1969 em Portugal Continental. Finisterra. 1969;**IV**(8):273-277

[18] Miranda J M, Carrilho F. 45 anos do sismo de 28 de fevereiro de 1969. Instituto Português do Mar e da Atmosfera. 2014.

[19] Buforn E, López-Sánchez C, Lozano L, Martínez-Solares JM, Cesca S, Oliveira C, et al. Re-evaluation of seismic intensities and relocation of 1969 Saint Vincent Cape seismic sequence: a comparison with the 1755 Lisbon earthquake. Pure and Applied Geophysics. 2020;**177**:1781-1800. DOI: 10.1007/s00024-019-02336-8

[20] Guevara LT, García LE. The captive- and short-column effects. Earthquake Spectra. 2005;**21**(1):141-160 ISSN: 8755-2930

[21] Guney D, Aydin E. The nonlinear effect of infill walls stiffness to prevent soft story collapse of RC structures. The Open Construction and Building Technology Journal. 2012;**6**:74-80. DOI: 10.2174/1874836801206010074

[22] Aparicio J A. 1947 Cádiz, la gran explosión. 1st ed. Cádiz: Cazador de Ratas;2017. 528 p. ISBN: 978-84-945867-5-0

[23] Murphy P. Terremoto de Lorca 11 mayo 2011. In: 4th National Congress of Seismic Engineering; 18-20 May 2011; Granada.

[24] Susanto A, Kasyanto H, Susilahadi S. Experimental study of 90° hook and standard hook at the end of reinforced concrete beam stirrup. IOP Conference Series: Materials Science and Engineering. 2020;**830**(2). DOI: 10.1088/1757-899X/830/2/022072